Moving Violations

Hagley

Hagley Library Studies in Business, Technology, and Politics
Richard R. John, *Series Editor*

MOVING VIOLATIONS

AUTOMOBILES, EXPERTS, *and* REGULATIONS
in the
UNITED STATES

Lee Vinsel

Johns Hopkins University Press
Baltimore

© 2019 Johns Hopkins University Press
All rights reserved. Published 2019
Printed in the United States of America on acid-free paper
2   4   6   8   9   7   5   3   1

Johns Hopkins University Press
2715 North Charles Street
Baltimore, Maryland 21218-4363
www.press.jhu.edu

Library of Congress Cataloging-in-Publication Data

Names: Vinsel, Lee, 1979– author.
Title: Moving violations : automobiles, experts, and regulations
in the United States / Lee Vinsel.
Description: Baltimore : Johns Hopkins University Press, [2019] |
Series: Hagley Library studies in business, technology, and politics |
Includes bibliographical references and index.
Identifiers: LCCN 2018046667 | ISBN 9781421429656 (hardcover : alk. paper) |
ISBN 9781421429663 (electronic) | ISBN 1421429659 (hardcover : alk. paper) |
ISBN 1421429667 (electronic)
Subjects: LCSH: Automobiles—Law and legislation—United States—
History. | Automobiles—Safety measures—United States. |
Automobiles—Motors—Exhaust gas—Law and legislation—United States. |
Automobiles—Motors—Exhaust gas—Environmental aspects—United States. |
Traffic violations—United States—History. | Traffic regulations—United
States—History.
Classification: LCC KF2209 .V56 2019 | DDC 343.7309/44—dc23
LC record available at https://lccn.loc.gov/2018046667

A catalog record for this book is available from the British Library.

Special discounts are available for bulk purchases of this book.
For more information, please contact Special Sales at 410-516-6936
or specialsales@press.jhu.edu.

Johns Hopkins University Press uses environmentally friendly book
materials, including recycled text paper that is composed of at least
30 percent post-consumer waste, whenever possible.

For Abigail, Henrietta, and Alban
and
With gratitude to the Baron and the Gypsy

CONTENTS

Moving Violations

Introduction

In 2009 the Insurance Institute for Highway Safety celebrated the fiftieth anniversary of its founding by crashing two cars into each other. The crash test pitted a 1959 Chevrolet Bel Air against a 2009 Chevrolet Malibu. If the 1959 car was unaltered, it would have weighed between 3,600 and 4,000 pounds, whereas the 2009 Malibu would have weighed between 3,400 and 3,600 pounds.[1] In other words, the 1959 car was heavier. But when the cars were crashed, the exact opposite seemed to be true. In the video of the crash, which is widely available online, the front end of the Bel Air almost seems to evaporate. A camera within the Bel Air shows the lighter Malibu nearly enter the passenger compartment, crushing the crash test dummy. Meanwhile, the Malibu's front crumple zones crumple violently, but the car's interior remains intact. Moreover, the dummy in the Malibu makes contact with an expanding airbag. Sensors in the crash test dummies suggest that a human driver in the Malibu would have suffered minor knee injuries. A driver in the 1959 Bel Air would have died almost instantly.

The safety differences between these vehicles are largely the result of government regulation. If the testers had also run emissions tests on the two vehicles, they would have found that the 1959 Bel Air produced a great deal more air pollution than did the 2009 Malibu. Furthermore, the 1959 Bel Air got on average 13.5 miles per gallon, whereas the 2009 Malibu got about 27.5 miles per gallon (if you average the Malibu's city and highway economy ratings).[2] These differences between the vehicle's emissions and fuel economies, too, largely result from law and public administration. We should not be naïve, of course. Tens of thousands of people still die in automobile accidents every year in the United States. Air pollution from automobiles is still a major problem in

American cities, and the country still consumes a staggering amount of gasoline, producing huge quantities of greenhouse gases. Yet even the greatest relativist must admit that, in general, automotive safety, emissions, and fuel economy have improved massively over the history of that technology.

This book is about the relationship between regulation and technological change, and it uses the history of automobile regulation in the United States as a way to study this relationship. The automobile makes an ideal case for such a study because innumerable rules, laws, and regulatory efforts have influenced the car, its users, and the entire automotive system. Federal regulations alone cover at least five different aspects of the automobiles—their safety, the emissions they produce, the noise they create, how much fuel they consume, and the repair costs they incur in low-speed collisions—and, of course, state and local laws deal with many other parts of these vehicles and the ways we use them. Moreover, the automobile has a history that spans much of the industrial era in the United States, since the Duryea brothers built the country's first internal combustion engine–driven vehicle in 1893.

The patterns of regulatory efforts focused on automobiles from the 1890s to the present have changed with the shifting political orders, mirroring broader patterns that are well known to students of US history. By tracing automotive regulation through these various periods, we learn not only about the history of this important technology but also about changing structures of government and expertise, the role of scientific and technical knowledge in regulation, and how a democratic nation dealt with risks that arose from technologies when that nation valued technological progress as a core aspect of its self-image.

In this book, I define both *regulation* and *technological change* broadly. I use the word *regulation* to refer to any effort to reduce a public problem. Such efforts range from bringing experts and concerned parties together to enacting formal legislation and government codes. From this perspective, regulation is not only public—or governmental—but also private. It includes the efforts of local, state, and federal governments, as well as the work of private groups, such as voluntary associations, professional and engineering societies, trade associations, and advocacy organizations. Taking such a perspective opens our eyes to the wide variety of methods people use to influence the world around them. Similarly, by *technological change,* I mean not only changes in technologies themselves—such as the car—but also transformations in how technologies are used and the systems that surround them— such as developing and altering roads, traffic signs and signals, the workings of gas pumps, and the seemingly endless stream of objects that make up the world of the automobile.

The first goal of this book is to counter a fairly widespread narrative about how technologies develop. We live in a society in which many people deny the value of government. Things would be better, they insist, if only government would get out of the way. Let the entrepreneurs and innovators take over. Regulations only inhibit us; remove them and watch as economic growth and vitality return. All of these ideas are quite common today; the reader has likely heard them expressed hundreds of times. But, in many cases, these ideas are also demonstrably false. They are based on weak assumptions, and they betray a deep, often willful ignorance of history.

Regulation has shaped the automobile itself from its beginning. There was no golden age before governmental power took hold. Before the emergence of federal regulations, state, local, and even private forms of control attempted to abate the worst aspects of the new world of automobiles. Furthermore, contrary to popular opinion, regulations have not hindered technological change, or "innovation," but rather have fostered and even generated it. Conversely, some thinkers have argued that government research and development is central to innovation, and so have asserted that government should conduct or fund even more research and development. In the case of automobiles, however, government research has not been a great success; it has been nowhere near as effective at generating innovation as regulation has been. Put simply, if we want to make claims about the relationship between government and technology, we must enter into a frank and sober assessment of it.

But how should one study the car's relationship with regulation? The topic can seem so large and unruly. In this book, I focus on communities: I examine how different groups emerged to deal with the hazards of the automobile and how they acted to curb those hazards. The makeup and mix of the groups that have made such efforts have changed a great deal over time. In the first years of the twentieth century, advocates for auto safety tended to work in citizens groups. While such efforts have persisted to some degree, over the course of the twentieth century we increasingly put the responsibility for automotive hazards in the hands of experts. This delegation of responsibility to experts fits a well-known pattern. Scientists, engineers, educators, lawyers, and other professionals took advantage of opportunities that arose around automotive risks and, consequently, shaped the auto world.

Some groups influenced the historical course of automobile regulations more than others. The auto industry itself, elected officials, and government agencies played dominant roles. But consumers, and groups and organizations that claimed to speak in the name of consumers, also had an enormous impact on the development and shape of automobile regulations. To date, most histories of technological regulations have focused on large-scale technological systems under central ownership, such as railroads and telegraph, telephone, and electricity systems. The fact that automobiles were owned and used by individuals created numerous dynamics that typically did not apply with those other technologies.

Since at least the 1920s, the insurance industry—which often claimed to speak for and aggregate the interests of consumers, and sometimes even did so—also played a continual, though constantly shifting, part in automobile regulation. Insurance executives led efforts to decrease auto accidents, injuries, and deaths in the 1920s and 1930s. The industry was an important funder of crash safety studies in the 1950s and 1960s. And since the 1980s, its nonprofit trade organization, the Insurance Institute for Highway Safety, has played a crucial role in publicizing information about vehicle safety and the results of crash safety tests. In all of these ways, the insurance industry is the clearest and most constant example of how private groups shape regulation—sometimes by influencing government action but often by acting outside the bounds of government altogether.

As the groups and communities focused on decreasing the car's risks changed, so did perceptions of what those risks entailed. Moreover, car hazards changed as the car changed and as the ways people used it changed. For instance, the 1901 Pierce Motorette, which, like the 1904 Oldsmobile Curved Dash, was steered with a tiller, came equipped with gas headlamps, called Neverout Insulated Kerosene Safety Lamps. The "safety" in the safety lamps referred to Neverout's patented system that was supposed to keep the lamps from exploding or turning into fireballs. If you were driving down a bumpy, unpaved dirt road in the pitch black of night in 1901 or 1902, having your car's lamp catch ablaze could be a real problem. Similarly, closed-body automobiles, which came to dominate the car market only in the 1920s, introduced new hazards to automobiling. With open-bodied automobiles, drivers could see in whichever direction they chose to turn their heads, though dust, mud, and insects could also fly into one's eyes or cover one's goggles. The closed-body vehicle inevitably created blind spots, which led to accidents. Closed bodies also shut drivers off from the environment, making it harder to hear oncoming hazards, such as trains. Part of the story in this book, then, is the story of changing realities and perceptions of risk.

On its own, this point about the changing nature of risks and the perceptions of them can seem facile. The insight becomes important, however, when we consider how people acted on their perceptions, how these actions influenced parts of the world, and how we continue to live with their formulations and products. For instance, the psychologists who created the first standardized driving tests believed in the notion of "accident proneness," the idea that a certain percentage of the population was predisposed to causing accidents. Accordingly, the psychologists created tests to weed out such people. By the 1940s the notion of accident proneness fell into disfavor among experts. Yet we still live with these driving tests, even if the idea that originally gave impetus to them have long since fallen away. Similarly, the mundane technologies that surround us everywhere and play crucial roles in our ordinary routines—electric traffic signals, for example—were once the subject of contest, heated discussion, and debate. While these debates are now largely forgotten, we still live with the objects and systems that resulted from them.

My approach, therefore, focuses on how groups formed, how they came to identify and perceive problems, and how they tried to solve them. Once people define problems, solutions present themselves. Throughout this work I trace where solutions came from. Often, they traveled from other industries or unexpected places. Many of the methods and technologies used to make streets safer during the 1920s, from electric traffic lights to driving tests, were first developed to foster safety in the railroad and streetcar industries. Likewise, much of the research that eventually went toward creating automotive safety standards in the 1960s was originally conducted by the US military, which sought to keep pilots safe when they ejected from jets. The methods that governing bodies developed to address automotive risks ended up influencing both our use of this technology and its ultimate form. These methods have stuck with us, whether they are the driver's education classes most of us suffered through in high school or the crash tests, shown in television commercials, that have given shape to structural elements in every car sold in the United States for the past fifty years.

None of these developments—the formation of groups, the perception of risks, the creation of new ways of doing things—happened in a vacuum. This book focuses on the United States of America, a capitalist country and a liberal democracy with a strong tradition of federalism. The fate of auto regulation was strongly tied to capitalist developments, the growth of industry, the emergence of consumerism, and shifting philosophies about how best to curb the negative aspects of capitalist production, including pollution, product safety risks, and depletion of scarce natural resources. Regulation developed against the backdrop of a political tradition that emphasizes individuals rights, avoids regulating corporations as much as possible, and prefers that states, rather than the federal government, devise and implement solutions whenever feasible. As we will see, individual states, such New York and California, have played influential roles in the development of regulation in the United States. To tell the history of automotive regulation, therefore, is to tell the history of capitalism and the history of constitutional democracy, with all of the hopes and vagaries that these histories entail.

Focusing on the history of US political development and organizational change over this period leads to a broad view of the relationship between regulation and technological change, a view I explore more deeply in the book's conclusion. Time and again, regulation has influenced technological change by prompting experts to think about specific problems. Regulation forms and reforms expert communities. As one would expect, this process has worked in different ways over time. For example, in the 1920s, Secretary of Commerce Herbert Hoover preferred what scholars later called an "associational" form of government, which avoided regulation via federal codes or legislation in favor of bringing experts together voluntarily in meetings, conferences, and working groups. Hoover's Department of Commerce played a key role in standardizing many parts of the automobile, such as brakes and headlights; the automotive system, including signs and signals; and even drivers, through psychology tests and eventually driver's education. Trained experts led all of these standardization efforts. Later, in the 1960s and 1970s, federal legislators passed a series of laws to force the automakers to address problems such as safety, air pollution, and fuel economy. Here, too, the solutions both drew on existing expert knowledge and, sometimes, pushed experts to generate new knowledge.

Most important, this book reveals that the structure of laws largely determines whether and to what extent regulations lead to the production of new knowledge. US automotive safety standards have rarely led to the creation of fundamentally new ideas but have largely drawn on existing ones, while emission control laws have been much more consequential. By the 1970s and 1980s, however, many individuals and groups had begun to have serious doubts about experts. Some argued that expert groups, particularly those ensconced in government, were self-interested utility maximizers who did little for the public good. Indeed, experts often harmed the public, according to this argument, by interfering with the proper working of capitalist markets, sometimes leading to fearsomely perverse results. Particularly with Ronald Reagan's election, leaders in the executive branch attempted to rein in regulatory agencies and the experts within them, as by building cost-benefit analysis into federal rule making. Since the early 1980s, automotive regulation in the United States has charted an uneven course. If this story is about how regulation shapes and reshapes expert groups and the knowledge they produce, it is also about attempts to control experts themselves by setting limits on how, when, and why they can exert influence.

The changing methods for reducing automotive hazards touch on a deep and profound shift in modern culture: the transition from a handmade, custom-crafted world of wood to a mass-produced, homogenized world of metal and plastic. The roots of the automotive industry lay in craft. When the Duryea brothers began producing the first cars for sale in the United States in 1896, they did so using craft methods that had once been applied to such objects as carriages and bicycles. These methods were based almost wholly on local knowledge, they were embodied in the skilled men who had long trained in such techniques, and they did not transfer easily from one place to another. The modernization of industry included attempts to surpass such craft methods of production. One way to modernize industry was to pool and transport knowledge of new production techniques. Today, someone can design a part or object on the computer and email that design to someone on another continent, whereupon a computer-controlled machine will spit the part out after making it with near perfect precision. Of course, in earlier times people did not have the technologies to do exactly this, but engineering and scientific societies, industry and trade associations, and government bodies all worked to aggregate knowledge of the best ways to do things. Much of this knowledge was eventually built into an organizational tool that took on increasing importance during this era: technical standards.

In the automobile industry, technical standards began as a way to simplify practices and increase coordination between manufacturers and suppliers. Under certain conditions, standards saved everyone money. Increasingly, however, standards also became the chief method of regulating technological risks. The automotive standards for steel and metal parts and wheels that emerged in the early 1910s can be seen as the first implicit safety standards, since they focused on resistance to breakage. The first explicit safety standards emerged in the 1920s and 1930s, however, and covered parts like headlights, brakes, and the glass in car windows. Standards also took pride of place in the wave of federal auto regulations that arose in the decade from 1965 to 1975. When US legislators and regulators tried to increase the automobile's safety and decrease its noise, pollution, fuel use, and the costs of its repairs, it was through standards, always standards.

The use of standards to decrease automotive risks was not without its ironies and perversities, however. Sometimes the structures of standards had unintended consequences. The creation of the sport utility vehicle (SUV), which increased road dangers and fuel use, was a corporate strategy to get around poorly made standards. Standards could also come into conflict with one another, leading to what engineers, economists, and others call trade-offs. For instance, some experts argued that fuel economy standards made cars less safe. Moreover, standards often came into conflict with core values of powerful interest groups and with some strains of American political thought. Increasingly, in the 1970s and 1980s, conservatives argued that standards levied an ungainly financial burden on struggling companies. They beat back proposed new standards and tried to eviscerate existing ones, sometimes leaving them unenforced. Perhaps most important, the use of standards to decrease automotive risks has by some markers been fairly successful, though auto corporations will always say the standards are too stringent and safety advocates and environmentalists will always say they are too weak. But where we have succeeded in shaping the technical aspects of the car, we have failed in curbing the often-increasing use of them, which is the ultimate cause of accidents and pollution. We control technologies but refuse to govern ourselves. For all of these reasons, the legacy of automotive regulation is mixed.

In the context of automobiles, the word *standard* has a pleasing and fortuitous ambiguity. When we talk about cars, we sometimes use the word to refer to technical standards, such as safety standards for airbags and other parts (a Google search for "automotive safety standard" gets over 42 million hits), but we also use the word to mean those features and technologies that come with the car when we purchase it. "Did that come standard?" we might ask of a friend's nice car stereo. In some ways, this second meaning of *standard*—the everyday, even assumed, shape that technologies take—forms an important basis of my story: how technologies travel from being special to being perfectly ordinary, what scholars call mundane technologies. As we will see, many parts of the car, from speedometers to electric self-starters, from energy-absorbing steering wheels to catalytic converters, from airbags to anti-lock brakes, have made this trip from being exceptional to being mundane. But it wasn't just the car itself that made this trip; many things around the car did too. When you look at a traffic light, you know that red means stop and greens mean go, even though this wasn't always the case. Road surface markings, street signs, traffic lights, and every other part of the automotive system had to be regularized and made dependable. Over the last 150 years, when we encountered hazards, our answer was to standardize the world.

Standardization efforts have gone beyond aiming at merely technical things to aiming at human beings themselves. Psychology tests, driver's education, crash test dummies—these and many other aspects of auto regulation assume that we can normalize human beings. In other words, we often try to standardize ourselves. Furthermore, attempts to decrease negative outcomes of automobile use often contained ideas about or perceptions of human nature and psychology, what scholars—somewhat pretentiously—refer to as *constructions of the subject*. Some auto regulation accounts of how humans work have been quite explicit, as in the case of the various reports and papers produced by the National Research Council's Committee on the Psychology of the Highway, which was formed in the early 1920s and existed in one form or another until the 1950s. Psychologists hoped to characterize the psychology of driving, to discover the limits of human capability, and to have engineers take these limits into account when they designed cars and roads. Other images of human nature were implicit, such as those of regulators who based safety standards on the assumption that drivers were fallible and unperfectible, or the images legislators and automakers had in mind in discussing how consumers thought of their vehicles. When we examine such pictures of human being, whether implicit or explicit, we see that attempts to create rules about technology are almost always ways of creating rules about one another.

⁓

This book is divided into four parts. The first, "Standards," examines attempts to deal with automotive safety from roughly 1900 to 1940 and serves as an overture for the rest of the book. The first chapter describes the rapidly changing circumstances around automobiles from the late nineteenth century, when cars were playthings of the rich, to the 1910s, when the mass production of automobiles dramatically transformed the safety situation and injuries and deaths mounted by the thousands. The emergence of this mass automobile culture took place within the context of deep and fundamental changes in American society, including the rise of big business, the increasing centrality of the professions, and the struggles of progressive political reform. The second chapter, "Standardization Is the Answer," examines the first national automotive safety movement, which emerged around 1920. Experts during this period attempted to make automobile use and many other social activities safer by promulgating a number of technical and other standards, including headlights, traffic signals, and driver's license tests. In some ways, these standardization efforts were highly successful, and they created, or at least reinforced, a tradition of seeing standards as the answer to public problems associated with automobiles. But they also had their limits, and deaths continued to increase on American roadways for decades.

The second and third parts of the book, "Safety" and "Pollution," mirror each other. Each part examines, first, the emergence of new sciences in the 1950s around the automotive hazards of crash safety and air pollution and early attempts to control these two problems; second, a major turning point in attempts to regulate these hazards in the 1960s, in each case culminating in an important new law; and third, how new regulations developed in the few years after these laws were passed as regulators tried to produce meaningful technological change in the auto industry. The parts end with reflections on why automotive safety and air pollution regulations played out so differently and had such widely varying effects on the development of new technologies.

In part 2, "Safety," chapter 3 describes the rise of impact biomechanics, the science that examines how physical forces affect human bodies and how to reduce forces and, therefore, injuries in accidents. The scientists, medical doctors, and others who gave birth to this science of crashworthiness did not see the automobile industry as an adversary or publicly call out the industry for making unsafe cars, however. Chapter 4, "From Movement to Government," examines a group of individuals, including Daniel Patrick Moynihan and Ralph Nader, who used the science of crashworthiness to push for government regulation of automobile safety. Their efforts ultimately led to the National Traffic and Motor Vehicle Safety Act, the landmark safety law that required all automobiles sold in the United States to conform to safety standards and created a new regulatory agency, the National Highway Safety Bureau (later renamed the National Highway Traffic Safety Administration). The law itself did not specify what those standards should be. Chapter 5 explores how regulators at the new agency attempted to define and set the first national safety standards. It finds that the auto industry fiercely re-sisted these efforts and in many ways won the battle, significantly weakening the proposed standards and setting a pattern of fairly weak US auto safety regulations, especially when compared to the accomplishments of automo-tive air pollution regulation.

Chapter 6 turns to the history of automotive air pollution to examine how experts discovered that automobiles caused photochemical smog and thereby launched legislators, policymakers, and others on a long, uncertain, mean-dering road in their attempts to control automobile emissions. By the late 1960s, legislators and citizens had grown sick of the auto industry's inaction on air pollution. Chapter 7 describes how Edmund Muskie developed the 1970 Clean Air Act Amendments, one of the toughest set of regulations in US history, and how an experienced and uncompromising civil servant, Eric Stork, came to head the Environmental Protection Agency's automotive pol-lution program. The eighth chapter explores a series of hearings on auto-motive emissions controls that the Environmental Protection Agency held in the early 1970s. Participants came to describe the hearings as a "seminar" that established the state of the art, formed new knowledge communities, and ultimately led to creation of the catalytic converter.

The book's last part, "Bureaucracy," describes the course of federal auto-mobile regulation since the establishment of the two primary regulatory agencies with jurisdiction over the issue, the National Highway Traffic Safety Administration and the Environmental Protection Agency. Chapter 9 exam-ines bureaucratic struggles that arose around the issue of fuel economy, in particular the Environmental Protection Agency's attempts to preserve cred-ibility and turf. Agency leaders and staff members developed their own in-terests and, in effect, became an interest group that influenced the agency's actions, though their choices were often meant to defend federal regulations themselves rather than fulfill personal self-interest. Chapter 10 recounts the rise of the federal deregulation movement of the late 1970s and early 1980s. The argument that civil servants and regulatory agencies acted in their own interest rather than for the public good undergirded at least part of the push for deregulation. Efforts at deregulation, particularly those made during the Reagan administration, sometimes had lasting consequences for automobile regulation, but just as often these efforts faced real limits, with subsequent presidents wholly undoing some of them. The final chapter describes the uneven, often pendulum-like history of automobile regulation since the late 1980s. Increasingly, the two major US political parties became polarized over the issue of regulation and the state's role in guiding the economy and technological change. The overall effect is one of indecision and uncertainty. Both parties tried to use government-funded research and development of technologies such as electric and hydrogen vehicles, but, as I argue, these efforts have been nowhere near as successful at generating technological change as regulation has. Finally, the chapter considers the case of automakers and other companies increasingly using electronics and computers, including the development of autonomous, or self-driving, vehicles, to examine the US government's uneven role in generating or influencing technological change in American automobiles.

In the book's conclusion, I put forward a general picture of the relationship between regulation and technological change and examine how this relationship changed over the course of four historical periods. The conclusion is much more social scientific in both tone and intent than the rest of the book, which is fundamentally a narrative history. In the conclusion, I draw together several bodies of historical and social scientific literature to produce a synthetic vision of technological regulation. Readers who have no interest in such theoretical reflections will, I hope, rest content with the chapters that precede the conclusion.

PART ONE

STANDARDS

CHAPTER ONE

The Auto World Gets Organized

In August 1902 an elite social club held a dinner on Long Island. No special business was planned for the night. Yet over the course of the evening, the influential men gathered there homed in on a theme—a new menace upending Oyster Bay's idyllic peace. Members stood and bemoaned "reckless operators of automobiles" damaging the "costly roads" around Oyster Bay. They "resolved informally to discourage the fast running of the machines." Oyster Bay had only recently constructed roads of macadam, a surface made of various sizes of crushed stone and often a petroleum-based binding agent, a new method of construction widely adopted in the closing decades of the nineteenth century. Local leaders hoped the roads would last for years. But experts claimed that drag racing could have rendered the roads into "masses of loose stone" by autumn and that only a rainy summer, which discouraged automobile driving, had prevented such an outcome.[1] The club's members resolved to create a new organization, the Highway Protective Society. Some locals protested that they did not want their tax dollars going to this endeavor, but the group hired traffic law enforcers on its own dime, in effect creating a private police force.

One of the club's directors was W. Emlen Roosevelt, whose cousin Teddy was president of the United States. Although the social club denied that Teddy played any part in its efforts, it "was believed to have his sympathy." Perhaps that was true. The president did not like automobiles. On the day he left the White House, Roosevelt reportedly refused to be driven away in a car, saying, "I came to the inauguration in this horse-drawn vehicle and I will leave in it."[2] More annoying for Roosevelt, his daughter Alice had taken to the new contraptions, ultimate symbols of wealth, from which Roosevelt distanced himself by cultivating a masculine, everyman image.[3] The car was associated with rich-boy racers, and Alice shook the gender norms of her day by driving unchaperoned with her female friends and setting speed records. For all kinds of reasons, then, automobiles may have given Teddy Roosevelt a headache. Yet, while Roosevelt was famous for his support of progressive and antimonopoly laws and regulations, he never considered directly regulating cars or their hazards. It was Emlen Roosevelt and others focused on shaping local conditions who took responsibility for curbing the problems of the automobile in the first years of the twentieth century.

This chapter examines efforts to control automotive hazards from 1893 to World War I. Nearly every aspect of automotive culture—cars themselves, the auto industry, the people using cars, and the individuals and groups trying to make them safer—went through profound changes during this period. American society was also going through its own significant transformations, including the advent of big business, the rise of professions, and the emergence of progressive reform efforts. An overview of these changes is essential for understanding the longer history of automobile regulation.

If one theme emerges, it is increasing organization. Early efforts concerning the automobile were ad hoc: automakers relied on craft methods of production involving little if any standardization. As a result, cars were expensive, often seen as playthings of the rich who terrorized local communities. Local and state governments, as well as voluntary associations like the Oyster Bay social club, attempted to address these problems, but their solution rarely took legislation in other states or localities into account, leading to a vast patchwork of rules and unspoken codes. All of this changed by the 1910s. Ford and other motor companies reorganized manufacturing to mass-produce vehicles. People other than the rich could afford cars, though mass ownership came with problems of its own, including ever-rising numbers of injuries and deaths. The auto industry also began to organize in new ways, as by vastly increasing the use of standardized parts, and responding to changes in liability law. Moreover, new institutions, such as the National Safety Council, emerged to tackle public hazards in unprecedented ways whose influence would be felt for decades. None of these developments aimed in any particular direction. They resulted from individuals and groups in specific situations trying to create solutions to problems as they saw them, feeling their way forward. Yet the changes they created came to form an essential bedrock for others' later efforts—organization built upon organization, regulation upon regulation.

The Emergence of the Automobile in the United States

By 1902 the automobile had come a long way from where it had begun a decade earlier. In 1893 Charles and Frank Duryea affixed an internal combustion engine to a horse carriage, making it the first vehicle of that kind in the United States. A year later, Elwood P. Haynes built a similar machine, and within a few years the feat of the Duryeas and Haynes had been repeated by Henry Ford, Hiram Percy Maxim, Ransom Olds, and Alexander Winton.[4] These first automobiles were truly *horseless carriages*. They had the high profile, the wooden frame and wheels, and the accoutrements of carriages, including gas lanterns for headlamps. Most were guided with tillers, not steering wheels.

The American automobile craze truly took off in 1895. In that year, two enthusiast magazines—*Motocycle* and *Horseless Age*—began publication in the United States, and the first automobile race was held in Chicago on Thanksgiving Day. Five cars started the 55-mile race. Only two finished. The Duryeas came in first. Races and the daredevil quest for speed became a constant fixture of early automotive culture. The Duryea brothers drew on the enthusiasm sparked by their win to begin producing automobiles for sale. In 1896 they opened their factory in Springfield, Massachusetts. The Duryea factory relied on labor-intensive methods of production that had long been used to build carriages and other contraptions. As a car was made, it stayed more or less in one place on the factory floor. Gangs of skilled workers moved from one vehicle to another. This way of making cars was slow and time-consuming. In its first year, the Duryea brothers' factory produced about a dozen automobiles, or an average of one per month.

Other automakers soon followed on the Duryeas' heels: Winton, Ford, Haynes-Apperson, Pope. Companies made electric and steam-powered autos in addition to those with internal combustion engines. By 1899 the periodical *Motor Age* estimated that one thousand shops around the United States were working on building cars and that "probably one hundred of them have been in operation for two years or longer without yet having advanced to the stage of manufacture, except in a very few instances."[5] The 1899 *United States Census of Manufactures* listed thirty automakers, which built about 2,500 vehicles that year.[6] By 1900 there were 8,000 automobiles registered in the United States, a staggering number considering how few cars had existed only five years earlier. But by 1901 that number had nearly doubled, jumping to 14,800. The industry boomed.

Figure 1.1. The Duryea brothers in one of their early automobiles, perhaps at the first North American auto race, in Chicago in 1895. Note the rudder steering and lack of headlamps. Their first horseless carriage had a top speed of about 7 miles per hour. National Automotive History Collection, Detroit Public Library

Early automobiles were *expensive*. Only the rich could afford them. After 1897, when the famous socialite O. H. P. Belmont imported a French car to Newport, Rhode Island, automobiling became a fad among the elites who summered there.[7] Harry Payne Whitney then filled an entire stable with European automobiles. Within the next year or two, their peers and friends— the Vanderbilts, the Astors, the Drexels—also imported vehicles. In the summer of 1899, local car owners took part in an automobile parade, a tradition also imported from France. They covered every inch of their cars with daisies, cat-o'-nine-tails, hydrangeas, and other vibrant flowers. Belmont's wife, Alva Vanderbilt, swung a "whip made of white and blue hydrangeas, with daisies" to spur on her horseless carriage.[8] As the wealthy car lovers drove through the crowd of onlookers gathered for the occasion, Belmont's servants transformed the lawn of his sixty-room summer cottage, Belcourt Castle, into an "obstacle park. Wooden horses attached to carriages, and

Figure 1.2. William K. Vanderbilt Jr. in the Steam Locomobile that won him first prize for speed in the Newport auto parade of 1899. The white objects on the flower-covered arbor just in front of and above Vanderbilt's head are stuffed doves. The Locomobile had gas headlamps, one of which can be seen to Vanderbilt's left. *Automobile Magazine* illustration, courtesy of Smithsonian Institution

sundry dummy figures representing policemen, nursemaids, and loungers, were scattered all over the place, after the manner of the French automobile driving academies, and among them all, a devious course was staked out by means of golf flags."[9] The crowd erupted with laughter when Belmont's car, which was topped with a bloom-covered arbor and a stuffed eagle, knocked over a few of the dummies. As night fell, each car, now adorned with garlands of electric "glow-lights," headed home. "Thus the procession of scintillating vehicles sped swiftly over the dark country roads, and at last glided into the sleeping town of Newport, like a veritable pageant of fairy chariots."[10] Automobiles were not for the common people. The average automobile cost $1,170 in 1903, when the median household income in the United States was below $500 a year.

Early Concerns about Safety and the Two Types
of Dangerous Drivers

Given the centrality of racing and the emphasis placed on speed in early automotive culture, it might seem surprising that the first auto-related fatality did not come until 1899. Nearly all of the cars sold during this period where open-bodied, and none of them had seatbelts, which would not be installed in automobiles for another seventy years. For drivers and passengers during those early days, to hit something was to be sent flying. But the cars' low top speeds meant that most crash victims were injured instead of killed. Henry H. Bliss was not so lucky. In September 1899 Bliss, a real estate broker, stepped from a streetcar at the corner of West 74th Street and Central Park West in Manhattan and was run over by an electric taxi cab. Bliss died the next morning, giving him the unlucky honor of being the first recorded person in North America killed in an automobile accident.

Opposition to and anxieties about the automobile emerged at about this time. In January 1900 the *Chicago Daily Tribune* wrote in an editorial titled "New Street Menace": "The collision between an automobile and a buggy on Jackson boulevard Wednesday as a result of the electric vehicle becoming uncontrollable suggests unwelcome possibilities."[11] Yet most conflicts during this period occurred between cars and horses, not cars and pedestrians in incidents that would come to dominate news coverage of the automobile by the 1920s. In the early 1900s, poems, images, jokes, and news stories featuring discord between horses and cars were so common that a cartoon in *Life* Magazine lampooned the genre. The cartoon pictured a horse-drawn wagon, flipped in a farm field next to a road, the vehicle's former occupants lying this way and that. "Did some automobile frighten them?" an onlooker asks. "No. They saw a cow," comes the response. The gag of the cartoon was that any such image would automatically lead readers to believe that it was commenting on the automobile.

During these early years, those concerned with automotive hazards basically worried about two types of dangerous drivers: the Auto Fiend and the chauffeur. These two popular images—which by no means represented every

Figure 1.3. This cartoon from *Life* Magazine satirically suggests how common images of this type were in popular culture. The cartoon's caption read: "'Did some automobile frighten them? No. They saw a cow.'" The image subverts common expectations that an automobilist would have caused the carnage. *Life* Magazine illustration, courtesy of Wisconsin Historical Society

driver on the road—reflected how car owners chose to use their vehicles.[12] Some owners chose to drive themselves; others had chauffeurs do the work. The Auto Fiend, an image common in popular culture during the first years of the twentieth century, primarily symbolized the earliest car owners: rich, young men who drove themselves because the thrill and danger of speed was, for them, the whole point of possessing an automobile. Books, newspapers, and magazines of the day featured jokes, stories, and other pieces of writing about the Auto Fiend, including a 1911 poem by Ernest Bowden:

> Look at the farmer! What's wrong with his horse?
> He or the creature is crazy, of course.
> Puppies and chickens, old women and brats—
> Send them all flying! And death to the cats!
> We are the people. Five thousand we've paid,
> For the right to make horses and women afraid.
> So away we go speeding, adventures to find,
> And leave all the fear-stricken mortals behind.[13]

THE AUTO FIEND
I'm the king of the highway and street,
And I scare every horse that I meet.
My chauffeur is in jail
When he isn't on bail,
But no cop can scare me off his beat.

Figure 1.4. The Auto Fiend dressed in the riding gear of the day—goggles, gloves, and a cap—was a central symbol of early automobility. The gasoline can was also a ubiquitous symbol of the automobile at a time in which there were no gas stations. This image also contains a representation of speed records, "2487 Miles Per Hour," and shows a trail of dead and injured humans and nonhuman animals, including a dog and a duck, that the fiend has left behind. He holds an accident register: 21 dead, 74 injured. Warshaw Collection, National Museum of American History, Washington, DC

The word *fiend*, which had earlier been applied to bicycle fanatics, has Germanic roots. It already had multiple connotations in Old English, referring variously to supernatural demons, human enemies or foes, or persons of "superhuman wickedness." In the United States the automobile arose in a society with a traditional metaphysical and moral worldview that included evil forces and people who fell prey to those forces. The young wealthy men who took to the car loved the thrills of their devil wagons too much and cared nothing for the lower classes they left in their wake.

In some ways, the problem was reversed when it came to chauffeurs. The word *chauffeur* was slippery during this period: it was often used as we use it, to refer to professionals drivers, but sometimes it was used as a catchall for drivers in general. The word's meaning eventually narrowed solely to professional drivers. Regardless, many wealthy car owners did employ chauffeurs as we think of them today. Cars designed to be driven by chauffeurs often reflected the class division of labor seen with horse-drawn coaches. The owners and other passengers would ride in plush, well-appointed, enclosed cabins, while the chauffeurs would sit on platforms open to the elements. Some of these vehicles included candles or oil lamps in the interior compartment so that the wealthy occupants could see one another while conversing on the way to and from evening gatherings. To bystanders watching such illuminated autos, it would have seemed as if a lighted living room was rolling by.

But who were chauffeurs? The answer to this question created the class tensions that led to chauffeurs' ill repute. Because chauffeurs were typically required to care for and repair automobiles, liveries, coachmen, and stable hands often lacked the skills necessary to fill that role. Traditional, trusted, well-known servants to the upper crust could not do it. Programs were opened in large northeastern cities, including Boston and New York, to train coachmen to be auto chauffeurs.[14] Yet few likely made it through these programs, and demand for such trained drivers outstripped supply. For that reason, wealthy owners had to rely on others. Chauffeurs invariably came from classes lower than those they drove.

In 1906, after two deadly accidents near Flushing, New York, the tony magazine *Town & Country* called for "the licensing and education of chauffeurs." The problem, the magazine complained, was that chauffeurs were paid too well, leading to all manners of lower-class people, "even bell-boys in hotels," to desire the job. "It takes an engineer on a railroad some time to qualify, and he must be a thorough mechanic, besides a man of good habits." The "automobile schools" that popped up around New York's Central Park, however, soon graduated new chauffeurs—"in a few weeks they are sent out with full responsibility to drive an engine." For the elite tastes of the day, according to *Town & Country*, these schools had become hotbeds of untrustworthy people. "The 'pupils' are not reassuring looking individuals. All day long they 'practice' with small automobiles, darting out of dark gateways, suddenly upon pedestrians without warning-bell or honk, make a speedway of the block, run hither and thither, and [are] quite oblivious to the rules of the road."[15] Concerned citizens demanded that governments regulate these schools and raise the bar for licensing.

The lack of trust in drivers and the sense of chaos that accompanied their machines played out in both the city and the country. Locals had time-honored traditions—and a habituated sense—for proper conduct on roads and in other public spaces. The habits of auto drivers often ran afoul of these traditions. Drivers' love of scorching down country lanes led to well-known tensions with farmers, which arose even in communities that had never before seen an automobile. The cars' reputation preceded it. One county clerk in Missouri wrote to the Automobile Club of St. Louis, "Gasconade Country sports neither a single automobile nor flying machine. Don't bring 'em here to scare our cows and horses and run over our dogs, chickens, and sech. The first 'honck' echoing over our peaceful hills will bring out a mountain howitzer at every crossroad. I therefore give you the danger signal—Honk! Honk!"[16]

Urban automobile problems were different than those in the countryside, though city dwellers certainly saw their fair share of speeding and scorching. When the automobile entered the city, it was thrown into a variegated mixture of existing modes of travel—horse-drawn carriages, wagons, and trams; electric street cars; bicycles; and, perhaps most numerous of all, people traveling on foot—all of which moved about with little or no rules. A film taken from the front of a streetcar in San Francisco in 1906 gives us some sense of the chaos that marked city streets.[17] Pedestrians crossed the street wherever and whenever they wanted to. Fast-moving automobiles weaved in and out of traffic, in front of heavily loaded horse wagons and slow-to-stop streetcars. One gets the sense that at any moment anything could go wrong, and often it did.

The First Safety Solutions: Traffic Regulation and Aftermarket Fixes

The first attempts to address automotive accidents and concerns about safety were fundamentally local, ad hoc affairs, involving little coordination between jurisdictions or between formal organizations. The most basic safety efforts during this period were the rise of traffic regulation and the emergence of aftermarket safety devices. Most important, the people who put forward these solutions were not trained experts in any meaningful sense.

Traffic reformers proposed rules that were little more than organized commonsense, and the automotive aftermarket was a riot of gadgets and devices created by independent inventors who had little connection to formal engineering societies.

Public officials responded to early automotive anxieties usually by enacting speed laws. Between 1899 and 1903, hundreds of state and local governments across the United States passed legislation affecting the automobile.[18] Some of these laws set speed limits as low as 5 miles per hour. Just as today, traffic rules were based solely at the state and local levels; the federal government was not involved, though some called for it to step in.[19] The person who was most responsible for changing the ruleless street culture was William Phelps Eno. Eno was born in 1858 into a wealthy Manhattan family. His father had an office on lower Broadway, which, Eno recalled, "must have been somewhere below the City Hall and near P. T. Barnum's Museum." From Eno's perspective, Barnum's carnivalesque spirit seemed to rule New York's streets. As a child, he witnessed a "traffic jam on Broadway" that "took at least half an hour to get out of." Most exasperatingly, the foolish moves of just a few vehicles could halt traffic completely "since neither drivers nor police, if there were any around, knew anything about the control of traffic."[20] Peregrination characterized Eno's education. He attended at least "fifteen schools," and he spent considerable time in Europe traveling with his family. After dropping out of college, he began working for his father's business, where he remained until 1898.

In that year, Eno turned his attention from business to regulating traffic. In some ways, he fit the progressive model of the socially engaged citizen of means. Like Jane Addams and other progressives, Eno sought to reform the city. In 1903 New York City adopted Eno's "Rules of the Road." The rules required slow-moving traffic to keep to the right, set a speed limit of 10 miles per hour, and most famously, mandated that drivers must make left turns by driving around the center point of the intersection, rather than cutting toward the near corner.[21] For several years, Eno himself paid to have the brochures printed listing the city's traffic laws.

While Eno's rules, especially the rule about left-hand turns, were adopted in several major cities around the world—including Paris—traffic regulation remained a basically local phenomenon during the first two decades of the twentieth century. Even when Eno's rules were adopted, they were often changed slightly. Eno hypothesized that such innovations were introduced into traffic codes because it enabled "somebody to get some political glory

out of having his name attached to it."[22] Early traffic laws thus reflected local color and preoccupations. For instance, beyond banning drunken driving, the Nashville traffic ordinance of 1914 mandated that no chauffeur "shall
use obscene or profane language while waiting on the streets or in the presence of his passengers."[23] This variety of traffic codes eventually created
great difficulty, as automobilists drove from one place to another with no
knowledge of local laws.

   Automobile lovers did not sit idly by watching these developments. During
the car's first decades, automobile clubs and associations sprang up around
the country to defend the interests of auto owners and users. The first
club, the American Motor League, began in late 1895. More influential organizations, including the Chicago Automobile Club and the New York–based
Automobile Club of America, did not form until 1899.[24] These groups were
exclusive, and their members were *rich*, including such luminaries as Samuel Insull, Alfred G. Vanderbilt, and Henry Clay Frick. *Motor Age* described
the Automobile Club of America as "an ultra-fashionable coterie of million-
aires, who have taken up the new and expensive fad of auto-locomotion and
banded themselves together for its pursuit and the incidental notoriety attributed to all the functions of upper swelldom."[25]

   The auto clubs and periodicals closely followed pending and enacted legislation and civil and criminal lawsuits involving the automobile. The penalties that drivers faced and their severity depended largely on local laws and
judges' beliefs and attitudes, which varied widely.[26] On occasion, the Automobile Club of America and other such organizations would step in to pay
the legal fees for defendants. Some advocates for the automobile originally
tried to claim that speed laws were an unconstitutional form of "class legislation," laws that applied to one group but others and, thus, were discrimi-
natory. In this radical vision, governments would not have been able to create
laws of any kind to regulate the automobile. The argument did not hold
water, however, and by 1907 even automobile journals were explaining to
their readers that governments could set speed limits.[27] With the class legislation argument set aside, partisans for the automobile had no option but
to fight for the most hospitable and permissive laws possible.

   From the beginning, new laws shaped the automobile and its attendant
technologies, starting perhaps with a device so basic that we can hardly
imagine a car without it—the speedometer. The first automobiles built in the
United States did not come equipped with speedometers. The 1904 Oldsmobile Curved Dash—made famous by the sexually suggestive tune "In My

Merry Oldsmobile"—did not have a speedometer, for instance; nor did it have anything resembling an instrument panel. Early drivers, if they wanted such devices, had to buy them on the automotive aftermarket, from the catalogs and specialty stores that sprouted up in the first years of the twentieth century. We have largely forgotten that drivers' desires for speedometers in the first decade of the twentieth century were driven by the enactment of speed laws. Speedometer firms competed with one another to make and sell the premiere devices. Advertisements for aftermarket speedometers played on drivers' fears of the police, laws, and speed limits. Indeed, advertisers even suggested that speedometers provided a means for challenging police officers who used pocket watches to "clock" the speed of cars. "The Mounted Officer with his primitive stop watch method of timing the Automobilist can seldom make a case. Guess work testimony doesn't go in court. The Officer means well, but his *method* is wrong," one ad suggested, playing up to auto owners' self-images as modern. Within a few years, the ubiquity of speed laws made speedometers a *standard* part of nearly all new cars. Changing laws, even of the most local variety, and changing technologies were deeply entwined from the start.

Attending to the conflict between early automobilists and others can make it seem as if car users did not care about safety; but of course they often did. Drivers wanted not only to avoid lawsuits and jail time but also to keep out of harm's way. For this reason, a thriving automotive aftermarket opened for products promising safety.[28] Sometimes the products on offer dealt with risks quite remote from today's worries. For instance, one catalog marketed something called the U.S.A. Liquid Pistol, a metal squirt gun. At first blush, the U.S.A. Liquid Pistol may not seem a safety technology, but the ad copy noted that it was "very effective in keeping off dogs."[29] If you were driving an open-bodied car that lacked doors and had a top speed of 8 miles an hour, angry dogs could be a real problem. Therefore, when we think about early auto safety—whether from the perspective of drivers, pedestrians, or policymakers—we must keep in mind that they did not share all of our concerns.

Early cars were stripped-down affairs that had few of the accoutrements that we expect in a car—no mirrors, no turn signals or brake lights, often no horn. Many did not come with a windshield. All of these attachments had to be purchased on the aftermarket and added to automobiles by owners. Early auto users accessorized for safety and customized for comfort. At best, the first cars had gas (acetylene) headlamps. The first aftermarket electric headlights were expensive and ran off batteries that were not recharged by the

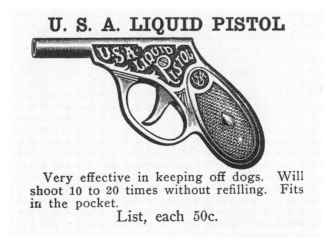

U. S. A. LIQUID PISTOL

Very effective in keeping off dogs.   Will
shoot 10 to 20 times without refilling.   Fits
in the pocket.
List, each 50c.

Figure 1.5. The U.S.A. Liquid Pistol was a metal squirt gun marketed in early automotive aftermarket catalogs. Drivers could use the gun to ward off angry dogs, a problem for riders in open-bodied automobiles with low top speeds. Some suggest that users filled the pistols with liquid acid. The meaning of *safety* has changed over time. Ballou-Wright Automobile Supplies Catalog, 1906. Courtesy of Oregon Historical Society, Portland.

engine, needed to be regularly recharged at home, and required a great deal of upkeep. The cars' pedals and running boards were often naked metal. If you wanted rubber covers on them to increase traction and decrease the risk of slips, you had to buy and put them on yourself.

Bumpers and fenders were also primarily aftermarket products that did not come standard on most vehicles. The first bumpers were marketed as safety devices. An advertisement for the Hercules Auto Bumper claimed, "It protects your lamps, radiators, and sometimes dangerous injuries. If you happen to hit a pedestrian, he will grab the fender before going down, very often saving his or her life; thus this device often prevents expensive lawsuits."[30] The front wheels on early automobiles did not come back to center on their own after turns as modern cars do; they had to be turned back by hand.[31] One aftermarket product, the Little Steersman, consisted of a long spring attached to car's undercarriage and each of the front wheels. It helped the vehicle right itself to center after each turn. The brakes that came standard on many automobiles were of poor quality and would burn out after one or two uses. They were not hydraulic, so they depended wholly on the driver's strength. The aftermarket developed some solutions to this problem as well. For instance, the Herz Safety Switch allowed drivers to cut the

HERCULES AUTO BUMPER

MAY SAVE YOUR CAR

Backed up by spiral springs to absorb shocks and collisions. Made of Shelby seamless steel tubing 3/16 in. diameter. It protects your lamps, radiators, and sometimes dangerous injuries. If you happen to hit a pedestrian, he will grab the fender before going down, very often saving his or her life; thus this device often prevents expensive lawsuits. It fits any car, arms being furnished in various lengths.

When ordering, state make, model, year and style of your car.

			Price	Our Price
No. 9007.	Plain finished.		$15.00	$4.00
No. 9008.	Brass polished.		20.00	4.50

Figure 1.6. Early automobiles did not come equipped with basic features, not even bumpers, which had to be purchased on the aftermarket. Bumpers were often meant to decrease property damage in mild collisions, but they also served as safety devices. Advertisers proclaimed that the Hercules Auto Bumper "sometimes" prevented "dangerous injuries." 35 Percent Automobile Supply Co. catalog (New York and Chicago, 1912). Courtesy of Division of Work and Industry, National Museum of American History, Smithsonian Institution, Washington, DC.

engine at the push of a button during emergencies, turning the engine itself into a brake.[32]

Focusing on the automotive aftermarket leads to several insights about the nature of technological change during this period. First, as the form of the automobile changed, some problems that the aftermarket attempted to solve went away. As more and more automobiles had closed bodies and the average speed increased, one no longer needed a squirt gun to keep off dogs. Similarly, for various reasons discussed later, auto brakes improved, rendering devices like the Herz Safety Switch mostly superfluous. Yet, at the same time, other changes in automobiles introduced new hazards. For instance, closed bodies created blind spots and made it harder for drivers to hear.[33] Mirrors were largely unnecessary on open-bodied cars, but an early advertisement for the aftermarket Owl Mirror, which could be added to the side of any vehicle, claimed to be "an absolute necessity for cars having large tops or closed bodies." Hazards were changing, unstable things.

In addition, many of the devices that could originally be obtained only through the aftermarket eventually became standard, assumed parts of new cars. It would be too much to say that the aftermarket was the core mechanism by which automakers came to know consumers' desires, but it is reasonable to assume that it played this role to some degree. At some point, car owners did not have to turn to the aftermarket for such features as windshields, rubberized pedals, speedometers, fenders, and mirrors. They came with the car, which soon became nearly unthinkable without them.

The aftermarket played one other role that had important implications for auto safety during the early period, namely, the improvement of driver and passenger comfort. Auto safety has always had a complex relationship with comfort. At the very least, discomfort can lead to distraction, which can have devastating consequences. By the 1920s, professional psychologists were hypothesizing that discomfort was leading to accidents. Others worried that if cars became too comfortable, they might encourage drivers to fall asleep, leading to other problems. Automotive catalogs teamed with goggles, masks, and veils to keep rain, dust, and insects out of one's eyes, mouth, and nose. They also contained many forms of automotive clothing, including heavy leather robes and jackets, the latter differing little from the jackets favored by motorcyclists in the post–World War II period.

Yet the quest for comfort could create its own hazards. For instance, early automobiles were used primarily in the warm months from spring to fall, though some brave automobilists did venture out in the winter.[34] For those who chose to go out in the cold, catalogs marketed auto robes, blankets backed with heavy leather and worn over the laps of drivers and passengers. But these robes could become tangled around control pedals, leading drivers to lose control of their vehicle. For this reason, the Burlington company marketed the "'Sho-fur' Automobile Robe" (its name a play on *chauffeur*), which contained a hole for drivers to put their foot through to safely maintain unimpeded contact with the pedals. For almost any need imaginable, the automobile aftermarket offered a solution.

Automotive clothing also affected authorities' ability to police driver behavior. In 1908 the jurist Herbert Billings Brown, who had stepped down from the US Supreme Court two years earlier, published an article in the *Yale Law Journal* titled "The Status of the Automobile," which summarized the automobile's place in law to that point. Brown was then in his seventies, and he had no love for the new machines. For users, Brown believed, automobiles "exhibited only their attractive features; to the outsider, only their repulsive ones."[35] The question was how to get control of drivers, and here Brown saw automotive clothing playing a negative role. As he wrote, speed laws had often "been found impossible to enforce" because "many of these machines largely resemble each other in construction and color; the faces of those who occupy or control them are often covered by masks, goggles, or veils; and they pass with such rapidity that it is impossible to discern their [license plate] numbers."[36] Brown leaves a strong impression of what automobiles were like for average observers and pedestrians: terrifying and puzzling.

Fordism and the Rise of Mass Automobility

The world of the automobile I have described so far was one of great inequality involving a division between wealthy, early adopters of the car and the rest of society. Nothing did more to undermine this world and create another than falling prices. With falling prices—and eventually the creation of consumer credit—came the opportunity for more-ordinary citizens to buy cars, which paved the way to mass automobile ownership. This trend was only reinforced by the emergence of the used car market, which promised even lower costs. No one did as much to create this new world as Henry Ford and the men who surrounded him. There are many biographies of Ford, and many more works that touch on the history of the Ford Motor Company. Only a brief summary is needed here to provide an understanding of the dynamics at Ford and the changes that followed.

The Ford Motor Company was founded in 1903. While early auto firms were spread throughout New England, the Midwest, and the mid-Atlantic, Ford followed a growing trend by locating in Detroit, Michigan, which had access to inexpensive natural resources, especially wood for the auto bodies. By the 1910s, Detroit had become the auto capital of the world. Ford began work almost solely as an automobile assembly business. At first, the company operated much like the Duryea brothers' factory had seven years earlier: teams of men moved from stationary setup to stationary setup, with parts being brought to each automobile under construction. In 1904 the firm built 1,745 automobiles, or only about a third of what the Olds auto works built that same year.[37]

By 1906 Henry Ford began to envision a new automobile, "a light, low-priced car with an up-to-date engine of ample horsepower, and built of the very best material," a car that would become the Model T. Ford's vision of an inexpensive automobile challenged the reigning notion of who automobiles were manufactured for. Because early automakers were mostly selling to wealthy customers, they had little pressure to decrease prices. They competed on quality and luxury. Indeed, from 1903 to 1905 the price of the average automobile *rose* from $1,170 to $1,784.[38] In 1906, the same year that Ford announced his idea, the president of Princeton University and future president of the United States, Woodrow Wilson, lamented, "Possession of a motor car is such an ostentatious display of wealth that it will stimulate socialism." The notion of a car for common people emerged as the US economy experienced a major transformation, the sharp rise of what would later be called consumerism.

Henry Ford was aided in his quest by several talented mechanics and machine tool experts who had spent their formative years working at firms like the Singer Manufacturing Company and International Harvester. Under the tutelage of Ford and his mechanics, the company created an ethos of constant improvement. In 1908, the company produced about 6,000 Model Ts, less than the 6,500 that Olds Motor Company had produced in 1905.[39] But Ford production improved and expanded constantly. The next year it nearly doubled its output, and by 1912 it made 78,611 Model Ts.[40] In 1913 the company instituted a truly momentous change by introducing the assembly line. That year, production more than doubled again—to more than 180,000 Model Ts—and it only expanded further from there. All the while, Ford dropped the price of the Model T. When the car was first introduced in 1908, it cost $850, but by 1916 it had fallen to $360. The company's style of manufacturing, which became known as Fordism, focused on three factors: the creation of interchangeable parts; the use of specialized machine tools, which enabled the company to increase output and "deskill" work; and reliance on the assembly line. Ford was extremely open about its production practices, which were written about and captured in pictures in magazines and journals, and within a few years, the practices had spread throughout the automobile industry and beyond.

Ford's Model T changed many aspects of the automobile. But in terms of the introduction of new automotive hazards, nothing about the Model T changed the situation so much as the sheer numbers of the things rolling out of the Ford plants. When the Model T was introduced in 1908, there were about 190,000 automobiles registered in the United States. In 1913 the company produced nearly that many cars on its own, and by 1916 it made *three times* that number. Of course, Ford was not alone; other companies were also greatly expanding output. In 1916 there were more than 3.6 million registered automobiles. The rise of mass automobile ownership fundamentally shifted debates about the vehicle's hazards. No more could automobile accidents be blamed on isolated Auto Fiends and crazed chauffeurs; if common people were to own automobiles, accidents would be a common problem. In 1905, 252 people died from auto-related accidents.[41] By 1916 that number had jumped to 8,200.[42] Consequently, the solutions that safety advocates pushed for after 1910 differed from those proposed from 1900 to 1910.

Examining a photograph of a Model T imparts other important insights about the nature of the automobile. Each Model T began as a chassis that moved down the assembly line, along which workers added all of its parts, systems, and devices. To look closely at a Model T was to see an automobile for what it was: a jumble of parts and devices with different functions, bolted together. In the late 1920s, stylists at General Motors and other firms would begin to smooth the automobile out, to pull the various devices into the body, to give the vehicle some sense of a planned, coherent whole. But the

Figure 1.7. A 1913 Ford Model T. Note the headlamp near the driver's door, reflecting a design tradition handed down from carriage manufacture, and reminiscent of the lamp on William K. Vanderbilt's Locomobile (see the photograph reproduced above). Division of Work and Industry, National Museum of American History, Smithsonian Institution, Washington, DC.

fundamental production question always remained the same: what parts and devices would automakers add to cars as they made their trip from the beginning of the assembly line to the end? How the automakers chose to answer this question had enormous ramifications for what hazards developed, as well as for many other aspects of the automotive world.

At least one other insight can be gained from contemplating the Model T. For many years, the common wisdom has been that "safety doesn't sell," and historians have asserted that safety was not a part of automakers' marketing. As we will see, neither of these claims is true. Perhaps it is more accurate to say that people both before and after World War II had different notions of the concept of automotive safety itself. The automakers in the earlier period did not sell the latter conception of safety, but they did try to sell some conception of it. Indeed, the selling of safety extended back to the industry's birth.

One of the novelties Henry Ford introduced in the Model T was the use of an alloy, known as vanadium steel, that was both lighter and stronger than other steels on the market. As a result, the Model T was considerably lighter than other automobiles on the market, especially the large, heavy luxury vehicles favored by some rich owners. The Ford Motor Company published a magazine, the *Ford Times*, to inform Ford dealers about new developments and to suggest sales pitches. Several articles in the magazine focused on vanadium steel as a selling point, and some emphasized its role in creating a safer automobile. They explained that heavier cars had more momentum than lighter ones and that the Model T could be "brought to a dead stop in one half the distance required for the ordinary car weighing double. When it's a question of life or death a quick stop is appreciated." The article marked an imaginary tally board: "That's safety point number one in favor of the light weight Model T." The materials excess in heavy cars also increased the bulk that "racks and ruins a car, sheers off bolts and generally subjects its passengers to danger," whereas lighter cars did not have such a "rough jarring" ride. "That's safety point number two in favor of the light weight T," the article proclaimed. It further argued that salesmen should make these safety qualities a part of their pitches: "If you've a customer who argues for heavy cars on the basis of greater comfort, durability, and safety, get him to ride in a Model T—select a hilly rough road and compare the results with those of a similar ride in any heavy car." From the very beginning, safety was one quality among many that automakers sold.

While the Fordist production of the Model T provided a paragon and guiding metaphor for truly mass production, the workings of Ford Motor Company alone cannot explain all of the dynamics of technological change in the automobile during this early period. Other firms played essential but different roles. One example is the Cadillac Motor Company, which became a part of General Motors in 1909. From the beginning, Cadillac was a manufacturer of luxury vehicles. In 1906 it became the first automaker to mass-produce closed-bodied vehicles. From this position as one of the key American luxury automakers, the company embodied a dynamic that to some degree still holds true: new devices often become a standard part of automobiles in the luxury market first. One famous example of this dynamic involves the electric self-starter. Until the 1910s, all vehicles were started with a hand crank, a dangerous act that often led to injury. The Olds Motor Company joke book, *Goop Talk*, defined "back fire" as a "disease of the automobiles similar to epileptic fit and caused by spark getting previous. Symptoms: Broken finger or sprained wrist and frequent use of word 'Damn' and similar expressions." Broken wrists were not uncommon when cranking went awry.

According to the mythology surrounding the invention of the self-starter, the head of Cadillac, Henry M. Leland, came to see the need for a self-starting engine after one of his employees died from complications caused by a cranking injury. Leland approached the mechanical genius Charles Kettering, an employee at National Cash Register, who earlier had invented the electric cash register and would go on to found the first automotive research and development lab, before working for General Motors. Kettering used some of the same basic principles inherent in his electric cash register to create the electric self-starter. The self-starter was introduced in the 1912 model year Cadillac cars, as were electric headlamps. The self-starter led to the general electrification of automobile functions as more and more devices were added that drew on the engine's electric current. The self-starter, electric headlights, and other novelties eventually diffused to other automakers. It was not until 1915 that Ford put electric head lamps on the Model T, and even then, its rear and side lamps were still lit with acetylene. Since that time, most innovations, including those that make cars safer, appear first in luxury cars and then spread out to other lines, until they become a part of the status quo.

The Emergence of Formal Organizations around Automobiles and Safety

As the automobile industry and automobile production matured, influential actors in the industry increasingly created formal organizations to serve automakers' interests. For instance, in 1911, representatives of the auto industry formed the Automobile Board of Trade, which was renamed the National Automobile Chamber of Commerce in 1913. The activities this organization conducted included political lobbying and disseminating industry statistics in a publication long known as *Automobile Facts and Figures*. Similarly, in the early 1910s, the Society of Automotive Engineers, which had formed in 1905, began important work in the creation of automotive technical standards.[43] Engineers had been forming societies and creating standards for decades before the practice took hold in the automobile industry. Earlier examples include the American Society of Mechanical Engineers, founded in 1880, and the American Society of Testing and Materials, founded in 1898.[44]

For centuries before the rise of these kinds of organizations, individuals had conceived and fashioned technological devices in local blacksmith shops and factories to suit local needs and tastes. Engineering societies, by contrast, were founded on the principle of sharing and aggregating best practices as well as state-of-the-art knowledge. Technical standards increased productivity and efficiency by enabling interoperability and the reduction of waste. The paradigmatic case of standardization in the United States was William Sellers's creation in 1864 of the screw thread standard, which guaranteed compatibility between nuts, bolts, and other threaded parts.

In 1910, spurred by a brief economic recession in auto sales, the Society of Automotive Engineers (SAE), established its Standards Committee.[45] Most of the committee's early work was dedicated to the standardization of metal parts, but the SAE also began to develop methods of destructive testing, techniques used to see how strong and resilient parts were. Destructive testing had a long history, extending back to the railroad industry.[46] Its entrance into the practices of automakers both brought them in line with wider engineering culture and had significant ramifications for the future development of automotive regulation. The historian George V. Thompson assessed the impressive acts of simplification and standardization the Society of Automotive Engineers accomplished in the early 1910s: "Thirty-five standard lock washers were selected out of more than three hundred designs previously used by industry. Standards on steel tubing reduced varieties from 1,100 to 150."[47] These efforts would only continue over the coming decades as automakers came to standardize nearly every part of the vehicle—an act that was logically necessary for car owners and mechanics to buy replacement parts through catalogs and auto supply shops.

The rise of Fordist mass production and automotive standardization were accompanied by shifting notions of safety, especially ideas about how to increase safety through coordination and organization. To some degree, Fordism and what became known as the "safety movement" shared a common heritage: they started as ways to improve and rationalize factory practices. In 1912 leaders from several industrial firms formed the National Industrial Safety Council, an organization dedicated to improving workplace safety.[48] The council's founding was at least partly motivated by the recent passage of workers' compensation laws around the United States, after years of pressure from labor organizers and others who were sick of workers receiving no help from employers after experiencing career-ending injuries.[49] Since these laws put additional financial pressures on companies for every worker injured, it made sense for them to find new ways of reducing workplace hazards. The safety movement in industry was grounded in training and in communication through posters, pamphlets, and the repetition of mottos, such as "Safety First," all of which was meant to engender a safety mindset in workers, foremen, and managers.

Within a few years, the National Industrial Safety Council had changed its name to the National Safety Council (NSC), and along with this name change came a broader mission. No longer content with focusing on workplace safety, the leaders of the NSC realized that the same logic that underlay all of their efforts could be applied to solving public problems. The NSC's methods, born on the workshop floor, walked out of factory gates to be applied to dangers everywhere—the streets, schools, playgrounds, amusement parks. While the NSC broadened its purview to deal with many hazards, its efforts often focused on the automobile.

One example of the council's public outreach was the children's book *Sure Pop and the Safety Scouts*.[50] The book featured the story of Sure Pop, a pigmy, a magical being from "Borderland" who resembled a leprechaun. "Sure Pop," a phrase with roots in the mid-nineteenth century, was used to declare certainty, most often in response to a question. Sure Pop's very name referred to the certitude, science, and predictability that progressive safety advocates sought. Through a series of adventures, Sure Pop taught children to avoid hazards, including street cars, household poisons, horseplay, fire, and electric wires. Two of the stories and several of the illustrations in *Sure Pop* featured the hazards of automobiles, especially from children's point of view as they attempted to cross or play in the street. Sure Pop, a kind of seductive and mythical stand-in for the safety experts who helmed the National Safety Council, believed that by teaching safety to children and making them safety conscious he would "be solving the whole problem [of safety] for

Figure 1.8. A youth narrowly misses being run over by an automobile. Illustration in the 1917 National Safety Council publication *Sure Pop and the Safety Scouts,* dedicated to "Six Timely Tips on Street Safety."

years to come." This vision would inform public and auto safety efforts for the following half-century. From the 1910s through the 1950s and even beyond, the National Safety Council was one of the most influential organizations in attempts to make streets safer and decrease automotive risk.

As the automobile entered a complex physical environment comprising horses, street cars, pedestrians, and cyclists, it also entered a complex legal world involving tradition, precedent, and often-conflicting laws. By the late 1910s, jurists began turning out legal treatises on the topic, and in 1911, Charles J. Babbitt, a Massachusetts-based lawyer, published a book on automobiles and the law that ran to over a thousand pages. Babbitt, like others, struggled to apply precedents in older laws and court decisions to the automobile by analogy. How was the automobile like a privately owned horse-drawn vehicle? How was it like an electric streetcar? While precedents sometimes worked to settle disagreements, court decisions involving the automobile also restructured the legal environment of the day, particularly in liability law.

In May 1911 a New York stonecutter named Donald C. MacPherson was driving his Buick runabout near Saratoga Springs when the car's left-rear wheel shattered catastrophically, flipping the vehicle on top of him, tearing his right eye apart, rendering it blind, and breaking his arm. MacPherson sued the Buick Motor Company, but liability law at the time held that he could bring suit only against the person or company that had sold him the product, which in this case would have been the car dealer. The lawsuit began an extended struggle that was resolved five years later when the case came to the New York Court of Appeals and Judge Benjamin Cardozo ruled in MacPherson's favor. At the heart of Cardozo's decision was the assertion that Buick "was not at liberty to put the finished product on the market without subjecting the component parts to ordinary and simple tests."[51] For the automakers, the most significant outcome of this ruling was an increased focus on materials testing and product inspection. Much of this effort fell on the shoulders of the Society of Automotive Engineers. In a sense, the *MacPherson* decision brought together the standardization efforts of engineers with the emerging safety movement: the engineers developed practices to ensure the production of safe, quality products.

In the immediate wake of the *MacPherson* decision, the Society of Automotive Engineers did not necessarily create any new technical standards, but its members were, at that moment, busy developing a literature on testing and inspection.[52] Perhaps they were motivated by the changing legal environment. Perhaps not. It may not have been wholly coincidental that in 1917, the year after Cardozo handed down the *MacPherson* decision, SAE president Russell Huff gave a speech at a society meeting in which he recommended formulas for determining steel composition and design, especially for "front axels, front-wheel spindles, propeller shafts, clutch-shafts, transmission drive-shafts, and rear-axle drive shafts," the breaking of some of which would lead to precisely the kind of accident that Donald MacPherson had experienced.[53] By 1919 E. H. Belden, the chief engineer of Willys-Overland Motor Company, was arguing that "the time has come for Government inspection of the materials used in car construction and their treatment." He went on: "A car designed to weigh 1800 lbs. should not be allowed to travel above a certain speed unless suitable materials properly heat-treated have been used in its construction."[54]

Yet *MacPherson v. Buick* also had a broader impact. As the historian of law Morton J. Horowitz has written, "The dramatic point of *MacPherson*, then, was to overthrow the traditional private law conception of duty in which one generally owed an obligation only to someone who was not a stranger."[55] Before around 1900, automakers had tailor-made cars for individual buyers, but thereafter they increasingly produced vehicles that had no known purchaser. This trend came to its highest fulfillment with the advent of Fordism and mass production. By the 1910s, people came to live in a world of things *made by strangers.* Cardozo's ruling in *MacPherson* held that these objects should be things that consumers could trust, and when they weren't, the buyer had every right to seek recourse from the maker.

By 1916, the year before the United States entered the First World War, the status of the automobile had changed considerably from its roots in the mid-1890s. The automobile had begun as thing for the rich, but by the 1910s, its appeal had broadened, as had its attainability. Henry Ford especially focused on cars' potential usefulness to the average farmer, a character he idealized as the foundation of American virtue. Farmers took Ford up on his offer and bought vehicles when they could. The auto's appeal to farmers was one sign of a general shift in American culture. For years, at least since the 1870s, the Good Roads Movement had languished, unable to secure the major changes in road construction its members desired. As one historian observed in the mid-twentieth century, "A primitive tax system, the doctrine of states' rights, the burden of expense which would fall upon states and counties for building roads through sparsely settled territories, and public indifference toward the whole question kept us stuck in the mud until well into the twentieth century."[56] It was not unheard of for horse carts to become mired in muddy roads during winter and not freed until the ground began to dry months later. By the mid-1910s, however, a new coalition of interests— farmers, auto and petroleum companies, progressive reformers, auto enthusiasts, and club members—had formed, and in 1916 the US Congress passed the Federal Aid Road Act, the first national law to fund highway construction in the United States. From that point on, the federal government played an active role in making automobiles and trucks easier to use.

From 1893 to 1916 the automobile's place in American society greatly changed, as had its hazards. Reactions in the late 1890s to the injuries and deaths caused by automobile accidents focused on the wealthy few who could afford such vehicles. Localities responded by passing speed laws, promulgating traffic codes that varied greatly from one place to another, and creating the new role of the traffic police. Some auto owners also worried about potential hazards, and they purchased aftermarket products to make their vehicles safer. Around 1910, however, the situation had begun to change: a new industrial order came to rule. The Ford Motor Company created new methods for mass-producing inexpensive vehicles, techniques that quickly spread to other firms. Mass-production of automobiles went hand in hand with the rise of formal organizations, such as the Society of Automotive Engineers. Increasingly, trained experts, including engineers and scientists, dominated the automotive world—a shift that fit a much broader change in North American culture during this period. A safety movement emerged that emphasized decreasing hazards by creating order and that relied on using public communications to foster a safety mindset. Federal legislators funded the construction of highways to ease travel and encourage the movement of people and goods; and judges, recognizing the changed world brought on by mass production, altered long-standing traditions in liability law to ensure that products were well made and trustworthy. Powerful individuals and institutions in American society had begun to rearrange the world around the automobile's potential, but it was not until the arrival of war that that potential was fully realized.

CHAPTER TWO

Standardization Is the Answer

Those who gathered for the National Safety Council's Twelfth Annual Safety Congress, held in Buffalo, New York, in 1923, were invited to see a new play, titled *Adventure: A Pageant of Life and Chance.* As the theater darkened, before the curtain was drawn, a narrator read out the prologue. "This is an ordered world, and man—each man—must find the order out," it began. The balance between safety and danger was a perennial source of human concern, but the problem was, the prologue insisted, that society was undergoing a profound transformation. As the narrator intoned, "Now the old order changes, and the new devises dangers ever more malign, and Man, amid the shifting orders that he makes, more and more loses his fight with Chance."[1]

The most dangerous of all the many forms this transformation took were the new devices and machines filling human life; individuals now found themselves "feeding strange flocks on still remoter hills." Troubles abounded. "So swiftlier gather round him unknown risks, and newer hazards circle like black wings around each new device enslaving Time; New perils wait by each unguarded bar where he sets his foot to pass the stream of space; and each wild energy that bears his rein leaves bloody hoof-prints in his field of life." The only question, then, was whether people would come together to create "new orders to find new prudences" and form "councils for the battle with Blind Chance." Would society organize to reduce these hazards? The prologue ended by asking the audience to make a choice: "Fate leaves the choice to us . . . Shall we choose Chance, or visioning farther, stanch her bleeding stroke, take Knowledge for our guide, and give the race far in the future deeper potencies?"

The author and theater director Thomas Wood Stevens wrote *Adventure* for the Education Section of the National Safety Council at the request of insurance executive Albert Wurts Whitney.[2] Whitney was both the head of the insurance trade association the National Bureau of Casualty and Surety Underwriters and the vice-president of the National Safety Council.[3] In 1922 Whitney convinced insurance executives to give annual grants to the Education Section, an arrangement that lasted until 1938.[4] As part of it, the Education Section was housed within the national bureau, and Whitney effectively controlled it. Whitney had been won over by the principles of progressive education. He believed that pageants were more effective than traditional methods of imparting knowledge because the best connection with learners was "made not through the intellect alone but through the more quickly-turning gate of the emotions as well and hence by means of the drama." He hoped *Adventure* would be performed in schools around the nation. "The play is not propaganda," its foreword read. "It is an attempt to show safety in its true relation to life. Only on such a basis can a successful movement be built."[5]

Adventure reflected one of Whitney's core obsessions: the early safety movement had rallied around the phrase "safety first," which shot into life around 1908 and began declining around 1919, at about the time Whitney turned against it.[6] Whitney thought the notion of safety first was absurd. He may have also have been motivated to attack the notion because physical education teachers—who were often made responsible for teaching new, applied topics in schools—believed focusing on safety would turn young people into "sissies."[7] Whitney wrote, "The fact is of course that safety is not the prime object in life. Exactly the opposite is true. The most important thing is adventure, and by adventure I mean a fresh, first-hand experience of life."[8] It was only because life was valuable and worth living that safety had any meaning. "Safety consists, therefore, quite as much in knowing how to face danger as in avoiding it." The generation and possession of knowledge were crucial.

If *Adventure* asked whether society would form "councils for the battle with Blind Chance," if it asked whether we should choose between "Chance" and "Knowledge," Whitney's answers were clear. Indeed, Whitney saw himself and his colleagues and the organizations they led as forming those councils and drawing together the best available knowledge to make the world safe—safe not for safety's sake but to enable human thriving. From the late 1910s into the 1930s, individuals such as Whitney and Secretary of Commerce Herbert Hoover would create new initiatives and new organizations to address the issues of public safety. Their goal was to convene experts—primarily engineers but also scientists, educational theorists, and others—to solve problems. Few individuals had more influence on auto safety during this period than Whitney and Hoover.[9] Automobiles were prime examples of those "strange flocks" leaving "bloody hoof-prints in the field" of life of which *Adventure* warned audiences. As commerce secretary, Hoover held the National Conference on Street and Highway Safety, the first national gathering to address the issue of automobile accidents, and Whitney, among his many auto-centered efforts, would go on to publish the first driver's education textbook, *Man and the Motor Car.* But what made these men so influential was that they created an entire movement—an extensive social network— of individuals and organizations working on automotive and other forms of safety. This chapter tells the story of that movement.

Many historians have identified the early 1920s as the time when concerns about automobile accidents formed into a national conversation.[10] A number of factors contributed to this emergence. The sheer numbers of traffic fatalities and injuries must have played a part. In 1916 an estimated 8,200 people died in auto accidents in the United States, but that number had more than doubled to 18,400 by 1923.[11] Note that the rise in deaths did not nearly match the increase in the number of cars on American roads, of which there were 3.6 million registered in 1916 but 15.1 million in 1923.[12] As much as the increase in accidents raised consciousness on its own, however, newspapers and other popular media sounded the alarm even more explicitly. Newspapers often ran prominent stories on violent automobile crashes. For example, the *Chicago Tribune* ran a daily column that collected stories of the region's traffic accidents and kept a running total of the deaths, which had reached 559 for the year in mid-December 1920.[13]

The mounting outcry spurred the evolution of the auto safety movement. Its members proceeded by *regularizing* the auto world, making it dependable and uniform, reliable and consistent. These efforts at regularization took many forms, but chief among them was standardization. Standardization became *the* dominant means for regulating technological risks, and it is the mode that continues to dominate to this very day. Automobiles were not the first area in which engineers and other influential actors used standards to address safety problems and other risks; they had successfully used standards to curb hazards in steam boilers, railroads, and other industries.[14] What was striking and new about standardization regarding automotive risk was the sheer number of problems that individuals and institutions tried to resolve through this method. In the name of safety, Americans standardized not only several aspects of the automobile itself (brakes, headlights, safety glass) but also signs and signals, traffic laws, road construction, accident report forms, driver's licensing tests, experimental apparatuses, psychological and vision tests, spaces for road tests, driver's education classes, ideas about how the public should use the street, and many other factors. Their efforts went well beyond the standardization of the technical to include the standardization of human subjects and ideas about normal and abnormal drivers. These acts of standardization typically involved the work of experts, and the resulting standards drew on expert knowledge and in some cases prompted the creation of new knowledge.

The auto safety movement relied on states, localities, and, especially, private organizations to reach its goals. By doing so, it drew on a long tradition of federalism in the United States. This is not to say that federal agencies were wholly absent. Often, they were central players, but they took different roles than they would later. The car became ensconced in a complex and dispersed system of laws, codes, and governments. The result was that a wide variety of groups and individuals—automobile manufacturers, courts, city governments, insurance executives, professional societies, charitable organizations, churches, schools, independent experts, and of course, individual drivers—took some responsibility for managing the risks that attended the car.

As concern increased, so did the search for solutions. A kind of marketplace of and for ideas arose during this period around the issue of auto safety. It attracted both "buyers" and "sellers." Policymakers, including the state motor vehicle commissioners, came looking for answers, and a multitude of actors rushed in to—often literally—sell their solutions. We see this dynamic again and again with different aspects of auto safety, as we have already with facets of headlights and their standards, such as the opportunities that garage owners and instrument manufacturers came to perceive. Many of the solutions that actors offered involved a large-scale transfer of knowledge from the railroad and streetcar industries, both of which had been trying to decrease the incidence of accidents for some time. The methods, techniques, and technologies that moved from these industries to the emerging domain of auto safety included ways of gathering accident statistics and psychological tests for drivers (discussed below), but perhaps the best example is the use of electric signal lights to manage traffic.

Along with organizations and institutions, the automobile itself and how individuals used it continued to change as well. Perhaps most important, the closed-bodied vehicle first came to dominate the automotive market in the early 1920s. Before then, most vehicles were still open-bodied, as the very first cars had been. Closed-bodied automobiles, originally machines for the wealthy who desired protection from the elements and who were not using their cars for primarily racing or scorching, were always available on the automotive aftermarket as well as from luxury automakers. But by the 1920s, use of the car had undergone considerable transformation. Owners came to desire cars that could be comfortably driven year-round as the automobile became firmly entrenched in practical, everyday activities like running errands. Changes in the automobile also gave birth to new risks, however, just as they decreased old ones. Closed bodies shut drivers off from parts of their environment, diminished visibility, added blind spots, and made it harder to hear approaching hazards, such as trains.[15] At the same time, closed bodies made it less likely that drivers would get dust in their eyes or choke on a bug, experiences that introduced their own hazards. With the rise of the closed body, the automobile took what historians consider its final shape. The car has continued to evolve to this day, but its changes since the 1920s have been incremental, not radical.

Still, in the late 1910s and 1920s, some sense of openness persisted about what technologies would be included in automobiles. For example, in 1919 voters in the city of Cincinnati weighed in on a ballot initiative that would have required all automobiles in the city to be equipped with speed governors, a technology that newspapers around the country frequently endorsed and that many managers of commercial fleets, from taxi companies to the US Post Office Department, installed to ensure that professional drivers hewed to speed laws and did not use their work cars for joyriding.[16] Voters struck down the measure by a wide margin, and the National Automobile Chamber of Commerce created a Safety Committee, in part to stave off any further efforts to mandate governors. Nevertheless, the Cincinnati incident shows that automobile's final form was not a closed matter in the late 1910s, nor was the world of laws, rules, and technologies that surrounded it. As actors stepped away from the rubble of World War I and laid aside its home front efforts, they confronted possibility.

Standardization before and after World War I

Formal organizations and standardization concerning automobile safety increased greatly after World War I—and in large part, because of war efforts—but these initiatives built upon foundations laid before the war,

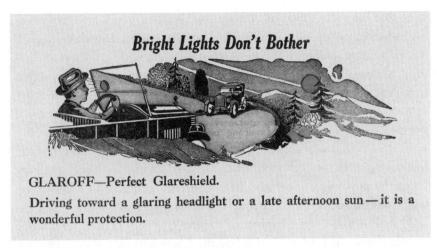

Figure 2.1. An advertisement for the Glaroff Amber Vacuum, an aftermarket product meant to protect drivers from glaring light, including from the electric headlamps of oncoming cars. Warshaw Collection, National Museum of American History, Washington, DC.

especially by engineering societies. As seen in the previous chapter, electric headlamps began as aftermarket products and did not come "standard" on automobiles until Cadillac first introduced them in 1911. Ford added them to its products four years later. These headlights introduced new hazards into the night. Street lighting was rare in the United States (compared to denser European cities). As drivers made their way over dark roads, their electric headlamps often blinded pedestrians and drivers of oncoming cars. This glare was such an issue that the automotive aftermarket offered at least one solution. The Glaroff Manufacturing Company of Sioux Falls, South Dakota, marketed a piece of amber glass that could be affixed to the inside of the windshield.[17] When cars with bright lights approached, the driver would lean over and look through the amber glass to avoid being blinded.

The headlight problem was part of a more general one: the glare that came with the spread of electric lighting throughout society. The Illuminating Engineering Society, founded in 1906, began tackling the problem of glare in the early 1910s. Illuminating engineers saw themselves as bringing light to darkness, literally and figuratively. Through the application of scientific knowledge about vision and engineering, they worked both to expand the application of electric lighting to all parts of culture and to solve the many difficult technical problems that came with this expansion.

In 1914 the society formed a Committee on Glare, which a society member called "one of the most complex subjects that have to do with lighting."[18] The committee comprised engineers, scientists, and medical doctors, including J. R. Cravath, the editor of *Electric World*, and Perley G. Nutting, at that time head of research at the Eastman Kodak Corporation and later founder of Westinghouse Electric Corporation's research laboratory.[19] Taking a broad purview, the Committee on Glare approached its topic in a number of domains and produced reports on a variety of topics, including street lighting, interior lighting, and automobile headlights.[20] Its members began talking about the possibility of creating headlight regulations and a standard on which to base such regulations, but they made little headway.[21]

From 1916 to 1917 the Illuminating Engineering Society had a Committee on Automotive Headlamps, which contained members from the earlier glare committees, and it began working hand in hand with the Society of Automotive Engineers.[22] Only in 1918 did the engineers settle on a headlight standard. Ironically, that 1918 standard was quickly outmoded by one created in Massachusetts. State authorities had conducted a study and found that glaring headlights were the primary cause of nighttime auto accidents.[23] Given this grim reality, the Massachusetts policymakers set a tough standard. The illuminating engineers found the tougher Massachusetts standard to be effective and adopted it as their own.

It was one thing for engineering bodies to create such standards. It was quite another to make them law. Worried about the growing automobile accident problem as well as the proliferation of new laws aimed at solving it, the administrators of motor vehicle departments in seven East Coast states in 1921 formed the Conference of Motor Vehicle Administrators to "promote uniformity, cooperation, and reciprocity in respect to the regulation and operation of motor vehicles."[24] State motor vehicle commissioners had typically begun their careers as tax collectors, charged with generating state revenues by collecting automobile registration fees.[25] They quickly found themselves in the unenviable position of trying to curb rising accident tolls. Although individual commissioners had the opportunity to influence the development of safety practices, they typically had no expertise in risk or safety. So they searched for solutions elsewhere.

Fittingly, in the early 1920s the Conference of Motor Vehicle Administrators adopted the Illuminating Engineering Society's (and, thereby, Massachusetts's) headlight standard and strove to build it into state legislation. But

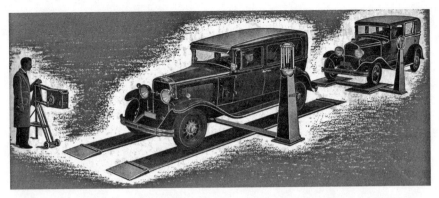

Figure 2.2. State and municipal governments mandated annual inspections of brake and headlights throughout the 1910s and 1920s, increasingly basing them on standards promulgated by organizations like the American Engineering Standards Committee. These laws opened new markets for testing equipment. In 1932 the Weaver Manufacturing Company trademarked the term "Safety Lane," used to describe annual inspections in some states several decades afterward. This 1933 Weaver advertisement shows brake, alignment, and headlight testing instruments that automotive garages could purchase as a way to profit from annual inspections. Scan courtesy Castle Equipment Company

these efforts met with limited success—a dynamic that dogged most safety efforts in this period, which depended heavily on the coordination of state legislatures and local lawmaking bodies. Many states adopted some part of the illuminating engineers' standard, but several factors, including state-level lobbying groups, regional political cultures, and politicians' egos, vitiated against total uniformity. Even by the 1930s, states had a wide variety of headlight standards.[26] Uniformity was the exception, not the rule. Viewed from another angle, however, the general movement toward state adoption of headlight laws was remarkable in that they gave birth to the annual automobile inspections that still exist in many places today. Perhaps never before had laws required consumer-owned technologies to undergo regular inspections. Within a few years, states around the nation passed legislation to that effect. By the mid-1920s, mechanics in rural Texas were being taught how to test and adjust headlamps.[27]

A significant result of these laws was that some headlights were excluded from the market. The state of California, for instance, promulgated its *Table of Approved Headlight Devices* in 1922, though the list did not include all of the lamps then available through catalogs.[28] An engineer on the city of

Chicago's Headlight Board estimated that the board had approved only about 75 of the 195 headlights it had inspected.[29] The Conference of Motor Vehicle Administrators approved even fewer lighting systems, 22, which if the Chicago engineer's estimate of 195 available headlights is to be believed, was a radical narrowing of the market.[30] These standards could wreak havoc for headlamp manufacturers, which could find their products banned from certain markets. But they also caused headaches for auto manufacturers. In 1925 the Conference of Motor Vehicle Administrators rejected the headlamps on General Motors' vehicles.[31] An interim deal was worked out while the company switched to better equipment, but the rejection caused stress as the head of research, Charles "Boss" Kettering, and other executives scrambled to find a solution.

Yet, if these kinds of regulations shut down opportunities for some companies, they opened them for others. In an advertisement in *Popular Mechanics*, the School of Automotive Electricity, a division of the Milwaukee School of Engineering, claimed that "the headlight laws alone—different in every state—means [*sic*] big pay for those who learn the *mechanical* requirements and how to conform auto lights to the regulations. Start now."[32] By the 1930s, watchers of the auto world knew that powerful groups were trying to shape state regulations in ways that enhanced their own self-interests.[33] Auto garages and the mechanics they employed were perhaps foremost among these. Inspection laws created a new kind of fat on which mechanics could feast. Auto supply stores also did well by such laws because, when lights were found defective, owners had to replace them. As a headline in the trade newspaper *Automotive News* put it, "State Crusades Inspire Sales of Headlights."[34]

Similarly, equipment regulations opened new markets for testing instruments. The first headlight standards involved labor-intensive tests that required testers to walk out hundreds of feet from the vehicle and use light meters to take measurements at different heights and distances. By the early 1930s, the Weaver Manufacturing Company and other firms had invented headlight-testing devices that could be placed in direct contact with a car's lamps to determine quickly whether they met regulations.[35] The Weaver Company—whose devices accounted for 90 percent of the inspection market—bundled headlight-testing devices with brake- and alignment-testing instruments into what it called the Safety Lane, which was marketed to auto garages. In some regions "Safety Lane" became a general term for having your annual inspection done.

Albert Whitney and National Standardization Efforts

Individual engineering societies, including the Illuminating Engineering Society and the Society of Automotive Engineers, led the drive to standardize headlights. But beginning in the early 1920s, standardization efforts were increasingly arranged by the American Engineering Standards Committee (AESC), an organization that formed in 1918 to coordinate the standardization efforts of more than one hundred different organizations, mostly engineering societies and trade associations.[36] Indeed, the organization did some minor work coordinating efforts between the Illuminating Engineering Society and the Society of Automotive Engineers to refine automotive headlight standards. The AESC went through several name changes over the years and exists today as the American National Standards Institute, or ANSI. The roots of the organization lay in attempts to create a standards-coordinating organization in the mid-1910s, but the AESC itself was a "war baby," as one of its leaders later characterized it. In its first years, the committee focused on industrial standards aimed at increasing efficiencies by reducing waste and increasing compatibility, but in the early 1920s it began working also on safety standards.

The AESC's turn to safety standards resulted largely from the arrival in 1922 of a new president, Albert Wurts Whitney, the insurance executive described at the beginning of this chapter. The physicist Paul G. Agnew, who acted as the secretary of the AESC for decades, later noted that Whitney led the organization during its "critical years" and exerted a "profound influence" on its "growth and development."[37] Before Whitney assumed the presidency, the AESC was already working on safety codes and standards, but its safety efforts expanded after he arrived. Its 1923 annual report highlighted some of the organization's safety efforts, including a proposed standard for traffic signals and a revision of the headlight standard.

In 1922 Whitney was leading the AESC, the National Safety Council's Education Section, and the National Bureau of Casualty and Surety Underwriters, and he used these positions and others to exert a huge influence on automobile safety during this period. For many reasons, it is worth taking a moment to trace Whitney's personal trajectory. Whitney had begun his professional life as a professor of mathematics. He earned a bachelor's degree in mathematics at Beloit College and then became a fellow at the University of Chicago.[38] He went on to professorships at the University of Nebraska and the University of Michigan before finally settling, in 1899, for some years at the University of California. Whitney appears to have been attracted to progressive education early on, and not surprisingly, given John Dewey's presence at the University of Chicago while Whitney was a fellow there. Whitney's pedagogical philosophy favored learning by doing. He turned from teaching pure mathematics to focusing on practical, business-oriented problems because he realized that he could persuade his students at the University of California to focus if his work at the chalkboard centered on matters they found immediately and obviously useful. When the University of California decided to offer a course on insurance and actuarial mathematics, Whitney was chosen to teach it.[39]

Whitney's work and focus on the mathematics of the insurance industry eventually brought him into the orbit of the business world in California. In 1905 he authored a report for the Pacific Board of Fire Underwriters, and he went on to write a series of papers and articles that greatly influenced insurers' practices across the country. He then undertook an important insurance study of the 1906 San Francisco earthquake for the San Francisco Chamber of Commerce. A few years later, Whitney was invited to work with the New York State Assembly's Merritt Committee in its investigation of fire insurance company practices.[40] Whitney's involvement with the Merritt Committee quickly pulled him into the speedily evolving policies and politics of insurance in New York, then the center of the US insurance world.

The Whitney family moved to New York in 1914, the year that Albert rose to head the casualty insurance trade association then known as the Workmen's Compensation Bureau, whose primary function was setting rates for workers' compensation insurance.[41] The rise of workers' compensation laws helped drive the safety movement. But Albert Whitney had grander visions for insurers' role in American society than one confined to workers. In 1919 he led the push to create the Education Section within the National Safety Council, and the insurance trade association paid for a staff and office space for its Safety Education Section beginning in 1922. In 1921 the insurance trade association was reorganized and renamed the National Bureau of Casualty and Surety Underwriters. Whitney used this moment to significantly broaden the organization's mission. Its new constitution set one of its goals as "the development and administration of plans for the prevention of accidents or other occurrences which increase hazards and adversely affect public interest."[42] Although Whitney would focus on safety of all kinds for the rest of his career, safety involving automobiles and the street always held a special place in his worldview.

Beginning in the early 1920s, the National Bureau of Casualty and Surety Underwriters came to focus increasingly on automobile safety. This shift reflected the ever-larger market for auto insurance. In a speech before the Insurance Society of New York in 1919, the insurance executive Eugene F. Hord recounted that since automobile insurance had begun humbly in 1898, "the volume of premiums has grown from nothing to approximately $150,000,000.00."[43] The core issue driving individuals to buy automobile insurance was tort law. Lawsuits stemming from accidents were already quite common and would only increase in number over the coming years. According to one study, 30 percent of the "new issues placed on the calendar" of the Supreme Court of New York County between October 1928 and April 1930 were "motor vehicle accident cases."[44] By 1919, personal injury costs from a single accident could already reach as much as $25,000.[45] Without insurance, automobile owners were on their own and could easily be bankrupted. Only about one-third of drivers were insured in the United States, but in cities like New York the number was higher, somewhere between one-half and two-thirds.[46]

All of this growth meant big business for insurance companies, but it also motivated them to reduce automobile accidents. In some ways, automobiles created new problems both for insurers like Whitney and for regulation. Whitney's trade association had traditionally sold insurance primarily to companies, and safety regulation had historically focused on controlling technologies owned by firms, such as steamship and railroad companies.[47] The automobile, however, involved insuring and regulating a technology owned and used by *millions* of individual consumers.

Whitney saw himself as aggregating and standing in for the interests of consumers. Throughout his career he emphasized that insurance was a business oriented toward the public good, which, he claimed, was why he was drawn to insurance in the first place. "What attracted me," he wrote, "was seeing business recognize the necessity, first, of co-operating for the permanent, common good, and second, of finding its own good in the line of service to the public."[48] Insurance's overarching value for Whitney was not that it socialized benefits after accidents but that it focused minds: "Not only . . . does insurance draw the fangs of misfortune by binding society together to bear each other's burdens, but it contains within itself a more potent virtue—the power to prevent misfortune itself as well as misfortune's effects."[49] Insurance "plays not merely the part of the distributor of benefits but the part of the conservator" of life and values.[50]

Whitney had already seen this dynamic play out in the worker's compensation movement. He believed the most important aspect of that movement was not that it provided compensation but that it had created the safety movement, which gave birth to the National Safety Council and greatly reduced workplace accidents. "The power of insurance for prevention," he wrote, "lies in the fact that it commutes an indefinite, unrealized hazard into a tangible money equivalent, thus bringing it clearly into view as an economic factor . . . and secondly that by concentrating many risks it makes possible the inauguration of comprehensive movements of great scope and effectiveness."[51] In the early 1920s Whitney saw himself bringing precisely these dynamics to bear on the problem of automobile accidents.

And it was in this capacity that Whitney first became interested in safety standards.[52] By the time he came to head the American Engineering Standards Committee, he was already deeply ensconced in the safety movement and had a sophisticated understanding of standardization.[53] It was natural, then, that he would move the committee toward creating safety standards.

Electric Traffic Signals

Under Whitney's leadership the AESC created multiple safety standards for several different industries, but when it comes to automobiles, one clear example is the standardization of electric traffic signals, one of the many technologies that moved from railroad safety to automobile safety. By the 1920s, railroads had been using electric signals for decades. As early as the 1910s, cities, including Salt Lake City (1912) and Cleveland (1914), had begun experimenting with electric signal systems for use in managing street traffic, but those systems had remained strictly local affairs started by local inventors trying to cope with their city's traffic problems.[54] Lester Wire, the Salt Lake City inventor, was a policeman and built his traffic signal apparently by putting red and green lights in a birdhouse atop a pole. Wire was likely not a member of any national professional organization, nor was he driven to try to market the devices. His signal did not spread to other parts of the country but remained within the Utah city where it began.

National discussion of traffic signals did not really emerge until the early 1920s, particularly after the first signals appeared in New York City. That city in 1921 installed traffic towers with colored lights at five intersections along Fifth Avenue, the same street that had been chosen as the test bed for Eno's traffic rules in 1903 and still the city's busiest. Deputy police commissioner Dr. John A. Harriss built the towers with his own money. They were

Figure 2.3. A traffic tower in the early 1920s along Fifth Avenue in Manhattan. The three colored lights (red, amber, and green), arranged laterally at the top of the tower, did not have the same meaning they do today.

"modeled somewhat similar to signal towers on railroads," Harriss explained. The Fifth Avenue Association, a group of local leaders who owned businesses along the street and who fretted about its appearance, held a contest for the design of the towers. The winning design was ostentatious. Twenty-three feet tall and made entirely of bronze, the towers featured neoclassical ornamentation, including four large eagles at the top corners. Each of the towers' four faces was equipped with a clock, a machine which by that point had been a symbol of modernity in the United States for seventy years. Importantly, the towers were manned. A traffic officer sat in each tower and looked through the windows in it to determine when to change the lights.

Everyone realized that drivers and pedestrians would require education about the meaning of the lights, just as police officers had needed to train street users in Eno's traffic rules twenty years earlier. The need for such education was particularly clear to railroad managers and executives, who knew its engineers had to be trained and retrained when new signals and other changes were introduced on the lines. At a 1922 meeting of the

Illuminating Engineering Society in which the New York City signals were discussed, A. H. Rudd, the Pennsylvania Railroad's chief signal engineer, drew on his experiences to make a case for education. Rudd "emphasized the necessity of not only arriving at intelligible standards [for traffic lights], but also educating the public as to the meaning and importance of observing them."[55] John O'Brien, the inspector in charge of traffic for the New York City Police Department, said that pedestrians posed the greatest challenge to police in persuading the public to follow the signals. People often commented on the many police officers the city had lined up "at the corners of 34th St and 42nd on Fifth Avenue," but their task was to "educate the pedestrians," O'Brien explained. "It is a tedious job and will require considerably more education along these lines, so you see, lights and education go together."[56]

New York City's traffic signals worked differently than the typical three-color signal does today, and in this way, the signals speak to the struggle for standardization that raged in the 1920s. The New York signals had three colors, as ours do, but the colors had different meanings. As Harriss described them, "The signals flashed from the towers indicate the following: *Yellow*: Traffic moves on Fifth Avenue and all cross traffic from side streets stops . . . *Red*; All traffic on Fifth Avenue as well as the side streets stops . . . *Green*; Traffic from side streets proceeds."[57] The lights would then go back to red before returning to yellow. Not only were the traffic signal technologies not standardized across cities—some systems had only two colored lights—but the *meaning* of the lights also varied.

The opportunities presented by the emerging market for street traffic signals further compounded the issue of nonstandardization. During the 1920s the electrical engineer for the city of San Diego, California, received more than forty brochures from at least fifteen unique companies, each of which was sending him mail and, often, banging on his door to see if he was interested in installing its system in the city.[58] Doubtless, other municipal engineers and decision makers had similar experiences across the country. The signal systems on offer came in a wide variety of forms, including two- and three-color systems and others that featured fourth colors, as well as a diversity of light arrangements. Street signals were meant to reduce hazards and improve traffic flow, but the lack of uniformity threatened to introduce its own dangers and inefficiencies.

Into this fray stepped the AESC. Three organizations—the National Bureau of Standards, the National Safety Council, and the American Association of State Highway Officials—formed a subcommittee within the AESC to address traffic signal standardization.[59] On May 14, 1922, AESC chairman Albert Whitney convened the first conference on traffic signals, "attended by one hundred delegates, representing more than forty different national organization."[60] The participants decided unanimously to standardize the signals, including "the coordination of color, form, position, and number of signals."[61] The ensuing standardization effort was massively successful. In 1936 the authors of the first driver's education textbook published in the United States could write unequivocally, "Universally a red color means STOP and a green color GO"—a statement that would have been untrue only a decade and a half earlier.[62]

The Ambivalent Role of the Federal Government

In the 1920s the federal government initiated a variety of activities designed to improve undesirable aspects of the automobile, efforts that influenced automotive design and markets during this period. None of these activities involved direct federal regulation, but they often shaped local and state regulations. And nearly all of them involved bringing experts to bear on pressing problems. The Department of Commerce, under the guidance of Secretary Herbert Hoover, led the way in the federal government's efforts. Hoover had begun his career as a mining engineer working in Australia and China. In 1908 he entered business for himself as a consulting engineer, based in Western Europe, and advised mine owners and managers around the world on improving production.

When the First World War began, Hoover organized the return of thousands of Americans from Europe to the United States. He then began coordinating the food relief effort in Belgium, which faced a terrible food crisis with the war's onset. Hoover's work as head of the Commission for Relief in Belgium brought him to the attention of the Woodrow Wilson administration, which appointed Hoover to administer the newly formed US Food Administration in 1917. In this role Hoover drew together an advisory council of business leaders, scientists, and politicians, an act that became a pattern for Hoover for the rest of his time in public life. The Food Administration worked to improve production, foster conservation and rationing, and facilitate the shipping of food both within the United States and to Europe.

After the war, Hoover made a first, unsuccessful bid for president, but having afterward supported Warren G. Harding's campaign for that office, he was offered a secretaryship in the Harding cabinet. Hoover chose to head the Department of Commerce, a decision that might have struck his contemporaries as odd. In those days, Commerce was seen as something of a backwater. Only a decade earlier, the department had been split off from the Department of Labor and Commerce. By the early 1920s, Commerce still lacked a strong sense of mission. Hoover later recalled that the department's administrators were said to spend less than two hours a day managing it, the rest of their time dedicated to such activities as golfing, hobnobbing, and attending parties.[63]

Hoover, however, had an ambitious and far-reaching vision for the agency and the role it could play in fostering economic growth. Encouraged by his experiences as an engineer and as a wartime administrator, Hoover pinned many of his hopes on the promises of standardization. Some critics at the time believed that standardization led to conformism, but Hoover opposed this view.[64] A proponent of free market capitalism and masculinity, Hoover was the author of *American Individualism*, published in 1921 while he was commerce secretary. He believed that American individualism was "the sole source of progress" in the nation. But, like Whitney, Hoover did not see standardization as opposed to individualism or consumer sovereignty; rather, standards enabled freedom and thriving, a belief in accord with a progressive vision of engineering widespread at that time. As Hoover wrote, "The man who has a standard automobile, a standard telephone, a standard bathtub, a standard electric light, a standard radio, and one and one-half hours less average daily labor is more of a man and has a fuller life and more individuality than he has without them."[65]

The first entity within the Department of Commerce to venture into the auto safety arena was the National Bureau of Standards (NBS).[66] Originally focused on the science of measurement, or metrology, the bureau's nearly constantly expanding staff moved into more and more fields of study and into playing new social functions, such as testing products involved in potentially fraudulent marketing claims between 1901 and World War I. This trend only increased after the war, as new leadership, both at the bureau level and in Hoover's Commerce Secretariat, encouraged bureau staff members to go further than they ever had before. Hoover and George Burgess, the bureau's postwar head, encouraged staff members not only to move into new areas but also to take a more active role in shaping markets. It was in this way that NBS staff members, like other actors discussed in this chapter, began moving into the developing domain of automotive safety.

Among the NBS's numerous divisions, the Heat Division was the office that worked most on auto safety issues.[67] The Heat Division had begun working on automotive issues during the First World War. At that time, the country had experienced several shortages of fuel, including gasoline, and this trend continued after the war. The National Bureau of Standards had led the formation of the Cooperative Fuels Research Committee, a public-private partnership that included the American Petroleum Institute, the Society of Automotive Engineers, and other engineering organizations. The committee aimed to increase fuel economy and design engines to use poorer-quality fuel by fostering cooperative research between experts in engines (automotive engineers) and experts in fuel quality (petroleum engineers). The committee made a number of strides during the 1920s and 1930s, including the standardization of some fuel and lubricant types.

But the division also moved into other areas, including auto safety, and it often did so in its role as one of the coordinators of the federal government's purchases. By the 1920s the US government was *the largest purchaser of automobiles in the world*. As such, it was concerned about the quality of what it was buying. The bureau's move into automobile safety was thus part of a more general shift in the agency. Its head, George Burgess, wrote a series of essays in the 1920s, several of which he delivered before the members of the American Engineering Standards Committee, outlining his plans for the use of technical specifications. Burgess hoped to coordinate the purchasing habits of the entire federal government by having bureau staff create standards of quality—embodied in specifications—and lists of producers who agreed to meet those specifications. Products would improve if government, using its purchasing power, demanded that manufacturers meet certain benchmarks.

Burgess set his sights even higher, though. He hoped that state governments, corporations, and private individuals would also make their purchases according to the specifications the bureau set. In this way, vast swaths of the American economy would demand higher-quality products, and markets would improve overall. The NBS would shape markets by shaping consumption, including, Burgess hoped, the choices of millions of individual consumers. The NBS published the nearly four hundred–page volume *National Directory of Commodity Specifications* in 1925.[68] Hoover was proud of these efforts. As he wrote in his memoirs, "A specification directory containing references to 27,000 items relating to 6,600 commodities was published, which served the buying agencies of the country as a guide in securing the best developed specifications. These specifications not only were used by the government and public institutions, but came into large use by consumers. As they called for simplified or standardized articles, they reinforced that whole program."[69]

Some businesses violently protested these specifications, believing that they were a violation of free market principles. But Albert Whitney, who knew both Burgess and Hoover, supported them wholeheartedly. As he wrote to Burgess, "It seems to me your plan of gradually building up lists of manufacturers who will agree to manufacture according to specifications is most clever and is likely to be very effective in producing results."[70] Specifications were "clever," he thought, in that they gave the government a way of shaping markets without creating formal regulations or legislating the matter.

The National Bureau of Standards Heat Division began working on automotive brakes in precisely this way. In January 1920—the earliest month for which the monthly reports of the Heat Division still exist in archival records—the division's head noted that Motor Transports Corps, the part of the US Army responsible for adopting and maintaining automobiles, had come to division staff members with a number of problems, one of which involved brakes and brake linings. Brakes on early automobiles were famously bad. The brakes on the Model T—introduced in 1908 but still produced until 1927—"shrieked like Banshees" and burnt out quickly because Ford used cotton rather than asbestos liners.[71] Moreover, until the mid-1920s, when four-wheel brakes were introduced, cars only had brakes on two-wheels. The automobile continued to change, and its risks changed with it. To address the Motor Corps' problems, the Heat Division developed a "standard test of brake linings."[72] By the middle of 1921 the division had created a specialized testing apparatus for brake linings.[73] The bureau received an increasing number of requests from other agencies in the government, including the US Post Office Department, for brake tests. Staff members began producing a list of brake linings that met the bureau's requirements, which it shared with other agencies. But the Heat Division also received a large number of inquiries from brake lining producers themselves, probably both because the producers were looking for ways to improve their practices and because their bottom line was being affected by the bureau's recommendations to other agencies—recommendations based on these tests.[74]

Heat Division staff began promoting their test and specifications for brakes and brake linings outside the federal government. In 1922 one staff member delivered a paper on the topic before the Society of Automotive Engineers.[75] Over the following years, the agency increasingly presented demonstrations, some of them public, to different states and municipalities that considered adopting the brake tests as part of their auto safety regulations, something that agency staff encouraged. By 1928 the NBS reported, "The type of decelerometer developed at the bureau is being produced commercially, and the bureau's advice is frequently sought by manufacturers interested in developing other types. It is estimated that nearly a thousand decelerometers of various makers are now in service. They are employed for brake testing by police officers, service stations, automobile clubs, etc., in some 30 eastern cities. Massachusetts and the District of Columbia have been pioneers in the adoption of the scientific method of brake testing."[76] Through the use of specifications and instruments such as decelerometers, the NBS influenced automotive regulation throughout many parts of the country.

The Department of Commerce affected automotive regulation in ways that went beyond the National Bureau of Standards' efforts, however. Indeed, more widely known than the bureau's work were the series of auto safety conferences that Herbert Hoover held in 1924 and 1926 and, while he was president, in 1930. For some time, various groups had been pressuring Secretary Hoover to do something about the traffic problem that was killing and injuring so many people around the country—including 19,400 deaths in 1924 alone. Early in that year, Hoover decided to hold a gathering to focus on auto safety. The organization that resulted came to be known as the National Conference on Street and Highway Safety. Its small core consisted of "only a full-time secretary, one stenographer, and a portion of the time of a publicity man."[77] The US Chamber of Commerce loaned the time of the secretary, A. B. Barber.[78] But the members of the conference's many committees carried out the vast majority of the work on a volunteer basis.

The National Conference on Street and Highway Safety reflected Hoover's general approach to governance, which historians later called associationalism.[79] Instead of pushing for federal legislation and regulation, Hoover preferred to bring together private interests to solve common industrial and public problems. He held conferences dedicated to difficult issues in the emerging aviation and radio industries, as well. Like the aviation and radio conferences, the National Conference on Street and Highway Safety was sponsored largely by various trade associations, including the US Chamber of Commerce, the National Automobile Chamber of Commerce, and Whitney's National Bureau of Casualty and Surety Underwriters.[80] But unlike the other conferences, the auto safety conference also received funding from the American Automobile Association, the National Safety Council, and, eventually, the National Research Council, groups that had deep industry ties but were not themselves "industry groups." Because the problem of automobile accidents touched nearly every part of society, the National

Conference on Street and Highway Safety brought together a wide range of interested individuals and groups: motor vehicle commissioners, publishers, insurance executives, police officers, taxi and streetcar business owners, safety educators, psychologists, consulting engineers, statisticians (from both industry and government), labor leaders, engineering professors, presidents of women's organizations, safety advocates, safety engineers, attorneys, physicists, heads of auto clubs and associations, government bureaucrats, and representatives of the National Congress of Parents and Teachers (later the Parent-Teacher Association). *Many* of the individuals and organizations mentioned so far in this book ended up participating in the conference, including (unsurprisingly) Whitney.

The subsequent Hoover conferences came to focus on uniformity, both in traffic laws and in signs and signals. As Hoover later recounted in his memoirs,

> In our work to promote [the adoption of a uniform traffic code around the country], we sent an automobile from New York to San Francisco and another from San Francisco to New York. The driver of each car had orders to follow scrupulously the laws of his own state and municipality. One of them was arrested eighteen times, the other twenty-two times, for violation of laws which differed from their own. The two together met with sixteen actual accidents and avoided scores of potential ones only because of their driving skill (and quick breaks from home-state laws).

The committee dedicated to working on uniform traffic laws had the largest and most geographically diverse membership in Hoover's traffic safety conference, in part because it had a different goal. Other committees aimed to aggregate expert knowledge on a given subject, such as the causes of accidents or the best way to collect accident statistics, but the committee on uniform traffic laws urged its members to push its uniform code on local and state governments. While the committees of the National Conference on Street and Highway Safety covered a wide range of topics, the conference's efforts came to focus, fittingly, on the uniform traffic code: the many groups and individuals involved in the conference could get behind this law as an act of standardization. As discussed below, this push for the uniform traffic code had a far-reaching but ultimately limited effect. Yet, between the work of the National Bureau of Standards and that of the National Conference on Street and Highway Safety, Hoover's Department of Commerce influenced the world of auto regulation in diverse ways in the 1920s and into the 1930s.

Standardizing the Driver

The discussion of standardization thus far has focused on technical issues and the law. But standardization efforts also extended to factors we often think of as nonstandard, such as human beings. A number of different actors and organizations sought to apply the standardization process to human beings, especially drivers, and their visions competed for resources and application.

The first group to make this attempt were professional psychologists. Like many individuals and organizations involved with auto safety, professional psychologists entered this emerging domain in the early 1920s, and they did so largely through the National Research Council (NRC). At President Woodrow Wilson's request, the NRC was founded as part of the National Academy of Sciences in 1916 to help ready the nation for entry into the First World War. In the war's wake, the council looked for new research projects and sponsorship.

In the early 1920s, the NRC's Division of Anthropology and Psychology began imagining a role for psychologists in the emerging area of auto safety. Psychologists envisioned a number of different roles for their expertise in this arena, including exploring the psychology of vision, signs, and signals and developing tests for drivers. Many practitioners, however, had experienced great success in developing and conducting intelligence tests during World War I, an effort that one historian has called "the most important early success of American mental testers."[81] Moreover, for the past two to three decades, professional psychologists around the globe had developed tests to sort careful, trustworthy streetcar motormen and railroad engineers from reckless, flawed ones. For all of these reasons, psychologists returned consistently to testing during this period and applied tools they had developed for other purposes, including the management of railroad accidents, to the new field of auto safety.

In 1924 the NRC psychologists contacted the organizers of Hoover's National Conference on Street and Highway Safety, who invited the psychologists to attend as members of committees that had already been organized. Two years later, however, the Hoover conference organizers asked the psychologists to form a new group, which became known as the Committee on the Causes of Accidents and was simultaneously a part of the National Research Council and of the National Conference on Street and Highway Safety. Walter Van Dyke Bingham, an influential professor of industrial psychology at Carnegie Institute of Technology, chaired the committee, which had a diverse membership, twenty-six in all, including Albert Whitney; George Burgess, the director of the National Bureau of Standards; and Lew Palmer, the author of *Sure Pop and the Safety Scouts* and a member of the National Safety Council.

Like many of the periodic committees and panels of its type, the Committee on the Causes of Accidents lamented that "reliable and detailed data on which to build a sound knowledge of underlying causes are very difficult to get."[82] Not much was known and more research was needed, the committee argued. While the committee's report laid out some tentative findings—most of which found that *drivers* were the primary causes of accidents—it mostly prescribed further studies and activities, especially new, better ways of gathering accident statistics. The psychologists believed that Hoover had promised them further support after publication of the committee's report, but that support was not forthcoming, so the Committee on the Causes of Accidents ended in disappointment for the psychologists. Eventually, the National Safety Council became the group that took up collecting automotive accident statistics on the national level, and that organization dominated this activity until the federal government took it up in the 1960s and 1970s. The NSC addressed the data quality problem by creating a *standardized* accident reporting form, which some state and local authorities adopted.[83]

The NRC psychologists also set about devising tests for commercial drivers, especially taxi drivers.[84] The psychologists initially considered developing tests for all drivers, but they knew that the public would rebel against having their driving privileges revoked based on the kinds of pen-and-paper and lab-based tests that the psychologists were creating. Businesses could be more restrictive about who they let drive their cars, so the psychologists started there. Much of their testing work centered on the paradigm of accident proneness, an idea popular in psychological circles during this period.[85] The notion of accident proneness held that a certain portion of the population was especially liable to cause accidents. In the world of auto safety, the accident proneness idea was supported by statistics that found that repeat offenders caused a large percentage of accidents. For about a decade, the NRC funneled research money first to Ohio State University and then to Iowa State College. Researchers at these institutions undertook a number of original-research projects, one of them involving the creation of one of the first driving simulators, used to conduct tests on drivers. This research was eventually published, in 1930, as *Psychological*

Principles in Automotive Driving, one of the first book-length treatments of its subject.[86]

Over time, however, NRC psychologists came to doubt the idea of accident proneness. Their suspicions arose from many causes, but one of the primary concerns was that some psychologists were using the idea to seek personal gain. They patented psychological testing instruments and tried to persuade localities to adopt them; they formed overly close relations with businesses, including some insurance companies; and they tried to shape local and state laws based on dubious "scientific" claims. In the mid-1930s, the NRC hired the psychologist Percy W. Cobb to conduct a study on accident proneness. Cobb traveled to Connecticut, the state with the best-quality accident records. Using those records as a statistical database, Cobb found that the standardized driving tests being pushed by some psychologists correlated with accident records only 35 percent of the time.[87] He showed these driving tests to be worthless for identifying accident-prone drivers, if such characters even existed. The notion of tests for accident proneness immediately fell into disfavor with leaders at the National Research Council, and by the 1940s the idea was discredited in professional circles. Mirroring the era's general exuberance for standards, psychologists had tried to create standardized images of human beings—"the man behind the wheel," as one psychologist put it. Although some of the psychologists' products stuck around, such as standard driver's license test forms, as a whole this effort failed.

These kinds of internal tensions and debates within psychological circles undermined the scientists' ability to get things done, but by far the greatest hindrance to the psychologists' work was their inability to find a steady, rich patron or any other source of funding for their work. As we have seen, one of the greatest sources of safety funding during this period and afterward came from the insurance industry. In 1925, NRC psychologists approached Albert Whitney to request funding for research projects, but Whitney turned them down. He excused himself by pointing out that his organization was already supporting a "large amount of safety work," including the "financing of safety education work, the investigation of benzol poisoning, [and] the establishment of safety work in local communities." Moreover, he could not be sure that members of the insurance trade association he headed would be interested in psychological research that was not directly tied to "preventative work." Even if more funds became available, Whitney argued, "there will be the serious question of whether [the] research . . . you suggest will be considered the best possible use of funds."[88]

What the psychologists may not have known was that, when it came to academic research, Whitney had other priorities. He was funding an alternative line of research that competed with the psychologists' paradigm of accident proneness. Whitney's research priorities arose from his own life trajectory, particularly his early attraction to progressive education, and these preferences came to influence the world of auto safety in key ways. As early as 1918 Whitney had told an audience that "the safety problem is fundamentally psychological. What needs most to be done is to make men think in terms of safety."[89] One might think that such a view would have predisposed Whitney to the psychologists' work, but he had a different view of human nature and how to improve behavior, one deeply informed by the writings of the pragmatist philosopher John Dewey.

Fittingly, when Whitney moved to New York, he began spending time at Columbia University's Teachers College with disciples of John Dewey, including William Kilpatrick and Harold Rugg. He socialized at the Teachers College's Scholia Club, an important hub of Deweyan thought. Whitney's excitement about developments at Teachers College led him to found a number of grants and fellowships there using National Bureau of Casualty and Surety Underwriters money. This financial support led to a series of eight studies, including the two dissertations-cum-books *Safety Education in the Elementary School: A Technique for Developing Subject Matter* (1927) by Ruth Streitz and *Safety Education in the Secondary Schools: Objectives and Materials of Instruction* (1929) by Herbert James Stack (Whitney's eventual biographer).

Whitney's own writings and the writings that he commissioned during this period have an undeniable Deweyan flavor. By the 1920s Dewey had turned his attention to the place of *habit* in human life. He believed that many, perhaps most, parts of our lives were not conducted deliberatively or intentionally but in a reactive, habitual manner. Habit for Dewey was not merely repetitive action but a recurrent reaction to a given stimuli. "Habit means special sensitiveness or accessibility to certain classes of stimuli," he wrote, "standing predilections and aversion, rather than bare recurrence of specific acts."[90]

Whitney picked up on this train of thought in his children's play on safety education, titled *How Knowledge Driveth Away Fear: A Morality Play for Children*. Whitney included a number of anthropomorphized fears in the play, including Fear-of-Crossing-the-Street, Fear-of-the-Fire, and Fear-of-Mice. Two children, Mary and Tom, labor under these fears until the character named Knowledge comes to dispel their fright and set them in right relationship to the world and also in right relationship to their senses. When the children ask the fears where they came from, Fear-of-Crossing-the-Street explains, "I came first to you children one day last spring when you were starting to go to your grandma's; you were dancing merrily along and looking at the blue sky and at a robin singing in the lilac bush in your grandma's yard, when a big red automobile came dashing by and shrieked its horn and almost ran into the telephone pole near the gate trying to get out of your way."[91] In this view, fears largely stemmed from mis- or overinterpreting perceptions and experiences received through the senses; it was a habit that had outlived its childhood usefulness.[92]

The goal of safety education as articulated in Whitney's play was to replace fear with knowledge. As Whitney wrote in the afterword to the play, "Gradually, knowledge and reason become able to assume control, the effect of fear remaining normally only in the form of certain habits, acquired in the earlier fear-regime."[93] Whitney believed that through reason "we have undertaken, for better or worse, to control nature and to control ourselves" and to create order.[94] "Safety," he wrote elsewhere, "is the condition that allows order to appear and here again safety opens the way to adventure."[95]

The philosophy of American pragmatism, led by such thinkers as William James and John Dewey, was in large part a response to and incorporation of the evolutionary theories of Charles Darwin.[96] In contrast to earlier philosophies that saw humans as God-made things, the pragmatists attempted to provide a naturalistic picture of human beings as biological entities with ideas, goals, uncertainties, doubts, and limits. Whitney, too, embraced evolutionary metaphors. Yet he went further and applied this perspective to the nature of technological change, which had implications for his understanding of safety education. Technology, like biological life, Whitney argued, was constantly varying. In this context, standards that stayed technology's development seemed backward, conservative, and regressive. But Whitney believed this was not the case. As he wrote in *The Place of Standardization in American Life*, "Standardization is thus the liberator that relegates the problems that have already been solved to their proper place, namely to the field of routine [or habit], and leaves the creative faculties free for the problems that are still unsolved."[97] The structure of this sentence exactly mirrors one of his sentences on safety, quoted above: "Safety is the condition that allows order to appear and here again safety opens the way to adventure."[98] Whitney worried greatly about the new risks brought on by industrialization and technological change, and he believed that order had to be created through standardization and education to enable human thriving, what he called adventure.

Yet, as discussed earlier, Whitney rejected the philosophy of "safety first."[99] Anyone who would utter the words *safety first* clearly did not understand the meaning of human life. "Safety for greater adventures" was Whitney's motto: the goal of safety was to enhance human existence, not to make life anemic in the quest to avoid hazards. While it is unlikely that Whitney's work influenced Dewey, Dewey made this same argument about "safety first" in his 1929 book, *The Quest for Certainty*: "'Safety first' has played a large role in effecting a preference for knowing over doing and making."[100] In other words, it drove people away from concrete existence into a fantasy world of pure thought where the hazards and accidents of life could be avoided.

Whitney's thinking about and support for safety education eventually culminated in putting resources toward developing driver's education. He helped compose the first driver's education textbook, *Man and the Motor Car*, published in 1936. Much of the book was written by Herbert Stack, one of the Columbia PhDs whose research Whitney had funded. *Man and the Motor Car* was a work obsessed with creating order. As with Whitney's earlier safety plays and other writings, the object of the driver's education textbook was to overcome habits based either on instinct or on sense-driven fears, and this aim meant creating a reasonable and disciplined regime for relating to one's senses. This focus on the senses included, for instance, how to read traffic signs and traffic lights and how to properly use headlights at night.[101] Yet the advice of *Man and the Motor Car* depended on a world where traffic signals and headlights were standardized. In this way, the two sides of Whitney's work on auto safety—technical standards and education—came together. The goal of driver's education was to standardize training but in a way highly conditioned on having a regularized world.

As we have seen, the professional psychologists' goal of creating exclusionary driving tests largely evaporated. Driver's education became the dominant way to deal with drivers and the hazards of human activity. Like John Dewey, Albert Whitney had a thoroughly optimistic vision of human capability and potential. Friends called him an "idealist," a "dreamer."[102] In the end, when it comes to driver's education at least, Whitney was far too optimistic. Driver's education surely improved the skills of drivers, especially young ones, but over the coming decades, deaths and injuries mounted. And it was only when later reformers began to question the centrality of driver's education as a means to safety that more thoroughgoing changes came.

Into the Depression

For several reasons, the auto safety movement died down after the coming of the Great Depression. Herbert Hoover entered office as president in 1929, and although there was one more National Conference on Street and Highway Safety, the topic did not receive the focus it had attracted while Hoover was commerce secretary. Franklin Delano Roosevelt, who became president in 1933, never focused on automotive safety, nor did he put much emphasis on the Department of Commerce, preferring instead to use the powers and capabilities of other federal agencies to carry out his plans. The field of auto safety required attention and care, but both the Hoover and Roosevelt presidential administrations devoted most of their attention to other endeavors, chiefly those related to pulling the country out of economic depression. There are indications that the more general safety movement, which included auto safety, fell on hard times during the Depression. Use of the phrase "safety movement" shot up in the early 1910s, when the National Safety Council was founded, but then plummeted in 1929.[103] Perhaps actors simply had different priorities in the midst of economic hardship. Perhaps safety was the kind of thing people focused on only when times were flush. Or perhaps they grew frustrated because they believed their efforts were not working. In 1924, the year of the first Hoover conference, 19,400 people died in auto accidents. By 1930, that number had risen to 32,900, and by 1941, the eve of World War II, it was 39,969. (Granted, in 1924, there were about 18 million cars registered; in 1930, nearly 27 million; and in 1941, 35 million; but such statistics were cold comfort to those concerned with reducing death and carnage on the roads.) What all of this meant, however, is that fewer experts focused on the problems of the automobile and fewer new ideas were being generated; organizational verve saw a decline as well.

The efforts of professional psychologists in the 1930s largely resulted in failure. Albert Whitney continued working on safety education and published *Man and the Motor Car,* but his efforts increasingly took place in his own, autonomous world of safety education, which had less contact with engineering societies and other groups than it had enjoyed in the 1920s. The National Bureau of Standards' Heat Division focused less on the automobile and turned its attention to problems in aviation; and the American Engineering Standards Committee—by that time renamed the American Standards Association—took up no new standardization efforts related to auto safety. The turning of engineering organizations away from auto safety standards raises the question,

Did engineers turn away from this topic because they believed that their work was done? Or did their attention shift for other reasons? Regardless, as a result of these various Depression-era developments, the social and organizational network that had grown up around auto safety slowly faded.

In 1930 the National Conference on Street and Highway Safety met for a third time. Beyond continuing the push for uniform traffic codes, the meeting's most significant accomplishment was publication of the *Manual on Street Traffic Signs, Signals, and Markings*. The committee that published the manual was led by the American Engineering Council. The committee contained the usual characters, including Sidney Williams and A. H. Rudd, the railroad signal engineer who had helped develop street traffic signals. Staff members of the National Bureau of Standards and the Bureau of Public Roads also took part.

To peruse the manual is to discover a number of signs that have retained their standardized form to this very day, including the octagonal stop sign, diamond-shaped curve sign, one-way sign, speed limit sign, and keep-to-right and no-left-turn signs. Of course, some of the signs in the manual—such as a square sign reading "Caution Cross Street"—are no longer common or have disappeared altogether, but the number of signs that have persisted since 1930 indicate this effort's great success. At the same time, other standardization efforts met with more resistance. A 1938 congressional report found that "no State has adopted the Uniform Vehicle Code [which Hoover's conferences had come to promote] in its entirety, although some have adopted a large number of its provisions."[104] The same report concluded that its analysis "clearly demonstrates an amazing lack of uniformity in laws and the methods provided for their enforcement. This chaotic nonuniformity prevails . . . in all major problems relating to traffic safety."[105]

Standardization of automotive safety equipment also broke down during the 1930s. Whereas engineers had standardized headlights and brakes in the 1910s and early 1920s, by the late 1920s, several automakers—Ford chief among them—began introducing safety glass into their vehicles. Other companies, most notably General Motors, resisted this trend for cost reasons. Engineers undertook no major initiative to make safety glass standard, and it did not come before the American Standards Association. Eventually it came "standard" in all new automobiles sold in the United States because a critical mass of individual states—driven by consumer outrage—required it.[106]

The future of auto safety lay in a notion implicit in safety glass—a design feature to make the automobile less dangerous during a crash—and glimmers appeared in the 1930s of a consumer safety movement demanding such changes. Publications such as the journalist J. C. Furnas's "—And Sudden Death" that described traffic accidents in horrifying detail whipped up consumer worries.[107] Furnas's article appeared in *Reader's Digest* in 1935, was reprinted thousands of times, and was even filmed as a noiresque Paramount Pictures melodrama a year later. Furnas compared getting into an accident in a 1930s automobile to "going over Niagara Falls in a steel barrel full of railroad spikes," car interiors were so full of knobs, handles, sharp edges, and other projecting features that gored the body during a collision.

One person who got a firsthand perspective of this violence and the resulting damage was the Detroit-based plastic surgeon Claire Straith. Straith put seatbelts in his family's cars in the early 1930s, and also installed cushioned crash pads on the dashboard. He tried to market the crash pads but seems to have sold few of them. More important for Straith, he met with automakers to convince them to remove dangerous protuberances from dashboards and doors. His pleadings met with some limited success. Chrysler, Studebaker, and the ill-fated Tucker company incorporated Straith's ideas, but for the automakers, these ideas often became "safety features" that could be touted whenever the companies introduced new models to the public. Within a year or two, these safety features would be gone, replaced by whatever new thing the companies thought would appeal to consumers.

For most safety technologies introduced, the companies would soon backslide into old habits. Faced with such practices, safety-conscious consumers had to follow the motto *caveat emptor*—let the buyer beware. In the late 1930s, the journal *Safety Engineering* ran a multipart series titled "Make the Automobile Safer!" Part 15 of the series contained a feature titled "Eliminate These Hazards!" that included removing "sharp radiator ornaments" and "hazardous door handles."[108] The auto safety advocate Ralph Nader would repeat these same criticisms nearly thirty years later. More to the point, *Safety Engineering* also ran an annual review of new cars on the market, rating them according to a set of criteria for safety features. One such review opened with the line "The safest car of the 1939, in our opinion, is the Chevrolet!"[109] The same review argued that "many hazards remain" and that "safety [was] sacrificed to style"; the latter idea would form a constant refrain for Nader and other critics in the 1950s and 1960s. *Safety Engineering*'s safe car ratings shared some characteristics with the commodity specifications published by Hoover's Department of Commerce. But whereas the Commerce specifications centered on devices like brakes and headlights that

were meant to increase the driver's mastery of the vehicle while not impinging on other drivers' performance (say, through glaring headlights), *Safety
Engineering*'s ratings mostly focused on making cars—both the interior and
the exterior—safer for collisions. This focus would form the basis of post–
World War II auto safety efforts.

But if the 1930s prefigured a new kind of safety movement, the decade also
witnessed the rise of automaker organizations that would resist such efforts.
In 1937 the Automobile Manufacturers Association, the chief industry trade
association, created the Automotive Safety Foundation, an organization that
sought to increase road safety throughout the United States.[110] Two years
later, the foundation's head, Paul G. Hoffman, the president of the Studebaker
Corporation, published *Seven Roads to Safety: A Program to Reduce Automobile Accidents*, a work that famously put all emphasis on driver education,
state motor vehicle administration, and proper road building.[111] While Straith
and the writers at *Safety Engineering* were beginning to advocate for redesigning automobiles, these ideas did not make it into Hoffman's work—and
they would not make it into auto industry publications for decades to come.

The tension between these two camps—safety advocates and automakers—
created a new dialectic that simply had not existed in the 1920s and early
1930s, when a wider variety of actors were involved and the network they
constituted focused on shaping a wider variety of aspects of the entire
world surrounding the automobile, not just the car itself. This network favored standardization as its chief means of influencing the world. While the
drive for standardization and uniformity encountered some limits—states
and localities often held tight to local variation—the sweep and extent of
standardization that arose between the two world wars was impressive. In
the 1910s, auto standards were limited to steel strength, steel tubing sizes,
wheel hub dimensions, and the like. By World War II nearly every part of
the automotive world, from brakes to driver's licensing tests, and from traffic
signals to accident reporting forms, had been touched by the standardization process. The dialectic between consumer advocates and the automakers
would eventually give birth to its own standardization revolution, but that
transformation would require social and organizational developments that
were still decades off. And it would require new kinds of science, which were
at least partly born amid the havoc of World War II. The birth of these new
forms of science are the theme of the following chapter.

PART TWO

SAFETY

CHAPTER THREE

The Creation of Crashworthiness

In 1956 the writer John Updike traveled to Upstate New York to do a bit of re-porting for the *New Yorker*.[1] The occasion for Updike's visit was the unveiling of a new, "radically safe automobile" sponsored by the Liberty Mutual Insur-ance Company and designed by the Cornell Aeronautical Laboratory. Eventu-ally known as the Cornell–Liberty Mutual Safety Car, the vehicle would be equipped with nearly every automotive safety technology yet created, includ-ing energy-absorbing materials and dashboard, recessed control knobs, extra mirrors to improve vision, "strap-equipped bucket seats," and experimental hydraulic steering controls. The interior looked like something out of a science fiction movie. Updike, however, was more interested in the characters who designed this object and who struggled to explain the ideas behind it. Edward Dye, the engineer in charge of the project, told Updike, "The human head is in its essentials like an egg. Both consist of an ellipsoidal, hard shell enclosing a semi-fluid. In the ideally safe vehicle, the passenger is packaged like fragile merchandise." Updike noted that Dye's "own head was squarish, balding, courteously inclined; a pink freckle in the center of his forehead, an inch above the brow line, marked him as a Brahman of industrial engineering."

Frank J. Crandall, a vice-president at Liberty Mutual, vouched for Dye's analogy: "Ed has done a lot with eggs. He built little airplanes for them that you could smack against the wall without breaking the egg." Dye murmured, "That was for cockpit research I was doing." Updike seemed to put little faith in the Brahman and the insurance executive, who showed him only a drawing of the safety car. He asked when he might be able to drive such a vehicle, and the men got defensive. "Understand," Crandall explained, "that our purpose is mainly to make the results of our research available to industry.

We want fewer injuries on the highway. Already, I think, we have made people conscious of seatbelts." Dye insisted, "We *are* building this vehicle," adding that they'd have a working prototype ready in eleven months. Updike ended his report by quoting Dye: "You'll be able to look at it, sit in it, handle the controls—everything except one thing. There are still too many imponderables in the steering mechanism. You won't be able to drive it." Throughout the piece, Updike's tone suggested these eggheads would not be transforming the American automobile anytime soon.

In some ways, Updike was right. It would be over a decade before any of the safety technologies in this experimental vehicle consistently came standard in new cars. Yet the very existence of the Cornell–Liberty Mutual Safety Car—which eventually *was* constructed and began touring the United States in 1957—marked how far thinking about safety had come since the 1930s. The core idea behind the safety car's design, an idea usually called crashworthiness, was that humans were injured and killed in most traffic accidents when their heads and bodies were thrown against hard or sharp objects in the car's interior. If their bodies could be slowed or the objects padded, or both, injuries and deaths would be reduced. This chapter traces the emergence of crashworthiness from the 1930s to 1960. As Edward Dye's own research trajectory suggests, crashworthiness research originally focused on aviation, particularly military aviation, before "spilling over" into the domestic sector and attending to automobile safety. This chapter examines how experts first began working on crashworthiness and how they formed a community around such research, before turning to how the auto industry responded to these new ideas. The automakers' track records were mixed. Leaders at Ford Motor Company were initially enthusiastic about safety engineering before backing away. General Motors fiercely resisted the very word *safety* and held that stance for years. Finally, the chapter considers how politicians, especially members of the US Congress, took up crashworthiness. These legislative efforts amounted to little, but in many ways, they set the stage for the significant transformations that came later.

Impact Biomechanics

When World War II came, the automakers stopped producing cars and turned all of their capacities to churning out war matériel. Tanks, jeeps, airplanes—all were built using the auto industry's mass production facilities, creating what boosters called the Arsenal of Democracy. The automakers' war efforts had huge consequences for auto production, however. The manufacturers made no cars during the war, and when they began producing automobiles again after it, they returned to prewar models. The 1940s were a lost decade for the evolution of the automobile—and for new efforts to curb the risks of the automobile. As the previous chapter shows, the auto safety and standardization movement that took off in the late 1910s largely fell into disarray and enervation during the Great Depression and the war. The exploration of auto safety did not come to a complete halt, of course. The psychologist Harry DeSilva published his magnum opus, *Why We Have Auto Accidents*, in 1942. Still, there were few substantively new initiatives during the 1940s. In 1946, after Roosevelt died, President Harry Truman held a national auto safety conference, but it just brought together many of the same actors from Hoover's earlier conferences.[2] Sidney Williams, the National Safety Council leader who had been involved in discussions of auto safety since at least 1920, led Truman's conference. In the end, the 1946 conference and a second one held in 1948 focused (still) on getting states to adopt uniform traffic rules and improving driver's education. No new technical standards or qualitatively new ideas were considered.

Although federal officials were focusing on old ideas, it is also true that the period between 1939 and 1949 saw the birth of a new science. Known as impact biomechanics, the field's researchers examined how forces affected human and nonhuman animal bodies. The creation of impact biomechanics and the notion of crashworthiness apparently mark an instance of multiple independent discovery. The basic insight behind these developments—that people could be protected from damage in collisions—was implicit in the work of Claire Straith and the editors of the journal *Safety Engineering*, which in the 1930s published rankings of automobiles according to their safety features. Still, neither Straith nor those at *Safety Engineering* developed their nascent insights into a program of systematic research.

Hugh DeHaven, sometimes called the father of crashworthiness, was perhaps the first person to undertake such research in a systematic fashion. Some historical accounts of auto safety have emphasized DeHaven's role in founding research on crashworthiness, but his later visibility was largely a matter of luck, self-promotion, and what the individuals and groups who later controlled auto safety in the United States did with DeHaven's research. Rather than arising from one source, the science of impact biomechanics emerged as researchers began work at different times in different places in response to different—mostly local—needs and concerns. This truism of local, idiosyncratic needs driving early research was no less true of DeHaven himself.

Hugh DeHaven was born in 1895 to a wealthy family in Brooklyn, New York, where his father worked as an inventor.[3] After attending two prestigious boarding schools, he spent one year each at Cornell and Columbia universities before attempting to join the US military. He was rejected on medical grounds, but DeHaven remained undeterred. He went to Canada, where he joined the Royal Flying Corps. While training as a pilot in 1917, DeHaven experienced a midair collision with another plane, and he fell more than 500 feet. The others involved in the crash were killed, but De-Haven survived with two broken legs and ruptured organs. Why had De-Haven lived? Although he could point to factors such as the seatbelt he had worn and the way the cockpit was designed, he could not answer definitively, and the question haunted him for the rest of his life.

After DeHaven healed, he returned home to the kinds of business his father had conducted for decades. He worked for a number of firms, including an automobile manufacturer in San Francisco, but his main calling was as an independent inventor. His profitable inventions enabled him to enter semi-retirement in Cooperstown, New York, by the mid-1930s. Settling into his newfound free time, DeHaven returned to his earlier thoughts about crash safety and began exploring the topic. His investigations took two forms. One was the egg drop, perhaps the guiding metaphor of the crashworthiness paradigm. DeHaven would climb to the top of a ladder and drop eggs from increasing heights onto rubber mats. He found that eggs—long a cultural symbol of delicate structures—could withstand falls from significant heights as long as they fell on energy-absorbing materials. DeHaven's other mode of investigation involved collecting news stories about apparently "miraculous" instances of people surviving falls from incredible heights. DeHaven's investigations led him to believe that such survivals were not so miraculous after all but rather the product of energy dissipation and absorption, an insight he believed could be used to redesign vehicles, particularly airplanes. He called this insight a discovery and began building a mythology around it.

DeHaven wrote airplane manufacturers about his ideas in the mid-1930s, not long after he undertook his first studies. Although some companies expressed a general interest in crash safety, little came from these initial contacts. Indeed, DeHaven's "discovery" languished until a new context provided it a ripe space for growth. That context was World War II. Still looking for a space to develop his ideas, DeHaven wrote to Eugene DuBois, chairman of the National Research Council's Committee on Aviation Medicine and professor in the Department of Physiology at Cornell Medical College.

Sensing merit in DeHaven's research, DuBois attached DeHaven to the Cornell Medical College and directed National Research Council funds to support research on crash safety there. Soon thereafter, the Civil Aeronautics Authority—predecessor agency of the Federal Aviation Administration—became interested in DeHaven's work, lending further financial support to what was soon known as the Crash Injury Research Project.

DeHaven used his new Cornell connections to expand his investigation of accident cases. The Civil Aeronautics Authority gave him access to its reports of aircraft accident investigations, but more important, DeHaven began contacting hospitals around New York City to request that they forward him detailed reports on any traumatic injuries from falls, crashes, or collisions that came through their emergency rooms. While DeHaven met with some resistance from doctors and hospital administrators who believed that his requests violated patient privacy, he was able to amass a considerable number of cases. In 1942, the same year that he helped initiate the Cornell Crash Injury Research Project, DeHaven published the article "Mechanical Analysis of Survival in Falls from Heights of Fifty to One Hundred and Fifty Feet," in the second volume of the journal *War Medicine*, a publication of the National Research Council and the American Medical Association.[4]

Focusing on eight of the cases DeHaven had collected, the article argued that the human body was not nearly so fragile as folklore held. Normal adult bodies probably could withstand considerable forces that needed only to be measured empirically. Moreover, even his preliminary findings suggested that changes in design would "enhance survival and modify injury within limits in aircraft and automobile accidents." Such modifications included reformation of "windshield structures, seats, instrument panels, and safety belts."[5] With this article DeHaven was the first person to publish on the nascent idea of crash safety, and its reflections on the potentials of redesigning vehicles set the parameters of the coming battle between safety advocates and industry. Yet DeHaven was not alone in conducting such investigations, nor was his "discovery" unique.

In 1939 Elisha S. Gurdjian and Herbert Lissner, two professors at Wayne State University in Detroit, Michigan, began dropping dried human skulls on steel plates and observing their impact and the resulting damage. They started inducing and studying brain concussions in living dogs soon thereafter. Gurdjian was a surgeon who, after receiving his MD from the University of Michigan in 1926, had begun practicing in Detroit in 1930 and had then become professor of medicine at Wayne State in 1939.[6] In the late 1930s,

Gurdjian realized that the problem he was studying—brain injury—required more than a surgeon's set of tools. His interest in such injuries grew out of his frustration as an emergency room doctor. No one had yet formulated a robust, systematic theory of concussion, and Gurdjian and others believed that existing theories lacked a scientific basis. Brain injury arose from forces, and the applied study of forces, Gurdjian reasoned, was properly the domain of engineers. Gurdjian inquired with the Department of Engineering at Wayne State as to whether anyone there might be interested in working with him. The engineer who replied was Herbert Lissner, who had received his master's in theoretical and applied mechanics at the University of Illinois in 1934 and had arrived at Wayne State only in 1938.[7] Lissner was a specialist in materials testing, more specifically the study of stress and strain in metals and other industrial products. Together, Gurdjian and Lissner possessed the expertise in engineering and medicine that other researchers in crash safety, such as DeHaven, lacked.

Both men had full-time day jobs, so Gurdjian and Lissner worked at night, often beginning at ten and not ending until three or four in the morning. They lacked facilities and equipment but found available resources at one of the Detroit automakers' facilities. "Taking anaethetized dogs into the auto plant at 10 P.M., past the plant protection, makes quite a story," one auto safety expert recalled.[8] The year 1939 was a propitious time for Gurdjian and Lissner to become interested in the study of forces and their influence on bodies. Materials testing had a long history in the auto industry and a much longer one in other industries.[9] Yet at the 1938 meeting of the American Society for Testing Materials (ASTM), the chief professional society for this field, engineers introduced two new methods for examining strain.

In the first method, testers applied a thin lacquer known as Stresscoat to the object under study. After submitting the object to intense forces, testers could examine it to see where stress cracks—now made visible in the Stresscoat—formed on the object. This method aided engineers in see where objects formed cracks that would lead over time to a material's weakening and ultimate failure. Gurdjian and Lissner applied Stresscoat to dried skulls and, after dropping them on various surfaces, used the cracks in the lacquer to determine the skull's weakest points.

After conducting Stresscoat studies, Gurdjian and Lissner applied the second testing technique unveiled at the 1938 ASTM meeting—known as the "bonded resistance wire strain gage"—designed to explore the physiological mechanisms of concussion. The strain gage relied on a basic electro-physical principle.[10] When a wire is stretched, its electrical resistance increases; when compressed, it decreases. Strain gages determine to what degree an object is expanding or contracting by tightly bonding wires to its surface. When the thing expands, the wire expands; the same with contraction. Testers use electrical instruments to capture the changing resistance of the wire. In 1936 researchers at the Guggenheim Aeronautical Laboratory at the California Institute of Technology (Caltech) had run up against a wall in their materials-testing projects. A twenty-five-year-old technician, Edward E. Simmons Jr., proposed a solution. Engineers had used changes in the resistance of electric wire to measure strain for some time, but primarily at large scales, such as across a bridge or dam. Simmons miniaturized the strain gage by weaving wire back and forth on a small piece of cardboard that could then be applied to nearly any surface. Simmons's strain gage became available commercially in 1938.

Many of Gurdjian and Lissner's strain gage studies involved gruesome animal experiments that would not make their way past university ethics committees today and that make for tough reading. In one experiment, they applied strain gages and related instruments to the study of concussion. First, they took a number of "mongrel dogs weighing 5 to 15 kilo" and, with the dogs "under intravenous nembutal anesthesia," they cut off the animals' scalps and carved away their masseter muscles.[11] After stopping all bleeding, the researchers "carefully dried and polished" the skull and affixed a stress gage to it. Then, Gurdjian, Lissner, and their associates would prepare the "pressure plugs." Pressure plugs relied on a dynamic similar to that behind strain gages: researchers stuck a wire into an enclosed, typically liquid-filled space. When the space's internal pressure increases, the wire shortens, thus decreasing its electrical resistance. The researchers drilled two holes in the dog's skull. They filled these holes with the pressure plugs, ensuring that the wires made "contact with cerebrospinal fluid and the brain." Then they hit the dogs on the head with calibrated radial hammers. The strain gages and pressure plugs indicated both the structural deformation of the skull and changes in its internal pressure as the hammer struck the dog's head. In some experiments, researchers hit the dogs on their heads with hammers every half hour, including in one case for six and a half hours, until the dogs died.

With the Stresscoat and strain gauge experiments, Gurdjian and Lissner initiated a long tradition at Wayne State of pioneering innovative methods for studying human (and nonhuman) limits. Lissner would eventually go on

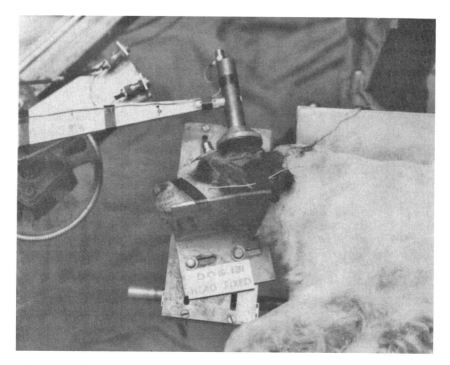

Figure 3.1. A living, anesthetized dog instrumented and positioned to receive a blow to the head from a hammer, in a study of concussion at Wayne State University. Nonhuman animals, both living and dead, and human cadavers were frequently used as testing materials in impact biomechanics studies. Stapp Car Crash Conference, Society of Automotive Engineers International, 1965

to cowrite the definitive book on strain gage methods, which focused on testing industrial materials but included "one of the more unusual applications" of strain gages "to bone, both human and animal."[12] These studies lead to construction of the Wayne Curve, which represented human tolerance to physical forces and which later greatly influenced federal regulation. Much of the research at Wayne State was supported by the auto companies in the form of both money and facilities. General Motors' Fisher Body Company and the Chrysler Corporation lent the researchers equipment for their early studies, and Ford and General Motors provided significant financing as well as collaboration in later research.[13]

In the 1950s, the researchers at Wayne State removed the cage from its elevator shaft in the eight-story medical school building, turning the elevator shaft into an impromptu laboratory.[14] They began dropping human cadavers and living nonhuman animals down the shaft. A pneumatically

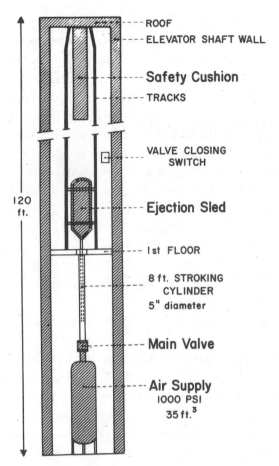

ROOF
ELEVATOR SHAFT WALL
Safety Cushion
TRACKS
VALVE CLOSING SWITCH
120 ft.
Ejection Sled
1st FLOOR
8 ft. STROKING CYLINDER 5" diameter
Main Valve
Air Supply
1000 PSI
35 ft.³

Figure 3.2. The Wayne State University Accelerator was perhaps a unique scientific instrument—an elevator shaft with its cage removed. The elevator shaft was a highlight of one of the Stapp Car Crash Conferences located in Detroit, with Stapp and other experts touring the facilities. National Academy of Sciences, Space Science Board, Man in Space Committee, *Impact Acceleration Stress* (Washington, DC: National Research Council, 1962), 64.

driven sled also allowed researchers to accelerate and decelerate the thrusting of test subjects (cadavers and nonhuman animals) up the elevator. In some experiments, cadavers were dropped upside-down so that their heads slammed into a heavy steel plate mounted at the bottom.[15] Many years later, when the author Mary Roach visited the facility, researchers fitted the cadavers in "Smurf-blue leotards" (purchased from an unwitting dance store) to keep all the various bits together during the impact.[16] Because of the "scarcity and non-uniformity of cadavers," as well as "the aversion a few people have to working with this particular test material," Wayne State researchers contributed significantly to the construction of dummies and other stand-in technologies.[17]

If impact biomechanics had remained a series of disconnected inquiries by the likes of DeHaven and the Wayne State researchers, its influence would have been limited. The individual most responsible for drawing this emerging field of study together, eventually into an extended network, was Colonel John Stapp. Born to Southern Baptist missionaries in Bahia, Brazil, in 1910, Stapp developed an interest in science early.[18] An old Brazilian castle served as both his family's home and the American Baptist College, which Stapp's parents ran. Strange sounds nightly filled the castle, which was rumored to be haunted. A deeper investigation, however, led Stapp to a nearby rum factory, where opossums were "sipping the mash, getting tanked up, and scampering over the college roof." Fascinated, he continued studying these animals and other local flora and fauna.

After studying zoology and chemistry, Stapp earned a PhD in biophysics at the University of Texas, Austin, then, in 1943, an MD from the University of Minnesota. During his internship at a hospital in Duluth, Minnesota, he had a conversion experience. "I had only seen pure scientists before, the prima donnas in universities working in their nitpicking ways at academic doodling to impress each other. Now for the first time I saw science and men of science working as a team, bringing everything to bear, the enormous facilities of the hospital, their own talents and devotion to the saving of human life."

In 1944 Stapp joined the war effort as a first lieutenant in the US Medical Corps. While touring the Aero Medical Laboratory in Dayton, Ohio, Stapp asked to see "something interesting."[19] He was shown early studies in flight biomechanics. Airplanes, especially emerging turbojets, subjected human beings to previously unknown forces. Test pilots and others were injured and often killed by cutting-edge technologies, as when they triggered ejection seats and slammed into still air. Medical doctors, biologists, and other scientists worked to understand and diminish the effects of these forces by conducting a number of novel experiments involving controlled airplane dives, compression chambers, and large centrifuges. Not long after this tour in Dayton, Stapp founded laboratories and initiated a series of innovative biomechanical studies—first at the aeromedical facility of Edwards Air Force Base in California, and then at the Aero Medical Field Laboratory of

Holloman Air Force Base in New Mexico—studies primarily focused on the needs and limits of jet pilots.[20] He moved up quickly in rank and stature.

The scientific instrument that became central to Stapp's work and eventually made him famous was the rocket sled, propelled down 2,000 feet of rail built on the flat desert lands at Edwards Air Force base. He also used the Northrop Aircraft Gee Whizz rocket sled.[21] His primary interest was studying the effect of rapid deceleration on the human body so that designers could improve ejection seats, safety harnesses, and other airplane devices. Stapp began experimenting with anesthetized chimpanzees and anthropomorphic test dummies, rocketing them down the rail at hundreds of miles per hour before bringing them to a dead stop within one or two seconds.[22] The many animals used in the research—mice, hamsters, cats, dogs, hogs, bears, and baboons—were kept in an area of the air force base called the zoo. After experiments, the hogs were barbecued and served at "group picnics" on the sun-drenched base.[23] It was a practice that Stapp had started in college when he would cook the meat of pigeons and guinea pigs he dissected for his studies. "If it breathed it had protein, and if it had protein, I ate it."[24] Stapp later said before the US Congress, "You wonder why I use hogs and chimpanzees? Well, man is somewhere between the hog and the chimpanzee. Some people are more like hogs; others are more like chimpanzees."[25] In one of the rocket sled experiments, the 185-pound anthropomorphic test dummy known as Oscar Eight-Ball broke free of the sled's safety harness, "slammed through the inch-thick pine windshield as if it were tissue paper, and soared 710 ft. down the track bed."[26] Even so, Stapp was determined to begin conducting rocket sled experiments on himself to prove that humans could take more force than common sense of the day assumed.

Stapp's first self-experiment took place on December 10, 1947, and he rode the rocket sled fifteen more times over course of the following five months, a feat he would perform nearly thirty times by the mid-1950s. His experiments took him to the limits of what any human being had ever experienced. He found, for example, that during extreme accelerations and decelerations his body's visual system was profoundly affected, likely from insufficient blood flow: "My vision became a shimmering salmon-colored field with no images. . . . It felt as though my eyes were being pulled out of my head, about the same sort of sensation as when a molar is yanked. . . . When the sled stopped, the salmon-colored blur was still there. . . . I lifted my eyelids with my fingers, but I couldn't see a thing. It was as though I was looking directly at the sun through closed eyelids."[27] Stapp's eyeballs would become

Figure 3.3. Military researchers load Colonel John Stapp onto a rocket sled. Stapp used the sleds on crash test manikins and living and dead nonhuman animals but also volunteered himself as a test subject to explore the limits of biological tolerance to physical forces. In his most famous and extreme test, Stapp reached speeds of about 632 miles per hour before stopping in a little over a second. National Air and Space Museum, Smithsonian Institution (NASM 00176366).

dark-purple bruises, and it took minutes for him to regain sight. He broke his arms and wrists more than once; he cracked ribs; and six fillings flew from his teeth. In 1954, now at Holloman Air Force Base, Stapp conducted his most famous and intense test on the Sonic Wind No. 1 rocket sled (also a Northrop product), reaching 632 miles per hour before stopping in a little over a second.[28] Afterward, he wanted to conduct experiments to go even faster, but the military ordered him to stop.

Stapp's experiments brought him fame. Magazines and newspapers put him on their front covers and pages. *Time* called him the "Fastest Man on

Earth." He appeared on the television show *This Is Your Life*, and Hollywood made a movie, *On the Threshold of Space*, partly based on his life.[29] Over the course of his career, he also won many awards, professional recognitions, and medals. Yet, while Stapp enjoyed this attention, he did not rest content. In 1955 he held the first of what would become known as the annual Stapp Car Crash Conferences, through which he would have the largest influence on the development of automotive safety. Although only twenty-six people came to Stapp's first conference, hundreds attended by the mid-1960s. Researchers came from universities all over the United States, including Harvard Medical School, the University of Minnesota, Tulane University, and the University of Michigan, as well as from abroad, and the Stapp Conference became the institutional home for crash safety research.

Stapp initiated the conference for a number of reasons. He realized early on that his research could have important implications for civilian domestic problems, especially auto accidents. A constantly repeated impetus for Stapp's establishing the conference was his realization that, as one essay put it, "almost as many Air Force Personnel were killed and injured in automobile accidents as in airplanes, so he added increased automobiles safety as one of his goals."[30] Stapp doubtless also knew that creating beneficial spillover from the military to the civilian sector would help ensure continued funding for his efforts. Stapp emphasized a culture of witnessing at the conferences.[31] Academic scientists, industry representatives, and law enforcement agents would come together to watch crash tests and experiments on cadavers and nonhuman animals.[32] Observing crash tests became a conference tradition. The armed services and industry, as well as government agencies such as the Public Health Service, financially supported a great deal of the research presented at the Stapp Conferences. In 1958, *perhaps* under the influence of lobbying by automakers, the congressional committee in charge of the military's research budget threatened to cut funding unless Stapp stopped his auto safety research.[33] Stapp halted his own research but continued the conference, made appearances before Congress, and supported auto safety in other ways.

While Gurdjian, Lissner, and others involved in the emerging impact biomechanics community became interested in auto accidents early on, most of the early studies in the field—and most of the money for research—concentrated on *aviation* and often specifically on *military* flight. The studies that DeHaven, Stapp, and others spearheaded took hold in a Cold War environment focused on developing new weapons systems, such as the jet fighter. Just as the auto safety movement of the 1920s had involved transfer of knowledge, including the use of signal lights, from the railroad industry to city streets, the science of crash safety developed primarily around aviation problems before being mobilized to deal with automobile accidents. Before 1950, nearly all instances of the word *crashworthiness* appeared in publications related to aviation, *not* automobiles.[34] In 1950 the Institute of the Aeronautical Sciences published one of the early papers on the subject, "A Note on Crashworthiness," by William I. Stieglitz, an employee of Republic Aviation Corporation, a manufacturer of military aircraft. Moreover, the crash test dummy, or manikin, which later became one of the chief symbols of auto safety, was also a product of the aviation world. One of the first known dummies, Sierra Sam, was invented in 1949 for the purpose of rocket sled experiments and other tests rooted in aviation safety.[35] The science of auto safety stemmed largely from the science of airplane safety.

The character of safety research changed and expanded throughout the 1950s as the groups began to focus more explicitly on automobiles rather than solely on airplanes. DeHaven's work is a good example. In 1951 Elmer C. Paul, a corporal in the Indiana State Police, began a novel program of accident investigation and data gathering.[36] The following year, Paul approached De-Haven to see if the Cornell crash researchers could scientifically analyze the data he had gathered, which eventually led to a formal relationship between the Indiana State Police and the Cornell program. As the previous chapter shows, accident investigation was not new in the 1950s: the National Safety Council had pushed standardized accident reporting forms in the 1920s and 1930s. What made the Indiana-Cornell program pathbreaking was that its standardized forms were focused not on what had *caused* the accident, such as driver distraction or poor visibility around a blind curve, but on what *injuries* had resulted from the accident. The program's forms included outlines of human bodies on which police would draw injuries. DeHaven and Paul worked with the new Traffic Institute at Northwestern University to develop a program that trained police officers to use these forms.[37]

Paul also made an important contribution to thinking about auto safety. Drawing on his accident investigation data, he reasoned that most people were injured or killed during auto accidents when their bodies were thrown upon protruding objects within the vehicle or when their heads slammed into dashboards, support columns, or other physical structures. In this way, Paul argued that people are not harmed by the "first collision"—that is, contact between their car and another car or, say, a tree—but by the "second collision" between their body and the car's interior.[38] The goal, thus, would be to design the auto interior in such a way that that this second collision was rendered less severe. Later, consumer safety advocates would take up the notion of the second collision in their argument for federal automotive regulation. Other metaphors, such as the idea of "packaging" those in automobiles, like eggs in a cardboard carton, or of placing them in a "cocoon," were also quite common among safety advocates.

In 1952 DeHaven and other members of the Cornell faculty formalized these automobile-focused research efforts by creating the Automotive Crash Injury Research Project. The initiative received additional collaborators from the Cornell Aeronautical Laboratory (later the Calspan Corporation). Started during World War II as part of the Curtiss-Wright Airplane Division, the facility became the Cornell Aeronautical Laboratory in 1946. The addition of the aeronautical laboratory meant the inclusion of engineers, leading to the medical-engineering hybrid that already described the research at Wayne State. Hugh DeHaven led the Automotive Crash Injury Research Project for two years, until he retired in 1954 and John O. Moore, formerly a researcher focused on flight safety at Republic Aviation Corporation, took over its management.[39]

DeHaven had an industry-friendly vision of safety research. Throughout his tenure with the project, he hesitated to go public with any findings for fear of not seeming objective.[40] When he held a planning conference for the project that included representatives from the automakers, he attacked the article "Are Car Manufacturers Killers?" which had just appeared in *Magazine Digest*. "Articles of this type," DeHaven declared, "are appearing in increasing numbers. For the most part, they are distinctly unfair in their reflections on automobile engineers and manufacturers who have not had any crash-injury data to show what is happening in their products."[41]

Moore and his colleagues carried out a number of auto safety research projects at the aeronautical laboratory, the most noteworthy of which was the creation of the Cornell–Liberty Mutual Safety Car described at the outset of this chapter, a proof-of-concept vehicle that featured at least sixty safety innovations. The car was primarily designed by Cornell laboratory staff member Edward Dye, a civil engineer who had gained a great deal of road-designing experience before working at Douglas Aircraft and then Curtiss-Wright during World War II, and Frank Crandell, a vice-president and the chief engineer at Liberty Mutual Insurance Company.[42] The insurance industry had plenty of economic incentive to participate in auto safety research. By 1930, auto accidents took up 30 percent of new lawsuits in many urban courts, and this was before states passed compulsory insurance laws.[43] Massachusetts began requiring that all drivers be insured in 1925, but other states did not follow suit for over thirty years. Yet in the mid-1950s the compulsory insurance issue reemerged in political discussions: New York state passed a compulsory insurance law in 1956, and North Carolina in 1957. The insurance industry was about to become financially responsible for a vastly increased number of accidents.

When John Updike visited Dye and Crandall in 1956, the two engineers explained that their design drew largely on features and principles developed for the aviation industry and for the football helmets that Cornell researchers had created.[44] As a safety award given to Dye later put it, "The project's purpose was to reduce the research from previous detailed [impact biomechanics] investigations and integrate the findings into a more understandable and usable form; to demonstrate that such [safety] designs could be engineered; and to make this information freely available to the automotive engineering community. An additional purpose was to create public awareness of safety features and their feasibility by means of a concrete example."[45] The Cornell–Liberty Mutual team unveiled the car to the public in 1957 and toured it around the country, with stops including the Smithsonian Institution. The vehicle came to have an important influence on the young lawyer Ralph Nader. Yet Liberty Mutual president Bryan Smith considered the unveiled safety car to be the culmination of the project's efforts, and he did not publicly challenge the automakers to adopt its design elements.[46]

Other areas of the country also saw important developments in crash safety science. In the late 1940s and early 1950s, the University of California Los Angeles (UCLA) became another influential home of crash science. The California legislature formed the Institute for Transportation and Traffic Engineering (ITTE) at UCLA in 1947. The institute's founding in part reflected the fact that the development of Los Angeles and other major California cities depended heavily on the automobile. Complaints about traffic in the City of Angels went back at least to the late 1930s. In addition, legislators believed that infrastructural maintenance had been neglected during the Depression and World War II. Thus, the ITTE was to comprehensively study California's transportation systems. In 1949 the institute expanded this scope to include crash safety research.[47] While several engineers at UCLA played influential roles in the development of auto safety beginning in the 1950s, two stand out for our purposes. The first was Derwyn M. Severy, the chief research engineer at the ITTE, who had piloted the F6 Hellcat during

World War II and remained a commander in the US Navy Reserve through-out the 1950s.[48] The other individual was one of Severy's students, Robert Brenner, whose career is examined in the next chapter.

Initially, the institute worked with local stunt drivers—the Joie Chitwood Auto Daredevils—who crashed cars for a living in front of stadium audi-ences. The UCLA engineers instrumented the vehicles and used external high-speed cameras to capture the collisions, but the results were less than satisfactory. (The use of high-speed cameras in the study of impact biome-chanics also seems to have been tried first in aviation.[49]) As the engineers later noted of the stunt driver tests, "While we learned a great deal and re-ceived splendid cooperation from management and drivers, we were even-tually forced to the conclusion that the demands of the audience and our research requirements were incompatible."[50] Like DeHaven, researchers at Harvard University, and others, the institute for a while focused on accident investigations, but the UCLA engineers found the information that could be culled from accidents too limiting because, for instance, "speeds at the in-stant of impact cannot be accurately determined [afterward] in the majority of accidents."[51] Over the following few years, UCLA researchers made a num-ber of fundamental contributions to the methods and instrumentation of con-trolled crashes, particularly in the use of dummies and high-speed cameras.

In this way, the science of auto safety involved the triangulation of three kinds of studies: accident investigations, direct experiments with bodies (human and nonhuman), and crash tests. Although DeHaven's examina-tions of accidents were initially important for suggesting that humans could withstand significant forces, the real role of accident investigations was to describe what kinds of injuries individuals sustained during crashes. Such investigations led to a statistical, epidemiological understanding of automo-tive injuries. The studies conducted at Wayne State shed light both on the relationship between known forces and injury and on the biological pro-cesses, such as concussion, that arose from injury. And, finally, crash tests, developed at UCLA and elsewhere, enabled researchers to examine how sec-tions of the automobile reacted to collisions and—via high-speed cameras—how dummies moved about the car during the seconds after impact. Over the course of the 1950s, these studies converged on points of relative consensus that later became the building blocks of federal auto safety regulation in the United States and elsewhere. And the Stapp Car Crash Conferences—and their published proceedings—were the primary spaces where these studies were brought together.

Industry Responses

The automakers and the Society of Automotive Engineers did not ignore the emergence of impact biomechanics, but developments in the field mostly took place outside the auto industry itself (though sometimes with industry funding). In the early 1930s, General Motors initiated a series of filmed crash experiments at its renowned Proving Ground, just before the steel-bodied car became the industry standard. The point of these experiments, which included both frontal barrier impacts and rollover tests, was less to protect human bodies within vehicles than to examine the durability of auto bodies.[52] GM's experiments were apparently short-lived because, as a later GM report concluded, "most of the early tests looked alike to the unaided eye and the gross damage on repeated tests was quite similar."[53] Therefore, "it did not seem necessary to conduct the tests on successive model years." The company returned to crash testing after World War II when, as one GM engineer put it, "changes in body design were evaluated much more carefully than had seemed necessary in the development of pre-war designs." GM engineers found rollover tests particularly useful in these postwar experiments and, consequently, lowered both the overall vehicle height and the center of gravity and improved tire treads to improve vehicle stability.[54] GM also ran a number of frontal crash tests during this period.

Interestingly, the GM engineers' findings were precisely the opposite of the conclusions DeHaven, Stapp, and other pioneering members of the impact biomechanics community had reached. For the GM engineers, "the catastrophic nature" of their crash tests "resulted in the belief that the threshold of serious and probably fatal injury is far below normal highway speeds."[55] Moreover, they believed that the tests showed that "the severity of impact at low speeds is surprisingly high." For these reasons they concluded that "it is impossible to provide secure protection during impacts of this nature by any amount of design modification, or any restraining devices that the average driver would be willing to use." Consequently, even through the early 1960s, "safety research" at General Motors primarily focused on, as the title of one paper put it, "roadside design for safety," though simulated crashes were sometimes used in research on guardrails and the like.[56]

In 1951, as part of a Society of Automotive Engineers panel titled "Packaging the Passenger," in which DeHaven also participated, GM's chief safety engineer, Howard Gandelot, presented his findings. In his paper, "Engineering Safety into Automobile Bodies," instead of emphasizing the potential of the emerging crash safety research, he offered a spirited defense of how far the automobile and the industry had come in safety since the late 1910s, particularly in the design of automotive bodies.[57] He outlined a history of design changes, from the closing of the automotive body, to the improvement of visibility problems inherent in early closed bodies, to the coming of the all-steel body in the 1930s. "The automobile industry, since its inception, has been noted for its progressiveness," he argued, and consequently new cars were much safer than older ones. Yet Gandelot expressed resistance to many of the new safety features being proposed by crash safety advocates. "Convenience for the motorist is an important consideration. It is better for the Industry that the merit of a safety feature be judged from factual determinations instead of its possible value to the advertising agency copywriter." And he held little hope for seatbelts. "Imagine trying to interest one's wife who is wearing a light-colored dress, fresh from laundry or dry cleaner, to fasten a safety belt or get into a shoulder harness that had been in use for even a short time and had only a slight appearance of soil."

Gandelot publicly opposed the inclusion of seatbelts in automobiles after the Cornell auto safety researchers published a report in 1954 showing that ejection from vehicles was a major source of injury in accidents and that seatbelts were the most plausible solution. He argued that "there is not sufficient factual information on the protective value of seat belts in automobiles to form any definite conclusions."[58] When a New York bank executive wrote to GM complaining that the his son had broken his tooth on the dashboard and suggesting that GM pad its dashboards, Gandelot responded:

Driving with young children in an automobile always presents some problems. As soon as the youngsters get large enough to be able to see out when standing up, that's what they do—and I don't blame them. When this time arrived with both our boys I made it a practice to train them so that at the command "Hands!" they would immediately place their hands on the instrument panel if standing in the front compartment, or on the back of the front seat if in the rear, to protect themselves against sudden stops. This took a little effort and on a couple of occasions I purposely pumped [the brake] a trifle when they didn't immediately respond to the command so that they learned quickly. Even now, when either of them is on the front seat, at the command of "Hands," they brace themselves. I frequently give these commands even when there is no occasion to do so, just so we all keep in practice.[59]

When the banker wrote an angry rejoinder to this unsatisfactory response, Gandelot wrote back, "A lot of people are hurt in bathtubs too, aren't they? Do you hear anybody demanding that they take the bathtubs out of homes?"

Not all companies followed GM's position on this matter. Ford took another tack altogether. In the mid-1950s, Robert McNamara, who would later play a large role in the Vietnam war as secretary of defense in the Kennedy and Johnson administrations, was head of the Division of Vehicles at Ford Motor Company. McNamara was one of the "Whiz Kids," a group of ten men who had moved from the Statistical Control unit of the US Air Force to Ford after World War II. A fervent believer in the power of numbers and statistics, McNamara was bothered by the annual count of auto injuries and deaths in the United States.[60] He was determined to make Ford's cars safer and to demonstrate that safety *could* sell, pushing back against the industry mindset captured in the phrase "Safety doesn't sell," variously attributed to longtime General Motors head Alfred Sloan and Ford executive Lee Iacocca.

McNamara was helped in this quest by Alex Haynes, an automotive engineer who had worked in aviation with Curtiss-Wright and General Motors during World War II, before joining Ford in 1945. Company president Henry Ford II directed Haynes to look into the emerging science of crash safety after hearing of the research coming out of Cornell.[61] In February 1954, eight months before Cornell released its report on door openings, Haynes and his Ford engineers were already at work on safety door latches as well as a collapsible steering wheel that would absorb some of the energy in a frontal crash when the driver was thrown forward.[62] By the time the Cornell report came out, Haynes and his colleagues were already crash-testing these new devices. They joined forces with members of the impact biomechanics community. With the Wayne State researchers, Ford engineers conducted cadaver tests on their energy-absorbing instrument panel designs, and they had Stapp run tests on their energy-absorbing steering columns with anesthetized hogs.[63]

The new safety door latches were an easy fix, and they were added to all Ford cars late in the 1955 model year. Redesigned instrument panels and steering wheels and columns proved considerably more difficult to change in the Ford production process. Instead of adding an energy-absorbing steering *column* that partially collapsed during an impact, the company opted for a steering *wheel* connected to the recessed hub by thin metal bars that would bend and absorb energy during a crash.[64] The company also began offering seatbelts as an option in 1955.

McNamara was determined to go beyond this incremental process of adding safety devices, however. Outside the company, he directed $200,000 to research at Cornell and convinced Chrysler to lend some support as well.[65] General Motors refused. Inside Ford, McNamara convinced executives to develop comprehensive design changes for its 1956 model year cars. Some of these changes, soon known collectively as the Lifeguard Safety Package, including the deep-dish steering wheel, safety door latches, and rearview mirrors with safety glass, became standard on all Ford vehicles.[66] Others, like front and rear seatbelts and padded dashboards and visors, were offered as options. Perhaps more important, McNamara persuaded others at Ford to make safety the center of the company's advertising for the year. The safety advertising campaign began in the fall of 1955. The well-known actresses Mary Martin and Ethel Merman, who had been featured in Ford's Fiftieth Anniversary television event in 1953, hosted a television special on Ford's safety package. The company released a series of advertisements soon thereafter with lines like "Coming Friday, the first major contribution to your driving safety—Ford Lifeguard Design."[67]

General Motors publicly attacked Ford's advertising campaign, arguing that its claims were not based on sufficient scientific research. The public could easily observe the tension between the automakers over safety, with Ford and Chrysler on one side and the industry behemoth GM on the other. An October 1955 article in *Bluebook* Magazine titled "Big Three Fight over How Safe to Make Your Car" stated baldly, "The battle erupted over seat belts, which Ford and Chrysler are pushing and GM is opposing."[68] General Motors, it argued, is "refusing to break the tradition that insists safety is a dirty word." GM executives also applied pressure on individuals within the Ford hierarchy, focusing especially on Walker Williams, Ford's vice-president for sales.[69] It is difficult to judge how popular the optional components of the Lifeguard Safety Package were with the car-buying public. McNamara still stood by it in 1957, claiming that it sold and that the public would buy safety; and safety advocate Ralph Nader later claimed that public response to the company's advertising campaign "brought a demand for more seat belts than the company could provide at first."[70] But this was not the perception within Detroit or even within Ford, whose car sales had weakened while those of General Motors increased. The quip "McNamara sells safety. Chevy sells cars" said it all.[71] Three months into Ford's safety campaign, leaders at the company caved under the pressure. McNamara almost lost his job and was forced into an extended vacation in Florida under the pretense of illness, while others at the company scuttled the safety message and retooled the company's entire advertising campaign.[72]

Government Regulation of Auto Safety

In the end, the importance of impact biomechanics did not lie primarily in how automakers voluntarily took up its ideas but in what it prompted legislators and policymakers to do. The symbolic moment when the new science of crash safety became a *regulatory* issue came in 1957 when the impact biomechanics community—in the persons of Edward Dye, Derwyn Severy, and John Stapp—made its first appearance before Congress. The community's initial public presentation was the working of Kenneth Roberts, a US representative from Alabama. Roberts was a conservative Democrat (eventually voted out of office after passage of the Civil Rights Act of 1964) who as a young, newly elected representative initiated a campaign in the early 1950s to reform refrigerator locks. Children occasionally climbed into abandoned refrigerators and were trapped inside, where they suffocated, as had happened in Roberts's district. Roberts was one of the chief sponsors of a bill—eventually passed as the Refrigerator Safety Act of 1956—that required refrigerators to be openable from the inside and that ultimately led to magnetic locks as the industry standard.

In 1953 Roberts married, and during his honeymoon he and his wife were involved in an automobile accident in which their car was badly rear-ended. "Roberts opened the trunk expecting to find their china and crystal wedding gifts shattered, but to his surprise, the carefully packaged fragile items were intact."[73] A year later, Roberts, along with several other members of Congress, was shot when members of the Puerto Rican Nationalist Party stormed the House of Representatives. While convalescing, Roberts thought about the fragility of life and dedicated himself to doing more to protect others. His thoughts returned to his earlier automobile accident; by then, perhaps, he had heard of the emerging work in impact biomechanics (*Collier's* ran a cover story on Stapp titled "Fastest Man on Earth" in June 1954, three months after Roberts was shot).[74] For a variety of reasons, others in Congress supported Roberts's traffic safety initiatives. Indeed, the US Congress at this time was dealing with a number of compensation cases for the families of individuals who had died in auto accidents while in US service or who had been killed in collisions with US military vehicles on American roadways.[75] Moreover, since Congress had some lawmaking jurisdiction over the city of Washington, D.C, discussions of automotive safety would periodically pop up in the context of regulating the district's roads.[76]

In 1956 Roberts formed and chaired the Subcommittee on Traffic Safety, appointed by the House Committee on Interstate and Foreign Commerce.[77] Over the course of that year, the subcommittee held a series of hearings in various cities around the United States, including Detroit. More than one hundred experts appeared at the hearings, including police officers, state motor vehicle administrators, engineers and executives from the automakers, and representatives from the National Safety Council. Most of the commentary, both from the congressmen and the experts, focused on traditional auto safety ideas, such as improving traffic flow and heightening police presence on roads, though the findings of impact biomechanics research certainly made an appearance there too. The hearings opened with Senator Paul Douglas of Illinois urging the subcommittee to consider the promulgation of new automotive safety standards, either through "voluntary action" and "patent sharing" by industry; by a public, nonprofit body such as "a health institute" or Cornell; or by a government agency.[78] Participants opined that a worst-case scenario for the automakers—and likely everyone else—was that each state would adopt its own auto safety standards. Few followed Douglas in making such strong recommendations, however. Interestingly, the automakers themselves conducted most of the discussion regarding the relationship between crash science and auto design. John Moore was the only biomechanics researcher to testify. The representatives from the automakers who spoke celebrated their accomplishments in internalizing the lessons of crash science, rather than emphasizing the potentials for further design changes. For example, Ford's Alex Haynes explained that the company had adopted safety latches after the Cornell researchers demonstrated that current doors flew open 50 percent of the time in nonrollover accidents and 75 percent of the time in rollovers.[79]

At several turns the industry also defended its traditional authority and design practices. For example, Senator Douglas and the other subcommittee members criticized the auto industry for marketing horsepower, speed, and performance and focusing myopically on sexy design, and they submitted to the *Congressional Record* magazine and newspaper articles that made the same critical arguments.[80] These disputations came in the midst of the postwar "horsepower race," in which the automakers competed to out-design one another for power under their cars' hoods.[81] Yet C. A. Chayne, vice-president of engineering at GM, argued that increased horsepower was not leading to increased speeds.[82] He went further, contending that "one of the safety factors that we consider absolutely necessary is adequate performance," because increased torque led to better acceleration and improved ability to merge with speeding traffic, thus "reducing accident hazards."[83] A year later, in 1957, Abraham Ribicoff, the governor of Connecticut, held a meeting at the annual gathering of the Automobile Manufacturers Association during which all of the automakers agreed to stop the horsepower race and to cease marketing speed and racing.[84] Yet most of the auto companies continued to support increasing engine size clandestinely, and five years later, in 1962, Henry Ford II formally pulled out of the agreement. Speed sold, or so the automakers believed.

These developments in Congress during the mid-1950s stood in stark contrast to what was happening in the executive branch. In 1954, President Dwight Eisenhower held the White House Conference on Auto Safety. The ideas featured there differed little from those presented at the earlier conferences held by Truman or even Hoover thirty years earlier. After the conference, Eisenhower issued an executive order creating the President's Action Committee for Traffic Safety, though action was not one of its strongest features.[85] Auto industry representatives dominated the committee, and GM president Harlow H. Curtice chaired it. The committee also included representatives of federal agencies, especially the Department of Commerce, which housed most federal transportation functions before the creation of the Department of Transportation in 1966. But those federal staff members involved in the committee's work took their directions from its corporate leaders. The committee was not a conduit for new ideas. When Louis S. Rothschild, a member of the Rothschild family and undersecretary of commerce for transportation, appeared before Roberts's 1956 hearings on traffic safety, he argued, "Design and engineering cannot solve the highway safety problem alone. Human psychology, which makes everyone careless at times and some irresponsible, is a fundamental contributor to accidents."[86]

A year after the first hearings, the Roberts Subcommittee on Traffic Safety held another round on the narrower topic of seat belts. By the time the hearing convened, twenty states and Alaska Territory had installed seat belts in all state-owned automobiles and had found universally that they reduced auto injuries and deaths.[87] The General Services Administration, the federal agency created in 1949 to manage federally owned property and purchasing, was drafting specifications to do the same with automobiles purchased by the federal government. Although little came from the seat belt hearings, the impact biomechanics community (Stapp, UCLA's Severy, Cornell's Moore and Dye) figured more prominently in them than previously, overshadowing the fewer representatives from the automakers. In 1958 Roberts held yet another hearing, this one with the name "Research Needs in Traffic Safety."[88] No industry representatives appeared there; instead, impact biomechanics researchers predominated. Stapp, Moore, and faculty members from UCLA, Harvard, Northwestern University, the University of Michigan, and other institutions involved in auto safety research argued that the federal government should increase funding for crash science. It was clear to Roberts that these researchers more closely aligned with his cause than did auto industry executives.

By 1959 Roberts was demanding federal automotive safety standards. Few members of the House of Representatives or the Senate heard or answered his call. Perceiving an intermediary path to his goal, however, Roberts drafted legislation that would have required the General Services Administration to create a comprehensive set of safety specifications for all vehicles that the federal government purchased.[89] His hope was that, because the federal government owned about 225,000 automobiles, any safety device it required in vehicles might become industry standard for all cars.[90] In other words, he took up the same strategy of using government purchasing power to shape the national market as Hoover's National Bureau of Standards had in the 1920s (see chapter 2). The impact biomechanics community, including Stapp and DeHaven, supported Roberts's bill, as did insurance executives, public health experts, and the National Safety Council.

In response, seven executives, representing all of the major and most of the existing minor auto companies, made a joint statement.[91] (The effort was almost certainly coordinated by the industry trade association, the Automobile Manufacturers Association.) While it emphasized the great strides in safety that the auto industry was making, even this group agreed in principle that the federal government had a right to set specifications for the products it purchased, as it had done for decades. The industry executives insisted, however, that safety performance standards already existed and that the government should not go off on its own. Furthermore, they contended—without presenting any evidence—that they were "devoting . . . all of our technical manpower" to addressing safety and automotive air pollution. Roberts pushed back by asking what percentage of the automakers' profits was going to such research and how it compared to research on reliability. But the automakers said they were unable to give such information, stating only that together the companies put $1.75 million toward safety, though the vast majority of these funds went to programs focused on drivers. In 1959 General Motors made $633.6 million in profits; Ford, $95.7 million.[92] Roberts's bill passed the House but died in the Senate.

The impact biomechanics community supported Roberts's General Services Administration bill on the floor of Congress, but following their habit, the researchers made few public proclamations on the matter and never actively criticized the automakers, which had frequently supported biomechanical research. In August 1959, while Roberts's bill was under consideration, John Moore, head of Cornell's Automotive Crash Injury Project, appeared on the New York City public radio program *Campus Press Conference*. The program featured the editors of local college newspapers questioning experts. Moore opened by likening Roberts's bill to US "pure food and drug laws," which ensured the safety of food sold in grocery stores and the vaccines given to children. He argued that "this legislation in essence says that disease on the highways is an epidemic, a national disgrace, and that there are certain methods of inoculating or preventing the likelihood of this disease, which we have such difficulty in treating."[93]

One of the students suggested that the best way to deal with the auto safety problem was to restrict driving. But Moore didn't see it that way. Moore made a passionate defense of the automobile, echoing many of the experts who had worked on safety during the 1920s and the 1930s. "The most essential instrument in maintaining our economy and society is the passenger automobile," he stressed.

> We have more automobiles now than we have income tax payers. We have more licensed drivers *by far* than we have income tax payers. But the truth of the matter is, if we were to move in with any *harsh* legislation to restrict the issuance of licenses to our driving population or to restrict the number of vehicles on our streets and roads, we would completely upset the apple cart of the American economy. The automobile industry, you must understand, controls America in a fashion that most people don't recognize. Twenty-five percent of the steel we make goes into automobiles; 37 percent of the radios manufactured in America every year go into automobiles; 60 percent of our rubber products each year go into automobiles. . . . I'm afraid, desperately afraid, that if we move in this direction, looking for solutions to this problem,

then we will upset so many vital American structures that this would be like opening Pandora's Box.

In Moore's view, the only possible solution was technological. It was the dominant view of the auto safety community, which never entertained curtailing American driving.

Yet, later in the program, one of the students asked Moore whether auto safety should be a federal matter or whether it should remain "in the hands and control of the various states."[94] Moore responded, "Well, first, let me put myself on record by saying that I'm a Southerner, as though my accent hadn't betrayed me before, and as a Southerner, I'm a fierce states rights man. I for one hope that we do not have to turn to federal legislation and control." He went on: "The only reason I am interested in the federal government's intervention here is that the federal government represents a body politic that is less likely [to cave to] lobbying influences than the municipal or states groups are subject to. These gigantic [corporations in] American society have a way of influencing local and even state level legislation to an amazing degree." Furthermore, he said:

I believe firmly that states with enlightenment and knowledge and the manufacturers that make the agent that causes this disease can solve this problem far faster than we can by federal legislation. *But* in [the] event that this disease, which threatens the wellbeing of my eight children and the risk of them [not] growing to adulthood and contributing back to American society for the investment we have made in those children if action is not taken by [state and local governments and the automakers,] then we have to look to the federal government. But I hope we don't have to.

As we will see, Moore overestimated the federal government's immunity to lobbying and influence peddling and underestimated the auto industry's resistance to change. The individuals who later brought federal auto safety standards into being had an altogether different attitude about the federal government's role in modern American society. They also had no illusions about the automakers' beneficence.

CHAPTER FOUR

From Movement to
Government Agency

On March 22, 1966, James M. Roche, president of General Motors, went before a subcommittee of the United States Senate to confess a sin. The corporation had hired private investigators to snoop into the private life of Ralph Nader, the young lawyer who a year earlier had published the book *Unsafe at Any Speed: The Designed-In Dangers of the American Automobile*. Roche partly defended this decision. The company had hired the investigators to see if Nader was bringing product liability suits against the company's Corvair automobile while also using his book to drum up public anger about that vehicle.[1] Pursuing both of these efforts simultaneously was illegal under US law. But the private investigators hired by General Motors made inquiries that went *far beyond* the company's ostensible purpose. For example, they asked acquaintances about his sexuality, mental health, and political allegiances, prodding particularly to see if he was a member of any far-left-wing political groups.[2]

For months, Nader had suspected that he was being followed and spied on. He believed his hunch proved true when a former law professor of his congratulated him on receiving a new job. Nader did not have a new job, but someone had come to the professor's office pretending to be from a firm that was thinking of hiring Nader and asking the professor questions about Nader's personal life.[3] The private investigators made a dramatic error, however, when they trailed Nader through a federal building where guards stopped and questioned them before taking their names. The names ended up in the hands of a *Washington Post* reporter who brought the whole situation to public attention.[4] *Unsafe at Any Speed* had garnered only moderate sales during its initial release in late 1965, but the *Washington Post* articles made it an overnight blockbuster. Roche's Senate confession guaranteed the nearly unanimous passage of the National Traffic and Motor Vehicle Safety Act of 1966, the first federal law to require the installation of safety devices on *all* automobiles sold in the United States. The discovery of General Motors' investigation was a contingency—it could have gone unnoticed—but it greatly influenced passage of the law. Yet the journey to that moment was a long and often uncertain one that depended on the efforts of many people.

By the mid-1950s, experts in impact biomechanics had established themselves as the go-to authorities on automobile safety, and for a while, that is where things rested. The years ticked by. The biomechanics community amassed knowledge. Automakers designed, built, and marketed cars more or less as they always had. By the mid-1960s, however, the picture was quite different. Consumer and environmental advocates and representatives of the US government—especially in Congress—pushed for change. If the auto industry would not transform itself, others would. The chief proponents of the transformation that occurred between 1955 and 1965 were not central members of the impact biomechanics network. Instead, they were experts of a different breed, including doctors, lawyers, sociologists, politicians, all of whom believed that the best way to save lives on the highway was to redesign automobiles fundamentally. Because the automakers seemed unwilling to undertake such redesign voluntarily, these experts asserted that regulation provided the only road to safety. Between the 1910s and the 1930s, state and local governments had passed laws regulating automotive technologies, especially brakes and headlights, according to the best knowledge and practices at that time. These earlier laws centered on technologies that increased and aided driver mastery. The federal laws that developed between the mid-1960s and mid-1970s, in contrast, focused on the automobile itself and, consequently, brought the federal government and the auto industry into conflict, really for the first time.

As with many other historical changes that have affected the entire United States, the auto safety movement that emerged in the 1950s and 1960s was strongly rooted in local and state governance; in this sense it resembled the science of smog that preceded it (discussed in chapter 6) and *contrasted* with impact biomechanics (see chapter 3). The movement's leaders primarily hailed from the East Coast, though individuals from the American South, Midwest, and West also played essential roles. Perhaps no other place played as central a role in the early safety movement as did the state of New York. The work of the impact biomechanics community in that state, including research at the Cornell Medical School and Cornell Aeronautical

Laboratory and Hugh DeHaven's tireless public relations efforts promoting the lifesaving potential of biomechanical research, formed a crucial precondition for the rise of the safety movement there. But, again, it was not experts from that community who pushed for change; they provided only the background research, the knowledge, and perhaps the inspiration for that push.

After these initial state-focused developments, the next phase took place at the federal level. Alabama congressman Kenneth Roberts and Connecticut senator Abraham Ribicoff pushed for increased federal involvement in automotive safety. In doing so, they came to rely on a more activist strain of auto safety advocates, including the epidemiologist William Haddon, the sociologist Daniel Patrick Moynihan, and the lawyer Ralph Nader. Largely because General Motor's botched investigation of Nader's private life, federal safety standards became the law of the land. Yet, after the passage of this law, the individuals involved had to make the difficult transition from movement organizing to governance, including the creation of a new auto safety regulatory bureaucracy.

The New York Safety Scene

In 1955 the liberal Democrat, diplomat, and presidential hopeful W. Averell Harriman was elected governor of New York, breaking a twenty-year Republican hold on the office. Harriman's staff undertook new social initiatives, including policymaking in the areas of consumer interests, juvenile delinquency, and poverty.[5] One of Harriman's chief assistants was Daniel Patrick Moynihan, a liberal thinker with political ambitions who was still finishing his PhD in sociology at Tufts University. Among other duties, Moynihan became the chair of Harriman's Traffic Safety Policy Coordination Committee—the first New York state government institution devoted solely to the issue. The committee quickly took a rather hard line on auto safety, in part because Moynihan surrounded himself, as he later put it, with "a group of epidemiologists, engineers, lawyers, and political scientists."[6] Many of Moynihan's colleagues were drawn from the New York crash science community detailed in the previous chapter. By 1959 their findings led Moynihan's committee to conclude that the automakers were not "seriously trying to design safer cars" and "that government regulation of industry may become necessary."[7] As a journalist influenced by the committee's findings wrote in *Popular Science*, "Certainly if safety glass or turn indicators can be made mandatory, there is no earthly reason for not making compulsory such a valuable feature as complete crash padding."[8]

Through the committee, Moynihan befriended William Haddon Jr., a young public health expert who, at the age of only thirty-one, was appointed director of the New York State Department of Health's Driver Research and Testing Center.[9] Born in Orange, New Jersey, in 1926, Haddon had received his undergraduate degree in engineering from MIT in 1949 and his MD from Harvard four years later.[10] In 1957 he earned a master's in public health, also from Harvard. Between earning his MD and his MPH, he worked as an intern at Philadelphia General Hospital and as a fellow at the National Foundation of Infantile Paralysis (later the March of Dimes). During that time Haddon became interested in injury as a public health issue, and he quickly began applying what became known as the epidemiological approach to safety. In addition, through his work with the National Foundation of Infantile Paralysis, Haddon formed a lasting relationship with groups focused on child health, a network that later helped diffuse his ideas to wider audiences. At the driver research center, Haddon's charge was to study "human factors involved in highway accidents"—a traditional way of saying behavioral analysis. Early on, Haddon published important studies of drunk driving and vehicle accidents.[11] But he increasingly fell under the sway of crash science research and, through New York safety networks, met the "father of crashworthiness," Hugh DeHaven (see chapter 3).

Haddon built on the so-called epidemiological approach to safety—which depicted accidental injuries and deaths as a "disease" that should be studied and remedied like any other. The roots of the approach had been laid earlier. In 1948 Fletcher D. Woodward, a Virginia-based ear, nose, and throat doctor, gave a talk titled "Medical Criticism of Modern Automotive Engineering" at the American Medical Association's annual conference.[12] As early as 1939, the medical association had formed the Committee to Study Problems of Motor Vehicle Accidents, but the group had focused on traditional topics such as driver's education and drunk driving. Woodward went further, arguing that automakers should follow the example of aviation medicine. He compared the automobile to an illness, calling it a "lethal and crippling agent." "Since there appears to be little likelihood of accomplishing radical changes in human nature in general and [in] exuberant youth in particular," he argued, "it would seem the part of wisdom to shift the emphasis, for the moment at least, to desirable alterations in the machine itself, rather than to place all emphasis on attempts to bludgeon 'old Adam' into safer driving practices."[13] Woodward drew inspiration from an artist's rendering in *Popular Science* magazine that depicted a hypothetical safe automobile.[14] The imagined car included "airplane safety belts," crash pads, and a collapsible steering column. In closing his paper, Woodward quoted a recent editorial in the Richmond, Virginia, *Times Dispatch*: "Disregard for human life has always been a measure of barbarism. If we are to become more civilized, we must place the value of human life above the prestige value of speed and appearance in motor cars."[15]

Little immediately came from Woodward's paper, which was published in the *Journal of the American Medical Association* a year later. But the medical community continued talking about cars and the carnage associated with them. In June 1953 the American Medical Association passed two resolutions calling on automakers to do something about safety.[16] A year later, the American College of Surgeons passed a similar resolution, which went so far as to suggest minimum g-force limits for seatbelts and other automotive parts. The medical criticism of automobile design was mounting.

Cornell's auto safety researchers often relied on medical metaphors when making a case for their work.[17] They likened gathering accident statistics in collaboration with Indiana and other states to classical public health–oriented epidemiological studies. As Walsh McDermott, the head of the public health department at Cornell Medical College, wrote to August Heckscher, editor of the *New York Herald Tribune*, "Understandably enough, [our] approach is also 'medical' [in that it relies on] the same methods used to study any wide-spread malady for which no cure is available. This . . . so-called epidemiological approach . . . consists of studying individual cases—then collecting them in meaningful numbers—then analyzing the material to see the various patterns emerge which lend themselves to some practical treatment."[18] For the impact biomechanics community, that "practical treatment" was redesigning automobiles. Haddon tapped into this line of thinking as he continued to develop an understanding of automobile accidents and their human toll.

In 1959, under Haddon's tutelage, Moynihan published the article "Epidemic on the Highways."[19] Moynihan raised the stakes for the new and growing auto safety movement by going directly after the National Safety Council, which he argued was doing the public a "serious disservice." As noted in chapter 2, the council had played a major organizational role in the auto safety network during the 1920s and 1930s. Its power persisted into the 1950s. Moynihan attacked the organization for focusing exclusively on "the individual responsibility for accidents."[20] By doing so, he contended, "the Safety Council shifts public attention from factors such as automobile design, which we can reasonably hope to control, to factors such as the temperament and behavior of eighty million drivers, which are not susceptible to any form of consistent, over-all control—certainly not by a bunch of slogans." Moynihan's essay garnered attention. In 1959, it was included in the *Congressional Record* in relation to Representative Kenneth Roberts's ill-fated bill requiring safety devices in all cars purchased by the federal government (see chapter 3).

Haddon remained at the driver research center until 1961, when he became the director of the New York State Department of Health's Epidemiology Research Program, a position he held until 1965. For Haddon the change in position meant more time dedicated to research and writing. In the early 1960s, he and his colleagues published a wave of new studies on a variety of auto-related topics, including drunk driving, automobile-pedestrian collisions, and vehicle design.[21] Beyond these empirically driven studies, however, Haddon also became the foremost *theorist* and *methodologist* of the day on the epidemiological approach to accidents. His most important early theoretical statement appeared in the 1963 essay "A Note Concerning Accident Theory and Research with Special Reference to Motor Vehicle Accidents." Haddon put forward a general theory of injury, the central insight of which was that all injury involved *energy*. Some injuries, such as suffocation and frostbite, interfered with the body's "normal energy exchange." Others involved "delivery to the body of amounts of energy in excess" of its "thresholds."[22] This latter type included "impacts of moving objects such as bullets, hypodermic needles, knives, and falling objects, and those produced when the moving body collides with relatively stationary structures, as in falls, plane crashes, and auto crashes." He also included burns (thermal injuries) and injuries from radiation in this latter category. Haddon believed that by bringing injury under a unified purview, public health experts could think more clearly about how to decrease the prevalence and severity of injuries. Injuries were not magical occurrences—a thought Haddon would return to later.

Unlike members of the impact biomechanics community, Haddon had the temerity to criticize the automobile industry and what Ralph Nader called the "safety establishment," which included the National Safety Council, the Automotive Safety Foundation, the President's Action Committee for Traffic Safety, and other groups that had been advocating safety-via-training since the 1920s and 1930s.[23] In a letter to Moynihan, Haddon paraphrased the famously caustic journalist H. L. Mencken: "No one will ever lose money underestimating the intelligence of American safety experts. With few exceptions, a dull and stupid bunch."[24] Haddon was typically more careful with his rhetoric in public, choosing to work with other safety groups rather than antagonize them. Yet Haddon's underlying critical spirit distinguished him from his predecessors, and it would play an important part in his eventual role in reforming automotive safety in the United States.

Haddon's growing influence in the auto safety field soon led to the 1962 conference "Passenger Car Design and Highway Safety," cosponsored by the New York–based Association for the Aid of Crippled Children and the Consumers Union of the United States, Inc.[25] Formed in 1899, the Association for the Aid of Crippled Children (now the Foundation for Child Development) was a product of the progressive era. The Consumers Union, the publisher of *Consumer Reports*, had organizational roots that went back to 1926 and had become the most influential group in the nation's growing consumer movement. The conference brought together experts in impact biomechanics, representatives from nongovernmental organizations and insurance companies, and staff members of state and federal agencies, including James L. Goddard, who was then focused on aviation safety at the Federal Aviation Administration but would soon become a fierce reformer of pharmaceutical safety as head of the Federal Drug Administration. The Big Three automakers sent one representative apiece. GM's representative just listened; the one from Chrysler gave a paper titled "Evaluation of Vehicle Signals and Signaling Systems," which fit the old safety paradigm of increasing driver mastery rather than focusing on crashworthiness; and the engineer from Ford bragged in his presentation about the company's safety advances since the early 1950s.

The perceived need for regulation echoed throughout the conference. In his paper, John E. Ullmann, a business professor at Stevens Institute of Technology, argued that auto safety presented a fundamental conflict between costs and values. Consumers might value safety, but safety devices increased costs for the automakers. The only hope, Ullman claimed, was to modify "the competitive environment" to encourage "greater industry agreement on the inclusion of safety features." The professor put the matter directly: "Unless there is an element of compulsion or the threat of it, manufacturers do not appear to have introduced [safety] features as standard equipment, which would . . . increase production costs."[26]

Daniel Patrick Moynihan ended the conference with his own paper, "The Legal Regulation of Automobile Design," which outlined just such a threat of compulsion. Safety regulations were unremarkable, Moynihan argued. Historically, they were seen as "an unexceptional exercise of the police power of the state and have rarely encountered . . . fierce opposition."[27] They were "a commonsense field of activity." Moynihan surveyed federal regulation of transportation safety in the railroad, aviation, and trucking industries (including limiting driver hours to prevent distracted or sleepy driving). "Experience in other forms of transportation," Moynihan argued, demonstrated that "more extensive government regulation is the most promising course" for reducing the carnage on American roads.[28] Moynihan was sympathetic to but critical of Congressman Kenneth Roberts's attempts to deploy government purchasing power to get safety devices into cars. Such an approach, he maintained, would add only minor safety technologies to already-designed and fundamentally unaltered cars. Moynihan preferred a different, more radical tack. He believed that the automobile "may be heading for a period of profound technological change," including even the introduction of electronic controls that would prevent a car from colliding with others or passing through a red traffic light.[29] "From this point of view," he stressed, "it would be wiser to leapfrog the safety-devices stage altogether and go directly to a program of extensive federal involvement in automobile design development, similar to the federal involvement in aviation design, although not, perhaps, so extensive."[30] Moynihan's vision went far beyond those generally shared by the auto safety community, and the federal automotive safety standards that eventually emerged fell far short of his hopes.

In 1964 Haddon, along with Edward A. Suchman, a professor at the University of Pittsburgh, and David Klein, a researcher at the Association for the Aid of Crippled Children, published *Accident Research: Methods and Approaches*. The work became, for many years, the most important book in the field, a kind of textbook. Drawing together a number of historically important studies dating to 1948, *Accident Research* included a running commentary in which the authors assessed both what advances those studies had made and where they had gone astray. The book aimed not just to review comprehensively the state of accident prevention knowledge but also to suggest a unified approach that others should adopt.

Haddon and his coauthors cast themselves as Enlightenment figures in the sense that they sought to replace tradition and superstition with *reason*. Haddon often referred to received notions of accidents, accident prevention, and fate as "folklore" and as "traditional, prescientific wisdom" or magical thinking.[31] Some of this rhetoric was questionable. The psychologists, physiologists, physicists, and other researchers who had worked on auto safety from the 1910s through the 1930s certainly saw their work as scientific and certainly claimed so. Yet Haddon included no papers from that period in *Accident Research*, perhaps because he thought the earlier work worthless, or perhaps because he overlooked them. After all, a whole generation sat between Haddon and these earlier researchers, and the world forgets. Regardless, in opposition to prescientific ideas of crash safety, which Haddon denigrated as merely "descriptive," he proposed an "etiological," or causal, approach similar to that used in modern medicine.

In spelling out this divide between the descriptive and the causal, Haddon employed analogies from bacterial and viral diseases. For instance, prescientific medicine relied on description when addressing the symptoms of fever and "wasting." By focusing on the symptoms instead of root causes, doctors improperly believed wasting to be an independent illness instead of a symptom of a number of diseases, including protein deficiency, amebiasis, and tuberculosis. For Haddon, the transition from description to etiology was the paradigmatic mark of modern medicine.[32] He insisted that if crash science, and accident studies more generally, was going to move forward, it had to make this transition. As Haddon exclaimed, "The notion of an accident is descriptive, not etiologic." The accident "has a long history and close relationship to notions of personal vulnerability and invulnerability." Traditional notions were "illustrated, for example by the themes of the stories of Job and Achilles. Redolent of the extrarational and supernatural and prescientific, they remind one of Malinowski's natives in their approach to the hazards outside the reef, which they did not understand, in comparison with those inside which they approached in more rational terms." In Haddon's view, science was waiting in the wings to make accident research modern.

To conceive fully of human injury as a disease, Haddon created what became known as the Haddon Matrix, a graphical device used to model injuries and create distinctions between them. To create this tool, Haddon again turned to a viral infection—this time polio—as an analogy to injury. The Haddon Matrix broke injury and illness into three moments, or "phases of social concern." The first phase dealt with the agent reaching the human host and ways to prevent it from doing so. With polio, of course, this phase arose from humans encountering the polio virus, and, Haddon noted, it was once avoided by various means, including keeping children away from swimming pools and movie theaters. In auto safety, Haddon called this the "precrash phase," and it was the proper domain of safety campaigns, driver's education,

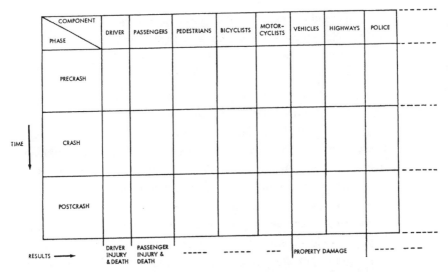

Figure 4.1. William Haddon conceived the Haddon Matrix as a graphical means to examine systematically different factors involved in accidents through three phases—precrash, crash, and postcrash. He believed that public health experts should think more clearly about how accidents could be avoided, how their effects could be minimized, and how recovery could be improved. The precrash, crash, and postcrash distinction was built directly into federal crashworthiness standards, with the standards being numbered in the 100s, 200s, or 300s respectively, depending on where they mapped on the distinction. Federal headlight standards are given the number 108, for instance, because they are involved in the precrash phase of avoiding accidents. William Haddon Jr. "The Changing Approach to the Epidemiology, Prevention, and Amelioration of Trauma: The Transition to Approaches Etiologically Rather than Descriptively Based," *American Journal of Public Health and the Nation's Health* 58, no. 8 (1968): 1431–1438.

licensing tests, road design, and standards for brakes, headlights, and other technologies of driver mastery.

In the second, or "crash," phase, the question was what to do once the agent and host had made contact. With polio, this included drug treatments. With auto safety, Haddon focused on how the automobile, especially its interior, could be transformed to lessen injuries during crashes. Over the following few years, Haddon predicted, automakers would bring about "the far better 'packaging' of human cargo, accomplishing with people what we long ago achieved with property." Finally, the third, or "postcrash," phase, in Haddon's words, "involves maximizing salvage, once damage has been done to the susceptible structures."[33] As chapter 5 explains, the Haddon Matrix came to deeply influence federal automotive safety standards.

Strikingly, the Haddon Matrix not only set aside issues of morality and politics but also viewed the human body as a "susceptible structure." With automotive crash safety, Haddon believed, the focus should be put on changing the automobile, not on shaping driver behavior. That is, the historical focus on what the Haddon Matrix called the precrash phase had proven a failure—or, at best, only one part of a multipart solution. Haddon believed that, at root, humans were fallible and beyond perfection or perfecting. Throughout his career, the simplest of his many arguments for focusing on technology was his analogy between auto safety and the dangers of early electrical systems. Instead of training people to use electrical systems safely to prevent fires and other hazards, Haddon pointed out, electrical engineers had created fuses to reduce the risks posed by both human error and technological failure.[34] Today we consider fuses natural, perhaps even inevitable components of electrical systems, but they are erstwhile engineering interventions that circumvented human fallibility and other potential problems.

Haddon's approach, which he called "technics in use," was deeply intertwined with his peculiar view of the division of research labor. In this perspective, only after scientists and doctors examined the root causes of human injury could technicians intervene and apply their lessons by designing technologies to negate these causes. Pure, abstract research would precede application. Haddon's view mirrored his times: after World War II, scientists and policymakers adopted a vision of innovation that has come to be known as the linear model.[35] This framework holds that science's attempt to gain knowledge about how the world works precedes the work of other fields, such as engineering, to apply that knowledge and remake the world to fit our desires. Whereas most advocates of the linear model focused on the "natural sciences," such as physics, chemistry, and biology, Haddon's version also included the social sciences. But he valued even the social sciences as ways of coming to know the world fundamentally, rather than for whatever practical knowledge they produced. Between his 1962 conference on auto safety design and the publication of his coedited volume on accident research, Haddon established himself by the mid-1960s as both an expert in crash safety and an authoritative critic of the auto industry.

Congressman Roberts's Ironic Victory

In 1962 Kenneth Roberts tried once again to get his bill, which required safety devices in all federally owned vehicles, enacted into law. Once again, the House of Representatives passed it. Once again, it died in the Senate. Of

Roberts's inability to get the law through the Senate, one Washington observer said, "Basically, it was a matter of lobbying. There are only one hundred senators, and the auto boys are very effective. They have about thirty people working on the Hill."[36] Roberts needed some way to overcome this resistance. His opportunity finally came in 1964 through the horse-trading conducted all the time in Congress. Senator Warren G. Magnuson (D-Wash.) was eager to see federal health benefits (in the form of US Public Health Service hospitals) extended to offshore fishermen in his home state. He asked Representative Roberts for help, and Roberts agreed provided that Magnuson would push through the Roberts auto safety bill. It worked. Looking back, Roberts said, "That shows the back-scratching process that goes on up here. Before that I had never got the bill out of a Senate committee."[37]

Roberts's victory came with some irony, however. The law passed only weeks before the representative was voted out of office. A conservative Democrat from Alabama, he was a part of the first wave of southern politicians who lost their seats after Lyndon Baines Johnson signed the Civil Rights Act of 1964. As we will see, however, the Roberts Act laid the foundation for all federal automotive safety standards that came afterward. Yet, perhaps because of the furor and hoopla that surrounded Ralph Nader's *Unsafe at Any Speed*, Roberts's contribution to auto safety has largely been forgotten. In an important sense, however, Roberts was, for a decade, the most important force in driving the nation toward federal auto standards.

The responsibility for creating these new auto safety standards fell to the staff of the General Services Administration (GSA). Accepting the task enthusiastically, these staff members aspired to influence auto safety well beyond the confines of the federal government. When Lawson B. Knott Jr., the acting administrator of the GSA, testified before Congress in 1965, he publicly expressed the wish that the agency's auto standards would "increase the inducement" for "the automobile industry to include as standard equipment" proved auto safety devices on all cars sold in the United States.[38] Moreover, the federal government came to think about auto safety out of concern not only for the health of citizens but also for the well-being of federal employees, the nation's civil servants. For example, in October 1960 the secretary of the interior told Franklin Floete, administrator of GSA, that employees of the Department of the Interior believed that seat belts were a necessity in government vehicles.[39] The secretary claimed that a number of injuries and deaths of Interior's civil servants in auto accidents could have been avoided if someone (automakers, postmarket suppliers, or government workers) had installed seat belts.

To aid the standards-making process, staff members of the General Services Administration put together an advisory council comprising thirteen federal agencies (including the Department of Defense), six automobile companies, and eight professional societies and trade associations, including the National Safety Council, the American Medical Association, the Society of Automotive Engineers, and the trade group the Automobile Manufacturers Association.[40] The impact biomechanics community played no major role in the process, however. Indeed, the presence of the National Safety Council bespoke the persistence of the old auto safety community established in the 1920s. Following the earlier paradigm, the safety council was especially focused on how to change and train drivers, not in reforming auto design. The GSA also faced real limits, however, particularly budgetary ones. As one critic of the auto industry later pointed out, the Roberts Act provided no "special appropriations" for the agency to hire "specialists and services of expert consultants."[41] Given that the agency could not hire experts and did not have internal expertise regarding auto safety, how would it create standards? The answer was that it largely depended on safety standards *created by the auto industry itself.*

This issue raises a historical puzzle that no known sources fully answer. Did automakers and their engineers and designers begin working on auto safety for their own reasons? Or did Roberts's proposed regulations more or less force them into it? By mid-1965 the General Services Administration had established seventeen safety standards for the automobiles it purchased.[42] Eleven of these standards were at least partly based on those created by the Society of Automotive Engineers. GSA's standards also contained a new design technique—the SAE two-dimensional fiftieth percentile male—which was not yet a formal SAE standard but was on its way to becoming one. The striking thing about these SAE standards is that, with the exception of the one for backup lights—which the SAE's Lighting Committee had created in 1947—all of them were established *after* 1959, the year that Roberts first tried to create safety standards for government-purchased cars. In other words, automotive engineers appear to have been creating safety standards largely in response to threatened government regulation. For example, the SAE created a manikin (or crash test dummy) subcommittee only in 1959, the same year Roberts first proposed his legislation.[43] What the economist John E. Ullman told those gathered at Haddon's auto safety conference in 1962 about manufacturers' reluctance to make safety features standard equipment in the absence of compulsion appears to have held true for formalized safety *standards* as well.

The GSA standards that were *not* based on their SAE counterparts say as much about the auto industry as do those that were. For example, although the automakers had used safety glass since the 1930s, the industry had not created a standard for it. Thus, the General Services Administration had to base its standard (515/8, as the GSA numbered it) on ones put out by the American Standards Association (the organizational ancestor of the American Engineering Standards Committee, discussed in chapter 2, which was later renamed the American National Standards Institute [ANSI]). Similarly, critics of the auto industry had, for at least a decade, been mocking its reliance on chrome, which often glared in sunlight, blinding other drivers.[44] Yet, although the federal antiglare standard used a Society of Automotive Engineers manikin to define where the average driver's eyes would be, the auto industry itself had no antiglare standard. The federal standard (515/13) relied on ones from the American Society for Testing Materials.

The General Services Administration crafted de novo only three of its own standards, covering bumper heights, the order of gears (PRNDL) for the gear stick on automatic transmission cars, and rearview mirrors. None of these three standards posed great difficulty for the automakers to meet, but the companies did not always speak with one voice. Alex Haynes, the Ford engineer who began working on safety issues in the early 1950s, vehemently opposed the bumper height standard, which aimed to increase the odds that cars in accidents would collide bumper-to-bumper, thereby maximizing energy absorption. Haynes argued that such a standard would place undo restrictions on auto design, pointing out that "our business includes styling," which was a "very necessary thing."[45]

For the most part, however, the automakers presented a coordinated front through the combined efforts of the Automobile Manufacturers Association and the Society of Automotive Engineers.[46] They beat back the federal purchasing specifications wherever they could. The companies hosted GSA staff members in Detroit for a three-day visit of facilities and proving grounds. Soon thereafter, the agency held a daylong conference on the pending standards that was dominated by auto industry representatives. Between the publication of the proposed standards and the final ones, several were weakened. Most significant, the final standard limiting instrument panel energy absorption was made half as tough as the proposed one.[47] Given that some standards had been attenuated and the fact that most of the others reflected auto industry predecessors, the federal specifications presented little threat to automakers. Ford president Arjay Miller said, "Although some reports may lead the public to believe that the GSA standards will be new, in most instances they are similar to or stem from our current engineering practice."[48] These standards were not going to induce serious change in the automobile industry or the cars it produced.

Safety advocates reacted caustically to the GSA standards. A Louisiana-based safety consultant, William I. Stieglitz, told the Senate Commerce Committee, which oversaw the GSA, that the standards were "utterly worthless" and enumerated standards that the auto industry had weakened, including raising the maximum threshold for the forces in head impacts.[49] In *Unsafe at Any Speed*, Ralph Nader wrote of the GSA specifications: "The final standards in general represented quite a triumph for the automobile makers. They obtained government endorsement of existing optional safety devices [as opposed to pathbreaking ones] and approval, by and large, of existing levels of safety."[50]

A year later, in May 1966, leaders of the General Services Administration decided that they wanted to revise and toughen some of their standards and create several new ones. The agency put together an advisory committee dominated by members of the impact biomechanics community, including Lawrence M. Patrick of Wayne State University; Robert A. Wolf, the head of the Automotive Crash Injury Research Project at Cornell; and Derwyn M. Severy, of UCLA's Institute for Transportation and Traffic Engineering.[51] In October 1966, the agency, under the guidance of this group, issued new and revised standards, all of which were tougher than the first wave of standards. The new standards included such requirements as padded seatbacks, which would protect the heads, knees, and legs of backseat passengers, and rupture-resistant fuel tanks. The automotive industry resisted these new standards passionately and loudly.[52] The Automobile Manufacturers Association and all of the major automakers condemned the new standards as too stringent for the industry too meet. Although the General Services Administration eventually established them, these new standards were once again weakened through the process of industry comments and suggested revisions. Yet, importantly, only by including members of the impact biomechanics community, who brought with them an aura of scientific authority and expertise, was the agency able to create standards tough enough to raise the industry's hackles.

Citizen Nader

When Senator Hubert Humphrey became vice-president of the United States after the 1964 presidential election, power in the US Senate was reshuffled. In November of that year, a relatively junior senator Abraham Ribicoff—former governor of Connecticut and secretary of health, education, and welfare under the late John F. Kennedy—became head of a Senate committee on executive branch operations.[53] Ribicoff began feeling around for issues to put before the committee. In the December 20 issue of the *New York Times*, he read a front-page review of *Accident Research*, the book William Haddon had coedited.[54] Ribicoff was known for having taken up the issue of auto safety when he was governor, though his solution at that time had been a traditional one, namely, initiating a police crackdown on speeding and unsafe driving, a policy that dramatically reduced deaths and injuries in the short run but had no long-term effect. The *New York Times* article now convinced Ribicoff of the wisdom of the crashworthiness approach. He called his chief of staff, Jerome Sonofsky, and told him to buy and read the newspaper. Ribicoff decided that his committee should pursue the auto safety issue, and he put Sonofsky in charge. Sonofsky called Moynihan and asked if he knew any experts on the topic. Moynihan suggested Nader. Sonofsky and Nader sat down for an introductory conversation that was supposed to last for twenty minutes but went on for three hours.[55] Ribicoff and Sonofsky had found their man.

The son of Lebanese immigrants, Ralph Nader grew up in the mill town of Winsted, Connecticut. At least according to his own account, Nader desired to be a "people's lawyer" even as a child. His father would occasionally take him to watch trials in the local courtroom. Nader earned a bachelor's degree at Princeton University in 1955 before going on to Harvard Law School. Multiple factors led Nader to become interested in auto safety. A frequent hitchhiker, he saw the aftermath of many auto accidents. Once, he was the first person to arrive at the scene of an accident where he found a young girl decapitated.[56] She had been thrown against the car's glove compartment door, which had popped open during the crash.

But Nader's impetus went deeper than mere personal horror or childhood ambitions. In 1956, while he was still a law student, the *Harvard Law Review* published a commentary written by the Chicago-based labor lawyer Harold A. Katz.[57] In his essay, "Liability of Automobile Manufacturers for Unsafe Design of Passenger Cars," Katz argued that tort law required updating. "Since the epic decision of Mr. Justice Cardozo in the *MacPherson* case," Katz wrote, manufacturer responsibility for product defects had become completely accepted to the point of being common sense. Even the auto industry did not "dispute its legal obligation." Yet there were real limits to how deeply such law had influenced the design and production of cars. The current state of auto safety, Katz maintained, "suggests that the motor companies, alerted to the potential risks involved [in defect lawsuits], have devised methods of production and inspection which have made defects of the *MacPherson* type relatively rare."[58] Such thinking, however, had not pressured the automakers to actively include safety technologies in their products.

Because the courts were not holding the automakers "liable for negligent design," Katz wrote, and given "general public apathy toward occupant safety, the manufacturers have in this aspect of design been in a 'rut worn deep by fifty years of unimaginative thinking.'" Given that "there is still no greater incentive for improved design than the threat of legal liability," Katz argued, it was time for automakers to be held responsible for lousy safety design. He took this phrase about the rut of unimaginative thinking from a speech that a medical doctor had made before the American College of Surgeons as that group was preparing to pass its resolution calling for increased auto safety. In other words, the medical-doctor advocacy for safety had influenced Katz. His essay, in turn, influenced Nader. "That was the beginning. That started it all," Nader later told an interviewer.[59] Soon after reading the essay, Nader contacted Katz. "I got a call from this fellow," Katz later recalled. "I'd never heard of him, but he was quite ecstatic. He told me he was utterly astonished and absolutely delighted by my article. He didn't have any prior notion of using tort law to reform the auto industry. The idea captivated him."[60] A decade later, Nader would be involved in the most fundamental redefinition of the notion of "defect"—and, therefore, of product liability—since the *MacPherson* decision.

Though otherwise an indifferent student in law school, Nader wrote an impassioned paper titled "Automotive Design: Safety and Legal Liability" for a seminar on medical law.[61] The paper earned him an *A*. He continued researching the topic of auto safety after graduating, and in April 1959 he published "The Safe Car You Can't Buy" in *The Nation*.[62] It presented a damning portrait of Detroit's indifference to automotive death and injury. Nader based his critique on the existence of the 1957 Cornell–Liberty Mutual Safety Car. Members of the impact biomechanics community not only had amassed the

The CORNELL-LIBERTY
SAFETY
CAR

SPONSORED BY
LIBERTY MUTUAL

PRODUCED BY
CORNELL AERONAUTICAL LABORATORY

Figure 4.2. A sketch of the Cornell–Liberty Mutual Safety Car, which played a central part in Ralph Nader's 1959 article, "The Safe Car You Can't Buy," in *The Nation*. The futuristic-looking vehicle included a reinforced body, stronger door locks, seatbelts, recessed knobs, and increased visibility. As the article's title suggested, the thrust of Nader's argument was that all of these safety technologies were already available but lay beyond the consumer's reach. From the Collections of The Henry Ford. Gift of the Family of Henry Austin Clark, Jr.

knowledge necessary to reform car design, Nader argued, but had also *already built* a vehicle that embodied these principles. The article's title thus reflected Nader's frustration that such a safe car was out of consumers' reach. Detroit had been unresponsive to the enlightened design of the Cornell–Liberty Mutual Safety Car. The essay began a pattern for Nader. From that point on, he drew upon the findings of the impact biomechanics field in order to level critiques at the auto industry that were much more polemical and pointed than anything biomechanics experts themselves had had ever offered.

Nader briefly practiced law after graduating from Harvard, but in the early 1960s he turned to other pursuits, including traveling the world and writing as a freelance journalist. During John F. Kennedy's presidency, Daniel Patrick Moynihan served as assistant secretary of labor, a position he used to initiate a number of studies on social problems, most controversially the report *The Negro Family: The Case for National Action*, which became known as the Moynihan Report. Not so well known, though, was that Moynihan had continued his interest in auto safety. One of his responsibilities was participating in the President's Action Committee for Traffic Safety, a creation of the Eisenhower administration that continued into the 1960s. Moynihan believed, as a biographer later put it, that "the executive director of the committee was a General Motors company man."[63] Using the excuse that auto accidents were one of the foremost killers of federal employees and workers throughout the country more generally, Moynihan initiated in 1964 a study of the auto safety problem. He began that study by hiring Nader, who hitchhiked to Washington, D.C., to lead the study. Nader finished a report for Moynihan in the spring of 1965 that ran to nearly 350 pages, though it was not this report that would come to have the greatest effect on auto safety. About this time, Ribicoff and Sonofsky contacted Nader, who moved from Moynihan's office at the Department of Labor to the Hill.

Nader was then writing *Unsafe at Any Speed: The Designed-In Dangers of the American Automobile*, which Grossman Publishers brought out in 1965. Nader's book is often compared to other powerful works of criticism from that era, especially Rachel Carson's *Silent Spring*, the foundational text of modern environmentalism. But *Unsafe at Any Speed* also fit within a small genre of books from the 1950s and 1960s that criticized the automobile industry for its failings and excesses. Unquestionably, the dominant popular culture of the day *celebrated* the car as the key technology of American liberty. As one historian has put it, "Postwar fiction, such as Jack Kerouac's *On the Road* and John Updike's *Rabbit* novels," as well as films like the *Rebel without a Cause*, equated "the ability to drive with the search for life's meaning."[64] Yet this celebration of the American automobile always met with resistance, especially from the highly educated. As Daniel Patrick Moynihan told those gathered at Haddon's 1962 auto safety conference, growing doubt about the industry's interest in auto safety coincided "with a period of great prosperity, great vulgarity, and considerable obtuseness in the auto industry. The mid-fifties was a period of increasingly bloated, overpowered, and overpriced automobiles, symbolized by the horsepower race and the tail fin, which at first sold in unprecedented numbers. Things got out of hand, however. Intellectuals grew disgusted and the public became annoyed."[65]

One such intellectual was the novelist Vladimir Nabokov, a Russian-born aristocrat who had fled the Bolshevik Revolution and, after living for nearly two decades in Germany, eventually fled the Nazis and emigrated to the United States. For Nabokov, car culture was the quintessential example of philistine consumerism in the United States. He satirized car culture mercilessly in his 1955 novel, *Lolita*. John Keats's best-selling book, *The Insolent Chariots* (1958), targeted the American automobile of the 1950s. A journalist and critic, Keats savaged the symbols of postwar affluence, including the suburbs, higher education, and industrial agriculture. But the overdesigned, overly chromed, tail-finned automotive monstrosities of the 1950s provided Keats with perfect examples of everything to be loathed. The automakers and their marketers played a similar role in Vance Packard's influential 1957 book, *The Hidden Persuaders*, which attacked the advertising industry for manipulating simpleminded consumers. In this way, Nader's *Unsafe at Any Speed* tapped into a larger cultural thread that held the auto industry up as the worst example of capitalism gone awry.

Moreover, critique of the auto industry during this time frequently went hand in hand with attacks on progressive education, the pedagogical theory undergirding driver's education, discussed in chapter 2. Progressive education, which had dominated American schools since the early twentieth century, experienced blowback in the 1950s and 1960s. The historian Richard Hofstadter denounced it in his Pulitzer Prize–winning book, *Anti-intellectualism in American Life*.[66] Nabokov also went after progressive education in *Lolita*, and Keats dedicated an entire book, *Schools without Scholars*, to the topic.[67] "What I find is that our schools pamper the jackasses, stuff the geniuses under the rug, and meanwhile envelope everyone in that fatuous diaperism they call life adjustment," wrote Keats.[68]

The critique of progressive education and the critique of traditional auto safety efforts came together in Edward A. Tenney's *The Highway Jungle*, published in 1962.[69] Tenney saw driver's education as a fraud rooted in progressive education's fundamentally unsound picture of human psychology. Driver's education had been around at least since the 1930s, and what was the result? Tenney asked. More highway deaths than ever. Furthermore, Tenney maintained, driver's education was simply a cover for automakers to continue doing nothing about safety. The traditional safety movement—what Nader and others called the safety establishment—"began with commerce, is presently controlled by commerce, and will in the future continue to be in commercial hands until the public acts on its own behalf."[70] The only hope was to reject driver's education in favor of meaningful automotive safety standards. Nader echoed these sentiments in *Unsafe at Any*

Speed: "The limitations of human beings in coping with the increasingly complex driving task, even under the most rigid law-enforcement or the most ambitious education programs, make it unrealistic to expect all drivers to control their vehicles perfectly all the time."[71]

In this way, Ralph Nader's *Unsafe at Any Speed* can be seen as the culmination of several historical and cultural strands that had emerged since World War II: it drew upon the scientific knowledge of the impact biomechanics community; it criticized the auto industry for its excess and carelessness, most famously in its condemnation of the Chevy Corvair—"It's one of the few cars I know that can do the *Bossa Nova* on dry pavement and the *Watusi* on wet"; and it set aside traditional ways of dealing with automotive safety, including driver's education rooted in progressive pedagogical philosophy.[72] Yet Nader's book also went further than any of these individual strands, further even than Harold Katz's notion of using liability lawsuits to reform the auto industry. Nader firmly believed that the only hope for the short term was strong federal action in the form of regulation.

Nader explained his philosophy on the matter in an address before the National Press Club in Washington, D.C., in December 1966.[73] At its core his understanding of auto safety was political. Nader did not adopt the rhetoric of Haddon and members of the impact biomechanics community, according to which the bodies of individuals were simply one more physical "structure" among all the others in an auto accident. Rather, Nader believed that the US government *had to act* in order to protect the inviolability of the human body. He approvingly quoted the American Romantic poet Walt Whitman, "If anything is sacred, the human body is sacred."[74] Whereas Whitman wrote his poem "I Sing the Body Electric" to celebrate the body in a conservative, Christian culture that diminished it, Nader sought in *Unsafe at Any Speed* to make the human form inviolable in a corporate-dominated consumer culture that casually destroyed people, through unsafe products, through hazardous pollution, and through dangerous labor practices. In this way, Nader was and is an archliberal; his arguments about justice were based on the ultimate value of the individual.[75]

Nader's politics arose from an allergic reaction to the theory of pluralism that ruled the liberal political philosophy of the day. At its core, pluralism applied to politics the economic theories of capitalism, which viewed "interests" as competing in the political realm as firms compete in "markets." Nader echoed John Kenneth Galbraith when he asked, "What avenues for reform, for the reduction of these [social] costs, are available to the citizenry when the diversity of competing, countervailing or penalizing centers of economic power is over-ridden?"[76] For Galbraith, large-scale buyers and sellers, including wholesalers and labor unions, could exercise such countervailing power by fighting back against corporate monopoly power, creating a process of struggle. Lamenting that Galbraith's idea now "reads like a quaint economic fable," Nader argued that corporate power simply overwhelmed these "pluralistic interests," ensuring that "the effective life-cycle of countervailing forces . . . is notoriously short." If "unions, government, different industries, big buyers, etc." could occasionally organize to oppose corporations, large businesses and their loyal legislators would quickly shut the opposition down. "Devolution sets in, and the system is praised as being 'pluralistic,' as if that description suffices to set our minds at rest."[77]

When Sonofsky brought Nader over to Ribicoff's Senate office in early 1965, then, Nader was intellectually prepared to take the next step in auto safety—helping to create federal safety standards that would apply to *all* cars sold in the United States, not just those bought by the federal government. As Ribicoff and his subcommittee colleague Robert Kennedy began to attract media attention for their auto safety work, the Lyndon Johnson administration joined the call for improved vehicles. The White House sent over a proposed bill that, in Nader's eyes, was weak, a "no-law law."[78] The president's proposed bill would have permitted the secretary of commerce to create federal safety standards but only *if* the auto industry did not move on its own. Nader saw the Commerce Department as an industry pawn and believed that making standardization conditional and discretionary would lead to inaction.[79] When General Motors' investigation into Nader's private life became public, the Senate passed a much tougher law than the one drafted by the Johnson administration. General Motors' Nader debacle also guaranteed unanimous passage of the legislation in both the Senate and the House of Representatives.[80] Signed into law by Johnson, the National Traffic and Motor Vehicle Safety Act of 1966 created a new agency, the National Highway Safety Bureau (later renamed the National Highway Traffic Safety Administration), thereby circumventing the Commerce Department. The law mandated that the new agency create automotive safety standards that would apply to all automobiles sold in the United States.

The National Traffic and Motor Vehicle Safety Act fell far short of Nader's hopes, however. Several provisions that he had written into the bill, including criminal penalties for corporate executives who produced and sold unsafe cars, were omitted from the final compromise bill.[81] Moreover, Nader had a far more radical and fundamentally different vision of the auto safety agency than the one enacted into law. While he believed that safety standards were important, drawing inspiration from Harold Katz's *Harvard Law Review* essay, Nader envisaged the new agency creating truth and knowledge that would be used in future lawsuits, especially product liability cases, that would force automakers to design and manufacture ever-safer products. It was a radical conception of tort reform. In fact, only a few of Nader's several congressional testimonies dealt with the issue of auto safety standards. He focused much more on the need to create an agency that would end corporate secrecy, the real threat to American citizens. He was greatly angered by a provision in the act that would have kept "any report or reports of any Federal agency, or officers, employee or agent thereof, relating to any highway traffic accident or the investigation [from being] admitted as evidence or used in any action for damages or criminal action"; the provision also forbade "any such officer, employee, or agent [from being] required to testify in such proceedings as to facts developed in such investigations."

For Nader, the fight against secrecy in favor of a policy of openness, a central pillar of the idealized norms of science, was necessary for a just society: "In the area of human safety, few practices are more deplorable than secrecy. Yet, the inducements to secrecy, absent legislative safeguards and legislative review, are numerous in an area where commercial interests have such a strong interest in the outcome of studies, investigations, and standards setting."[82] Nader contrasted the plan to keep the federal auto safety agency out of court proceedings with the work of the Civil Aeronautics Board (CAB), whose "reports, photographs, diagrams, maps, charts, and personnel, training and proficiency records of the [airplane flight] crew, and history and maintenance records of the airplane" were all admissible in the courtroom.[83] Furthermore, CAB employees could testify about the "facts" of a case, though they could not share their "professional opinions" about a crash's cause. Thus, Nader's hopes were dashed that the new auto safety agency would play a role in tort cases that would enable auto industry reform. The law and the safety standards that eventually resulted from it disappointed him greatly, and in the months and years that followed, Nader, in his drive for justice, would increasingly turn against individuals who had once been his allies.

Attenuation from Movement to Government Agency

When Lyndon Baines Johnson signed the National Traffic and Motor Vehicle Safety Act of 1966 into law, he compared automotive deaths and injuries to disease. "For years, we have spent millions of dollars to understand and fight polio and other childhood diseases. Yet until now we have tolerated a raging epidemic of highway death—which has killed more of our youth than all other diseases combined. Through the Highway Safety Act, we are going to find out more about highway disease—and how to cure it."[84] Fittingly, when it came time to pick the first administrator of the new agency, the National Highway Safety Bureau, the Johnson administration turned to Daniel Patrick Moynihan's colleague and friend the medical doctor and epidemiologist William Haddon. The administration rejected the consumerist roots of automotive safety as embodied in Ralph Nader's arguments. Against this highly politicized background, William Haddon's medicalized vision of auto safety must have recommended him as a safer alternative. Haddon had not discussed politics in his early writings. In his rendering, the automakers were not the enemies of automotive safety. The enemy was both older and more diffuse; it was superstitious ignorance, the treatment of safety issues in a "prescientific" way.

Upon taking charge of the new agency, Haddon soon found that the job demanded more than his medicalized approach to safety provided. The heads of federal agencies often play public roles, including interacting with the White House and Congress and making public speeches, that pull them from their supposed jobs. In such instances, administrators have to rely on their subordinates to carry on their agency's work. This dynamic proved true for Haddon, who was rarely in his office. He therefore turned repeatedly to one man, his deputy administrator, Robert Brenner, who, far more than Haddon, shaped the agency in ways that endure.

Born in 1922, Brenner earned a bachelor's in mechanical engineering from the University of Pennsylvania in 1943. For the remainder of World War II he worked as a designer at Lockheed Aircraft, then moved to American Canning Company, where he was a plant engineer. In 1948 Brenner entered a master's program in engineering at the University of California Los Angeles. After receiving his degree in 1949, he joined the faculty of UCLA's Institute for Transportation and Traffic Engineering, where he stayed until he became Haddon's deputy. In 1960 Brenner became the secretary of the Highway Safety Committee of NRC's Highway Research Board, chaired by

Charles W. Prisk, the deputy director of the US Bureau of Public Roads' Office of Highway Safety. Although Brenner had been a member of and made presentations before a number of national organizations, this was the first time he held a leadership position in one. His rise from that point was quick. In 1962, the same year that he received his PhD in engineering, Brenner became a member of the National Institutes of Health's Accident Prevention Panel, which had formed through a special grant from NIH's Division of Grants and Research.[85]

Brenner took the main leadership role at the National Highway Safety Bureau early on. The Johnson administration created a group within the Department of Commerce to plan the future Department of Transportation, which would be the parent agency to the safety bureau. Brenner, not Haddon, was a member of this planning group. He participated in the day-to-day activities of the months-long project, thereby exerting a major influence on agency's the organizational structure.[86] Although Brenner no doubt consulted with Haddon about the plans, Haddon took no part in the quotidian affairs. Brenner was able to build his own conceptual understanding of auto safety into the agency's genes, and some of these understandings differed from Haddon's own philosophy.

For example, although the Haddon Matrix reflected an encompassing view of the auto safety problem, Haddon was, in fact, focused rather single-mindedly on changing automotive design during the 1960s. But, for Brenner, the key metaphor and conceptual model was always the system—that is, everything involved in the problem: cars, drivers, roadways, signs, and signals. In this way, Brenner was very much a product of his times. The paradigm of systems reigned supreme in engineering culture and in many intellectual circles, including those at the RAND Corporation and in studies of system dynamics by the computer scientist and systems engineer Jay Forrester.[87] Brenner went quite far in his systems thinking. In his UCLA dissertation, he had built a computer model of the road network as a thermodynamic system made up of multiple interconnected subsystems.[88]

When designing the agency's organizational structure, Brenner and his team members turned once again to this organizational motif and built system into its conception. As their proposed mission statement reads: "The end results of concern—the safe and efficient movement of people and goods to achieve defined social goals—depend upon several mutually complementary factors comprising a coherent total system. The [agency] deals with the central factors which include (1) the vehicles, (2) the traveled way, (3) human factors in handling vehicles on the traveled way, and (4) the storage of vehicles and goods in transit."[89]

Some aspects of Brenner's thought and action distanced him from Haddon. Brenner placed much higher value on the contributions of engineers and practitioners of highway and road management to increasing automobile safety.[90] As he wrote, discoveries useful in ensuring highway safety "were made by people who had the skill and experience to understand a problem and recognize a solution, even when it appeared by chance."[91] For Brenner, in contrast to Haddon, this essential experience came from familiarity with the problems and dangers of the road, the proper domain of engineers.[92] In many ways, Brenner had a more conservative vision of reengineering the overall automotive system in the name of safety, a perspective that could feel like a throwback to the safety establishment of the 1920s and 1930s and was certainly far less aggressive than the views Haddon espoused in private. Brenner also had nothing close to Nader's adversarial approach to safety regulation. Moreover, Brenner had other means by which to shape the agency. After Haddon left the agency, Brenner became the acting administrator through 1970. Then, from 1971 to 1986, he became the agency's chief scientist, an important, if largely honorific, position in the organization. Brenner's influence on the agency ran both deeper and longer.

⌒

The road to federal auto safety standards required the efforts of many different individuals drawn from several different professions, including medicine, engineering, and the law. Yet many staff members of the National Highway Safety Bureau (not counting secretaries, janitors, and other essential support staff) were members of the Society of Automotive Engineering. As we will see, most of the agency's first auto safety standards were based on preexisting SAE standards. Engineering culture has thoroughly dominated the agency since its inception. While the agency arose from the politicized discourse of consumer advocacy, over the years, it has favored a reserved spirit of ostensibly apolitical technical authority. Whereas Haddon favored the "pure" research of doctors and scientists, the agency as guided by Brenner always emphasized the more practical approach of engineers. Doubtless, many factors encouraged the agency to develop in this direction. Legislators emphasized auto safety standards, and engineers were perhaps best equipped for this mission. They may also have simply fit the prevailing understanding of the agency's division of labor. But, in the end, understanding Brenner's role in the agency helps to explain a seeming paradox—that the agency did not develop even in its early stages according to the more aggressive regulatory visions of its first administrator, William Haddon, to say nothing of the man most centrally associated with this moment of auto safety, Ralph Nader. As Robert Brenner attended to the tedium of planning the bureau's work and pursuing mundane details, he bred into the National Highway Safety Bureau a strong, conservative engineering culture that would often disappoint consumer and safety advocates, as when it created the nation's first safety standards, the subject of the following chapter.

CHAPTER FIVE

The Limits of Federal Automotive Safety Regulation

When Lyndon Johnson signed the National Traffic and Motor Vehicle Safety Act of 1966, the auto industry's opposition to there being *any* regulation at all collapsed then and there. The automakers could do nothing but go along with the legislation and try to minimize its effect on their autonomy and profits. Yet, if the industry's resistance to legislation faded, the energy behind it simply moved elsewhere, namely to weakening the federal auto safety standards that would arise from the new law. Congress had not defined these standards. The new agency, the National Highway Safety Bureau, had to set them through the federal rule-making process. This process was defined by the Administrative Procedure Act of 1946, a law meant both to bring some rationality to executive branch rule making and to beat back the rule making by fiat that some had perceived during Franklin Roosevelt's presidency.[1] The Administrative Procedure Act required agencies to publish proposed rules in a public forum. Interested parties then had a chance to comment before the agencies published a final rule, which, the law's framers hoped, would take the public's views into account.[2] In matters of federal regulation of business, this comment process became a zone for industry challenges. It took on an adversarial hue much akin to a court of law. Companies would hire experts to criticize the government's proposals, especially to call into question the scientific and technical facts on which the proposals were based.

From 1966 to 1967 the newly created National Highway Safety Bureau designed and promulgated the first twenty, nationally binding safety standards. Bureau administrator William Haddon, Deputy Administrator Robert Brenner, and the civil servants who constituted the young agency brought with them an eager, reformist spirit. They sought to change the auto industry and make the roads safer for all. Their work did not go unopposed. Many of the agency's staff were new to government service and thus were naïve and inexperienced in the art and struggle of regulation. Agency leaders seemed to overestimate the industry's willingness to adopt safety standards voluntarily. Or at least they did so rhetorically in their publications, perhaps in an effort to spur the industry to transform itself. One report claimed, "There is every evidence that manufacturers will soon be competing actively in the safety field on a voluntary basis, and will on their own introduce numerous safety features as a stimulus to sales."[3] This prediction eventually came to partial fruition, but not for another twenty-plus years. In the meantime, the agency headed into a fight, one for which it seemed ill-prepared.

The first federal auto safety standards covered many aspects of the automobile, including tires, brakes, headlights, windshield wipers and defoggers, the arrangement of the gear-shifting lever, materials on car bodies that glared, and steering wheels. The most important standard of all for advocates of auto safety, however, was Standard 201—titled "Interior Occupant Protection"—and this chapter tells its story. (I refer to this standard simply as Standard 201.) In some ways, Standard 201 was just another in the first wave of standards. It was published in the same way, on the same days. Nothing would have marked it as special to the casual observer. But William Haddon, Robert Brenner, and the crash safety community cared deeply about Standard 201 because it most fully embodied the notion of crashworthiness that they had fought so hard to make law. To the degree that the auto industry rejected this notion, the standard was a philosophical battleground. As the *New York Times* observed, "The rule is considered by many the most important of the twenty [new standards]. Its purpose is to minimize injuries resulting from the 'second collision' that occurs when the occupant of a car is thrown forward by the impact of a crash."[4] Moreover, Standard 201, both literally and symbolically, penetrated the most private relationship between the automakers and consumers, the vehicle's interior—the basis of multiple visions of luxury, comfort, and contentment. Standard 201 exposed the most severe moral, technical, and ideological differences between the regulators and the automakers.

The auto industry opposed Standard 201 on numerous grounds. Viewed from one angle, its resistance arose from a clash of cultures. Regulators and automakers had different visions of engineering practice. The automakers claimed that the safety bureau had an insufficient understanding of traditional industry routines, while regulators argued that the auto companies were hidebound. Auto manufacturers protested, for example, that they could not meet Standard 201 within the allotted time. They were at pains to explain that, while politicians and bureaucrats lived by the Washington calendar, automakers had their own schedules. In part, their remonstrations were essentially attempts at education. The car companies had to explain to Washington how their industry worked and how it was organized. No doubt, to support their cause, automakers highlighted the longest cases of new-car development. Chrysler claimed, for instance, that it took the company three years to develop a new car. The safety bureau protested that much of this concern over lead-time was mounted to obstruct the development of effective safety standards. As the agency wrote, "The intent of the Congress in passing the legislation was precisely to encourage design changes, not to freeze design. At some point, arguments on lead time changes from a legitimate problem of industrial engineering to covert attempts to freeze design to the economic advantage of the manufacturer, but to the detriment of safety objectives."[5]

Viewed from another angle, then, the automakers' protests were strategic efforts to undermine the very ground of the new safety standards. In the 1920s, multiple interests convened in the name of Hooverian associationalism—sometimes spurred by agencies such as the National Bureau of Standards—to create so-called consensus standards.[6] Of course, some interests at the metaphorical table were always more powerful than others, and some groups had an outsized influence on a standard's final form. Still, the process of consensus standardization brought expert communities together. In contrast, the federal rule-making process pitted one group of experts against another. It gave rise to dynamics very different than those we have seen thus far, and it thereby shaped the automobile in different ways. The irony of Ralph Nader drawing on industry-sponsored science to make his case for auto safety came back to bite the young agency when some influential figures in the impact biomechanics community sided with the automakers over the regulators. Standard 201, as originally planned, contained a somewhat radical vision for technological transformation, but that plan ended in failure. In its final form after rule making, Standard 201 was a weakened thing, and that weakening set a tone for American auto safety regulation that has persisted ever since.

Drafting the First Federal Auto Safety Standards

When the staff members of the new National Highway Safety Bureau arrived on their first day of work, they found themselves responsible for creating the nation's first federal crash safety standards. The agency originally proposed twenty-three standards. But it still lacked sufficient evidence for some of them, including rules for tires, head restraints (almost always head rests, though automakers technically could have used other technologies to meet the test), and windshield wipers for small vehicles. So the agency was forced to delay issuing these standards, whittling the total to twenty. Several of these first standards, which covered many aspects of the automobile, had been around for decades, either as voluntary consensus standards used by the industry (which the Society of Automotive Engineers called Recommended Practices) or as standards adopted by state and local governments, such as the headlight and brake standards examined in chapter 2. The effect, then, was that several preexisting standards—which had authority only in certain states and localities—became truly standardized and for the first time mandated for nearly every motor vehicle sold in the United States.

As earlier chapters show, standards making is always a complex process involving many actors. When governments are involved, legislative mandates often add to this complexity. In the case of auto safety standards, the 1966 traffic safety act required the National Highway Safety Bureau's standards to be based on "already existing" standards. The spirit of the requirement was to keep regulators from creating excessively strict standards ex nihilo and to ensure that new standards would be reasonable and achievable. The requirement forced agency staff members, however, to draw on a wide variety of sources to cobble rules that would approach their hopes. As agency staff members wrote: "These twenty standards . . . were based on a total of forty-eight existing standards. These include nineteen from the General Services Administration (GSA), one from the Interstate Commerce Commission (ICC), twenty-two voluntary practices recommended by the Society of Automotive Engineers (SAE), one Swedish National Road Board Standard, one of the Uniform Vehicle Code, one of the Post Office Department, one of the U.S.A. Standards Institute, one from State laws, and one from the National Bureau of Standards (NBS)."[7]

The General Services Administration standards established as part of the Roberts Act (described in chapter 4) ultimately did the lion's share of the work. As the safety bureau report noted, "The Agency was able to accomplish as much as it did substantially because of the fact that the industry worldwide had been served notice by the earlier General Services Administration standards . . . as to the likely direction of future Government vehicle standards."[8]

Standards making at the agency faced additional limits. For example, standards had to be based on performance, not design. That is, they had to set criteria for technology performance, not which technologies should be used. For example, a design-based frontal collision standard would have specified that automakers mount foam padding of a certain thickness on all instrument panels. The performance-based frontal collision standard that actually came into being specified a test for determining with how many g-forces a head would hit an instrument panel or other surface, and limited how many g-forces the head could reach. Automakers were free to use whatever solution they chose, including foam padding, to pass the test. The rule mandated no technology.

Yet the line between performance and design was often murky, sometimes hopelessly so. From the get-go, the safety bureau stated that it would set performance standards while maintaining that some of these standards would inevitably affect design in determinable ways. A report on how the agency developed the standards approvingly quoted something Senator Warren Magnuson had said: "We are also pleased that the House agreed to the restoration of Senate language for the definition of 'motor vehicle safety,' recognizing that safety is related to design. Performance standards issued under the act are expected to affect" automotive design. As the safety bureau report summarized this idea, "There are no precise legal or other criteria for deciding when a performance standard begins to have design overtones."[9]

Standard 201 both set requirements for interior features, such as window cranks and control knobs, and mandated a test for surfaces that the driver's and front passenger's head might strike in a frontal crash. The test worked like this: A standardized dummy was placed in the front seat of the automobile. Engineers then bent the dummy forward at the hip to identify which surfaces its head would strike, including the instrument panel and the front roof beams. The identified surfaces were then removed from the car. Sometimes engineers removed a sample of the identified material and mounted that on a frame, but at other times they simply removed the entire instrument panel and mounted that. Either way, the mounted material was placed in front of a large pendulum equipped with a "head form," a rounded surface meant to represent a human head. Engineers placed an accelerometer, which measured g-forces, inside the head form. The pendulum was pulled back, and the head form was slammed into the mounted material, such as the instrument panel, at 30 miles per hour. The standard limited how many g-forces the accelerometer inside the head form would experience in such

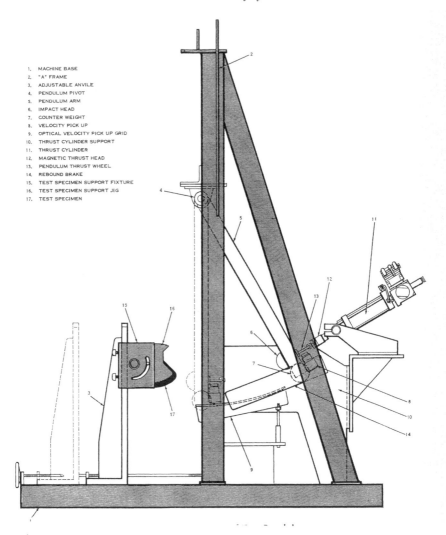

1. MACHINE BASE
2. "A" FRAME
3. ADJUSTABLE ANVILE
4. PENDULUM PIVOT
5. PENDULUM ARM
6. IMPACT HEAD
7. COUNTER WEIGHT
8. VELOCITY PICK UP
9. OPTICAL VELOCITY PICK UP GRID
10. THRUST CYLINDER SUPPORT
11. THRUST CYLINDER
12. MAGNETIC THRUST HEAD
13. PENDULUM THRUST WHEEL
14. REBOUND BRAKE
15. TEST SPECIMEN SUPPORT FIXTURE
16. TEST SPECIMEN SUPPORT JIG
17. TEST SPECIMEN

Figure 5.1. The Society of Automotive Engineers' standard J921 test pendulum, designed at Chrysler, drew on longer traditions of material testing. SAE Recommended Practice J921, SAE International, 1967

an impact. By reducing these g-forces, the standard decreased the head trauma and brain damage drivers and front passengers would experience in frontal crashes.

Standard 201 was based on six General Services Administration standards, all related to the vehicle's interior.[10] The standards that dealt with impact areas stemmed from two SAE Recommended Practices—J826, which used various devices to figure out seating accommodations, and J921, which used an "impact pendulum" test to determine the "safety characteristics of padded instrument panels and other passenger compartment components."[11] Engineers used J826 to determine the location of head impact zones inside the vehicle, and then used the pendulum device described in J921 to smash an accelerometer-equipped head form into those impact zones. The J921 pendulum test was often referred to as the lollipop test because observers thought that the head form looked like a lollipop when it was placed on the end of the shaft attached to the pendulum.

The Chrysler Corporation played an influential part in constructing both of these SAE testing processes. The SAE had for a number of years tried to standardize seating arrangement practices, including an attempt by the SAE Body Activity Committee in 1959.[12] Until the SAE established Recommended Practice J826 in 1963, however, Chrysler, Fisher Body (GM's auto body division), and Ford Motor Company all had different methods for determining internal configurations of their vehicles' body. In 1961 the SAE Body Activity Committee made another push for standardization and created a Manikin Subcommittee. The core innovation in "interior dimensioning" came out of GM, which took the best features of each company's practices and combined them into a single manikin and methodology, which became J826.[13] Recommended Practice J921 had a similarly long gestation. Chrysler's Rubber and Plastics Laboratory developed the precursor of the pendulum used in J921 in 1956–1957.[14] When the SAE Body Engineering Committee formed the Passenger Compartment Energy Absorption Subcommittee in response to the passage of the Roberts Act, Chrysler's Engineering Office redesigned the pendulum. As the General Services Administration picked up J921 as part of Roberts Act Standard 515/2, the industry helped shape the federal standards.

Agency staff admitted from the beginning that they had tightened certain aspects of the General Services Administration standards to create a more stringent interior-occupant standard. They added a "knee and leg" impact requirement to the already-existing "head" and "hip" impact criteria. They lowered the acceptable g-forces that the crash dummy could sustain, and they "changed the wording of the requirements for protrusions," items such as door handles, radio knobs, and window cranks that projected into the passenger compartment. Staff members stipulated that the test should be run for both a 95th percentile adult male (i.e., a manikin that only 5 percent of adult males were larger than) and a 5th percentile adult female (i.e., a manikin that only 5 percent of adult females were smaller than). Since men tend to be larger than women, if a car could safely accommodate both a large male and a small female manikin in a crash, it would be safe for the wide range of human figures, making exceptions for people who had forms of either dwarfism or gigantism.[15] The measurements of these manikins were based on anthropometry research published by the Public Health Service.[16]

Fighting Standard 201

On October 8, 1966, the National Highway Safety Bureau published an Advance Notice of Proposed Rule Making in the *Federal Register*, declaring that the agency would issue initial standards by the statutory deadline of January 31, 1967. The law required that interested parties be allowed "to participate in the making of [the] standards."[17] By November 1, the parties had sent 3,105 pages of technical commentary on the proposed standards. In the following month the agency prepared formal proposals for the initial standards that in theory reflected the input of the interested parties. The safety bureau published the Notice of Proposed Rule Making on December 3. The issuance allowed interested parties to submit comments on the proposed standards until January 3, and automakers and other parties submitted 4,525 pages on these proposals, an increase in heft of almost 50 percent over the round-one commentary. The agency then had "slightly more than three weeks" to sort through these documents and publish the initial standards by January 31. Things moved very quickly.

Because of the strong opposition to Standard 201, the Commerce Department decided to hold a "rulemaking proceeding pursuant to U.S.C. 553 (formerly section 4 of the Administrative Procedure Act)."[18] The agency appointed Russell A. Potter as presiding officer of the proceedings.[19] Little is known about Potter. He had worked at the National Transportation Safety Board's predecessor agency, the Civil Aeronautics Board, since at least 1947, and probably before then because he chaired a hearing that year, not a task ordinarily left to agency newcomers.[20] Potter's life was cut short in December 1968 when he and two others were lost at sea during a sailing trip to the Virgin Islands.[21] It is perhaps for this reason that Potter did not eventually rise high enough in the bureaucracy to be more visible in the historical and biographical record. Regardless, by 1967, when Potter presided over the hearing on Standard 201, he had accrued at least twenty years' experience regulating complex technological risks—albeit in his case airline safety.

The safety bureau had not yet created standard rule-making procedures, especially ones for dealing with controversial issues. So the new agency still had to devise not only technical standards but also its own routines. Potter laid out a process for dealing with the automakers' challenges to 201. The companies could file objections until May 1.[22] Then, three days later, Potter would hold a hearing in which he would compile and simplify objections to prevent duplication and wasted effort. One week after this conference, the agency would furnish a rebuttal to the objections, and then, on May 23, would hold a hearing allowing for cross-examination of both "direct and rebuttal" testimony in Washington, D.C. The parties could, within ten days, file further briefs on what was said at the hearing. Finally, the presiding officer would issue his recommendations as quickly as possible, and interested parties then had ten days to submit further objections. The rule-making proceeding was a reasoned process that included extensive scientific and technical information, but it moved at breakneck speed. Scientists and engineers, especially those within the agency, were often expected to turn around analyses of complex technical issues within days. Safety bureau members tripped more than once during the process.

The prehearing conference on May 4 bordered on chaos. The agency had received hundreds of pages of submissions and had three days to look them all over. Not only was the standard under examination inherently complex, but the submissions also brought up a plethora of other issues. The first objective of the hearing, therefore, was to settle what the issues were. The participants and the hearing's presiding officer handled this task by agreeing that the Automobile Manufacturers Association's submission summarized the major points and would form the basis of future discussions.[23] This decision strengthened the trade association's position within the rule-making process. The petitioners had adopted its submissions as the basis of their claims, and the hearing officer had chosen the trade association's materials as a rubric by which he would organize and interpret the proceedings. Thus, the association and its submissions took a central, intermediary role.

Throughout the process, agency staff members held that Standard 201 should remain intact because it would reduce deaths and injuries, it was "based on existing standards and relevant available motor vehicle safety data," and it was "reasonable, practical, and appropriate."[24] Cast in the language of the traffic safety act, the three terms *reasonable, practical*, and *appropriate* became flash points of disagreement.[25] The safety bureau had two nonagency auto safety experts, Derwyn M. Severy and Colonel John Paul

Stapp, testify on behalf of its position. (Stapp is discussed in chapter 3, Severy in chapter 4.) To counter these two agency expert witnesses, GM alone submitted written direct testimony from thirteen employees. The company also submitted a letter from Lawrence M. Patrick, a professor of engineering at Wayne State University, who co-led the impact biomechanics lab that Elisha Gurdjian and Herbert Lissner had earlier organized (see chapter 3). The automakers enlisted and demonstrated overwhelming expertise. Although the highway safety movement had persuaded a majority in Congress to sign on to its cause through the traffic safety act of 1966, when it came to dealing with detailed technical analysis and argument in the rulemaking process, the movement's proponents in the safety agency were seriously outgunned.

The industry frequently complained that Standard 201 was not "clear and objective," which typically went to the regulation's wording and definitions. The automakers defended this complaint with several arguments. First, the regulation did not live up to the exactitude demanded by the auto industry. Second, since Standard 201 was cobbled from several standards, it posed added complexities and vague aspects not contained in any of its constitutive regulations. Third, the standard sought to implement new stringencies that were not yet standard in the industry and thus had not yet received all the refinements standards typically received in such circles as the SAE committees. More particularly, engineers and firms had not signed onto the new parts of the standards as they would in a typical SAE standards-setting process. To the degree that the auto industry's lawyers' endeavors were obstructive, obscurant, and sophistical, the standard's language made a perfect battleground.[26]

Most of the criticisms regarding Standard 201's lack of clarity and objectivity centered on the definition of terms, including the "H-Point" (the *H* referring to both the hip on an adult human and the hinge on a manikin or crash dummy) and the various "impact areas"—head, pelvic region, and knee and leg. Debates about seemingly superficial issues often masked deeper disagreements about the standard's very justice. For instance, should the standards require the automakers to innovate, not just in terms of the automotive safety technologies but also in terms of the tests used to determine compliance? Grappling over the H-Point was a case in point. Industry representatives consistently maintained that the H-Point and the impact areas should be defined in terms of two-dimensional graphical methods—what was sometimes called "graphic equivalency"—for designing automotive inte-

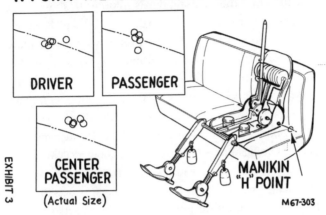

"H" POINT MEASUREMENT VARIATIONS

Figure 5.2. Federal Motor Vehicle Safety Standard 201, Interior Occupant Protection, was based primarily on the use of this manikin to identify which surfaces the driver's and passengers' heads could hit during an accident. Researchers then removed the identified surfaces and materials from the car and used a standardized test pendulum on them in a laboratory. National Highway Traffic Safety Administration microfilm, National Archives and Records Administration, Washington, DC.

riors, not in terms of a three-dimensional, "real-life" crash test. Two-dimensional methods enabled designers to work out problems at the drawing board *before* building any prototypes, which would then be crash tested.

General Motors argued, for example, that the graphic equivalent "is necessary to permit designing compliance into the car from the very beginning of the design process—the drawing board. The use of a graphic equivalent is also necessary to provide a precise and repeatable method for continually checking design at all phases in the automotive design and production cycles."[27] General Motors, like many other automakers, used the two-dimensional manikin prescribed in SAE Recommended Practice J826. Designers first worked out issues, such as spaciousness and placement of control elements within the driver's reach—door handles, window cranks, stereo knobs—in two dimensions. The automakers' real fear was that the standard would require them to build up every potential vehicle into a working model, only to find out that the car did not meet the space requirements. Such a discovery would add huge lags into the car's lead time and create an expensive pile of wrecked cars. Thus, automaker engineers and designers wanted two-dimensional methods to solve these issues beforehand. At the same time, automakers may have played up the difference between actual industry practices and the new standards in order to show that regulators were clueless and their mandates absurd.

The National Highway Safety Bureau preference for a standard based on crash tests instead of two-dimensional graphical methods doubtless reflected its scientific and technical network. The agency, after all, was a segment of the crash safety community. It crashed cars. It did not design them. This lack of knowledge about the ins and outs of the auto industry would continue to plague federal regulatory efforts. Still, members of the agency effectively shot back, what they cared about was the real-world object. Two-dimensional design and analysis were valueless unless they translated into a safer car, and the only established way to determine the car's safety was to run it into a cement wall or use pendulums to smash "heads" into parts.[28]

Questions over whether Standard 201 was "practicable" were inextricably tied to whether it set design, as distinguished from performance, criteria. What constituted a design standard versus a performance standard was a matter of interpretation. The chief industry complaint about practicability had to do with "protrusions," "bezels," and other interior objects, including window cranks, door handles, and support beams. The automakers submitted numerous blueprints and photographs demonstrating the absurd effects Standard 201 would have on interior auto design. Some submitted photos of redesigned window cranks showing that Standard 201 would make the head of the crank so rounded and oblique that it would be almost impossible to grasp. Similarly, they argued that Standard 201's requirement that these aspects of the vehicle's interior not be able to withstand more than 90 pounds of pressure (so that occupants would not be impaled or otherwise injured on them) was ridiculous because people often used things, such as window cranks, to close vehicle doors. If someone shut a 201-compliant door this way, the window crank might snap off in the person's hand. All of the written submissions ridiculed these requirements.

The agency's witnesses pushed back against these criticisms. Derwyn Severy wrote that the industry's assertions about protrusions suggested a "completely callous attitude toward the injury producing qualities common to protuberances."[29] Stapp claimed that some of the auto industry's representatives had gone to "grotesque intellectual effort to belabor the definition of protrusion" in order to make it seem incoherent, while "not one engineer suggested a better term."[30] Industry witnesses were trying not to help establish a better standard but only to destroy the proposed one.

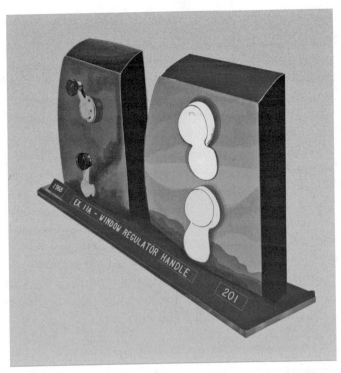

Figure 5.3. To highlight their point that the initial federal automotive safety standards would lead to unworkable design features, the automakers created absurd-looking parts, such as these window handles. National Highway Traffic Safety Administration microfilm, National Archives and Records Administration, Washington, DC.

General Motors and the Automobile Manufacturers Association recruited Lawrence M. Patrick to testify regarding the proposed Standard 201's human tolerance requirements—that manikins not exceed 80 *g*'s for 1 millisecond during testing. Patrick was a Michigander, born in Highland Park (home to Henry Ford's famous factory) in 1920. He did all of his secondary schooling at Wayne State, earning a bachelor of science in mechanical engineering in 1942 and another in aeronautical engineering in 1943, before receiving a master's in mechanical engineering in 1955.[31] After he became a professor at Wayne State, his work focused on human tolerance studies. General Motors supported much of Patrick's research in a "joint venture" with Wayne State, though Patrick did not reveal this relationship in any of his testimony.[32]

At the time of Patrick's testimony, the major upshot of research at Wayne State University was something known as the Wayne curve, a graph of the human head's tolerance to deceleration over time. The statistics underlying the curve came from merging two sets of crash data: cadaver tests and animal tests. Since most of the cadavers were elderly, their heads had a lower tolerance to forces than the "average" human head; since animals had harder skulls than human cadaver heads, their use was thought to balance the equation. Still, because of cadavers' frailty, most researchers seemed to believe that the Wayne curve was a conservative estimate.[33] All of this meant that, after the findings of the Wayne curve were eventually built into federal regulations, the image of the average adult human at the heart of federal safety standards was a statistical aggregate of animal and cadaver data—or, put more baldly, a mix of elderly cadavers and dog skulls.

The Wayne curve helped the industry's cause in two ways. First, it presented a clear, easily grasped representation of the issue, which helped win over non-experts, including Potter. Second, it suggested that the standard—80 g's for no more than 1 millisecond—should be lengthened, since the graph demonstrated that the human head could sustain 80 g's for longer intervals. The Wayne curve caught the agency flat-footed, which probably says something about its preparation and organization. The agency's two witnesses, Stapp and Severy, found the curve convincing and sided with the industry over the agency itself on the low g-force standard. Severy suggested that research at UCLA supported Patrick's 3 millisecond criteria as a "reasonable interim compromise," though he did insist that the 1 millisecond standard should be applied within a "2-year period."[34] Stapp argued that "the 1.0 millisecond value . . . was reasonably chosen by the Agency on the basis of the existing scientific evidence and especially in the interests of public safety."[35] He also claimed, however, that no one, including himself, had yet established with any certainty or scientific rigor how much force the human head could sustain. Ultimately, the agency bent the knee, feeling forced to admit that the limit should be loosened for the time being to 80 g's for 3 milliseconds, instead of 1 millisecond.

Staging 201: The Automobile Manufacturers Association's Theater of the Absurd

Potter, the automakers, National Highway Safety Bureau staff, and various witnesses gathered for the hearings on May 22 and 23, 1967, in Detroit, Michigan, and May 24 and 25 in Washington, D.C. For anyone who attended, the hearings must have been a strange show. For years, scholars studying science and technology have examined how scientists and other technical experts use rhetoric and dramatic techniques to bolster and solidify their authority in public.[36] The drama that representatives of the auto industry staged at the hearings had the exact opposite intention: it was meant to undermine the National Highway Safety Bureau's authority. The automakers' attorneys and technicians attacked the agency by staging Standard 201's absurdity.

The hearings were in many ways a face-off between Lloyd N. Cutler of the law firm Wilmer, Cutler & Pickering and Robert M. O'Mahoney, the safety bureau's legal counsel. The two men could not have been more different. Cutler's firm, cofounded a half-decade earlier, lobbied for the Automobile Manufacturers Association, among other auto-related interests, in Congress and the executive branch.[37] He was not highly regarded in safety and consumer circles. Ralph Nader once said of him, "People shouldn't forget that over the years he has represented the crassest positions of the auto companies, the drug companies and others in their anti-consumer and anti-environmental activities."[38] Cutler's influence and power grew in Washington. From Jimmy Carter's term onward, Cutler consistently acted as a presidential advisor. As his *New York Times* obituary noted, he "shuttled between a lucrative private practice and the White House."

O'Mahoney was an attorney who had dedicated his life to public service. Before working for a time in the federal government, he served as the deputy attorney general of Indiana. Later, O'Mahoney worked for the General Services Administration, where he addressed automotive issues. For instance, he played a role in that agency's efforts to purchase natural gas–powered low-emissions vehicles—yet another effort designed to use the federal government's purchasing power to foster technological change.[39]

The hearing began on a seemingly positive note. Cutler said, "We were pleased to hear counsel for the Bureau state . . . that a purpose of this proceeding is to produce information."[40] Immediately, however, the agency's counsel began to suspect that all was not as rosy as the automakers initially claimed. O'Mahoney stated for the record that automakers did not treat the agency's representatives with the deference that Cutler suggested. When agency staff members asked to see the materials Cutler brought for the hearing, he allowed them to see only one item, a gray book that held pictures.[41] The other materials he would unveil slowly, dramatically, as he made his case. Agency staff members found Cutler's performance shocking. As O'Mahoney exclaimed, "I must say I was incredulous then. I am incredulous now. I don't believe that."[42]

Reading the transcript of the hearing makes one yearn for a video recording of the event. At one point, while GM attorney Edward C. Adkins was questioning GM engineer Edwin Klove, Klove stood up and moved to a vehicle that was in the hearing space.[43] There, he demonstrated how a test manikin would be articulated to determine head impact areas in the proposed version of Standard 201. The transcript reads like a description of a vaudeville act. As Klove struggled with the bulky and ungainly manikin in the tight space of the passenger compartment, Adkins asked, "Would you . . . demonstrate its lack of space for doing the things required in the establishment of [the] head impact area found in the standard as a relevant definition?" Apparently, several people had crowded around the car in order to see the demonstration. Potter turned and asked, "Mr. O'Mahoney, would you like to come and look at this?" O'Mahoney asked, snarkily, "Is the dummy being used to demonstrate the non-existence of a dummy? Or the lack of capacity of that particular dummy which doesn't conform to Standard 201 to do what 201 requires?"[44] Potter responded calmly, trying the move the matter forward; Adkins tried to continue his questioning; and then Potter called a five-minute, off-the-record recess.

Minutes later, after the hearing resumed, Klove continued to demonstrate the standard's absurdity and the physical difficulty it posed. Adkins asked, "Are your getting tired yet?" Klove answered, "I am getting out of breath."[45] He continued the performance and reached a point where he was not able to manipulate the manikin in the required way with only two hands. The dummy's feet had to remain on the floor, while someone moved its body. O'Mahoney asked, "Can the gentleman standing on the side be of assistance to him?" Adkins suggested, "This is not a three-man procedure, is it? It is one operator for the dummy." O'Mahoney protested, "There is nothing in the standard about how many people [must] do it."[46] The point of all of Klove's maladroit wrestling with the dummy inside the car, of course, was that the standard was unrealistic, unrepeatable, and impracticable. Samuel Beckett could hardly have staged a better scene.

Slowly, over the course of the day, even a dozy member of the audience could have deduced that the automakers' testimony and especially their introduction of new exhibits were well-choreographed shots at making Standard 201 look foolish. O'Mahoney's anger began to build. If he was merely irritable during the dummy-wrestling demonstration, his temper would later flare unmistakably: "Mr. Potter, if it is helpful to shorten this proceeding, we will stipulate that it is the position of apparently General Motors and apparently the Automobile Manufacturers Association that their interpretation of the standards [is] to design idiotic things they have been showing us. But I don't see any point in going on with this."[47]

GM engineers Klove and George Ryder brought in redesigned interior features, such as window cranks, that they suggested met the standards criteria. Safety bureau members described the redesigned window regulator handles as "hopped-up." GM's outlandish design for such devices clearly betrayed the company's attempt to make the standard look foolishly impracticable. Agency staff members argued, in effect, that these artifacts "had politics."[48] That is, the cranks and other features were like Rube Goldberg devices, not in terms of complexity but in terms of their parodic intent to undress the human inclination to overdesign the world. If the proposed Standard 201 had a range of possible design interpretations, engineers and designers at GM took their reading to a logical, comic extreme.

There was nothing neutral about the bulbous, dome-like, silly-looking, seemingly ungraspable handles. In their very effort to be free of any sharp, pointed component that might injure a passenger, the GM-designed handles made their point. When O'Mahoney challenged this line of criticism, Klove maintained that "he knew no better solution."[49] Representatives of Ford Motor Company and the executive vice-president of Nissan Motor Corporation USA, Soichi Kawazoe, who testified on behalf of the Japanese Automobile Manufacturers Association, made similar statements.[50] Kawazoe declared, "I can assure you, sir, that this (the Japanese design) was strictly done in Japan by our engineers according to their interpretation of the standard. I particularly want to say, especially, [that] I had a big argument with Tokyo about showing this crazy-looking screwhead."[51]

These absurdist displays of how the automakers would have to redesign their cars' interiors finally drove O'Mahoney over the edge. "I must say," he declared, "as a show—and that is what this is—for this assemblage, without any advance notice that this was the sort of thing you were going to do, it seems to me it is terribly violative of the spirit of this proceeding. It is a sort of like a sideshow. I think it is unfortunate." Potter responded, "Mr. O'Mahoney, I believe that the demonstrative testimony is necessary and serves a very useful purpose in illustrating what the end result of the standards will be." O'Mahoney shot back, "Mr. Potter, that is just the problem. They don't demonstrate what the standards will be. They demonstrate what these people choose to make the end results of the standards by a rather extreme interpretation of them."[52] When Potter would not hear his complaints, O'Mahoney suggested that, at the very least, "I would like an instruction, Mr. Potter, from you, to the parties who are conducting this show, that they should conduct it in a manner befitting the dignity of a serious pursuit of fact involving the safety of the citizens of this country."[53]

By the end of the first day, it was clear that O'Mahoney and the National Highway Safety Bureau had been run over by the automakers' traveling theater troupe. As those testifying for the auto companies continued to discuss how the proposed Standard 201 would hypothetically affect interior design, O'Mahoney pleaded uselessly: "Mr. Potter, I came here ready to listen, to hear, and to see. I am still ready to do that. But I am not interested in hypothetical problems, only real problems. Neither is the administration for which I work. We want real problems, not hypothetical problems, not assumptions but problems."[54] Potter continued to believe, however, that the automakers' testimony had identified real issues, and he would take those into account in his final act as hearing officer, writing the Recommended Findings report.

Revising Standard 201: Potter's Recommended Findings and the Agency's Response

On June 22, 1967, Potter issued his Recommended Findings report. It devastated the National Highway Safety Bureau's cause. Potter divided his analysis into four parts, with the first three addressing the requirements of the law—the need to be clear and objective, the need for practicality, and the need for proved safety. Potter's fourth point treated the issue of whether it was "reasonable to require compliance by January 1, 1968."[55] He argued that it would not be fair to require the automakers to comply by this date, because it did not give them enough lead time.

Potter also asserted in his report that the 95th percentile male and 5th percentile female criteria for manikins should be jettisoned: "People are a variety of sizes. Long-legged men may be in the 95th percentile group for standing height though their sitting height is below the average."[56] In this view, the size definitions were completely arbitrary, or what scholars call "socially constructed." Potter concluded, mincing no words, "Therefore, the definitions of the head impact area, and the knee and leg impact area based on use of manikins are vague, misleading, inexact, and completely unworkable for design purposes."[57] Moreover, Potter was overwhelmingly moved by General Motors' material demonstration of how automakers' parts would need to be redesigned if Standard 201 was promulgated and enforced. He called the testimony of Edwin Klove and George Ryder on this point "the most informative and persuasive of all evidence introduced throughout the proceeding."[58] Potter therefore declared that the safety bureau should give up its requirement on protrusions and other interior features.

The agency issued a rebuttal to Potter's Recommended Findings, pointing to many errors and "erroneous concepts" the report contained. The agency argued that whereas "the Standard should have enjoyed a presumption of regularity, instead the report consistently balanced the evidence and found that the weight of the evidence was with the petitioners."[59] Staff members claimed that the agency's administrator had intended the public hearings and submissions as a means to amend and improve the Standard, but Potter had asserted the opposite, viewing the hearings as a review of Standard's justice. In the agency's view, "the presiding officer's complete and puzzling failure to grasp this rather simple instruction" had led him to produce a document that was of "little or no utility in achieving the objective of the Administrator's hearing order."[60]

As to how Potter had come to his basic position on the purpose of the hearing, safety bureau staff members speculated that "Bureau Counsel recognized this misunderstanding on the part of the presiding officer early in the proceeding and made frequent, but obviously unsuccessful attempts to turn the proceeding away . . . from the adversary character to which it was allowed to descend albeit with active encouragement of counsel for the Automobile Manufacturers Association and other United States manufacturers, most particularly counsel for the General Motors Corporation."[61] But O'Mahoney's argument was largely futile. Earlier, the *New York Times*— after discussing how Haddon believed that Standard 201 would stand up to scrutiny—had written, "The hearing commissioner [that is, Potter] will be given final authority on whether to grant the manufacturers' demands for modification, Dr. Haddon said."[62] It seems that the bureau in its rebuttal was trying to back itself out of a deal.

The final steps in Standard 201's path to its final form are as closed to us as the initial formulation was. No archival records document any meetings subsequent to the date the National Highway Safety Bureau submitted its rebuttal. Perhaps Haddon, Brenner, and others met and strategized, hoping to beat Potter's recommendations. Perhaps they resigned themselves to their fate. In any case, on August 13, 1967—a little less than three months after the hearings—the *New York Times* carried a story on Standard 201, "Safety Standards on '68 Cars Eased," noting that "requirements for extra knee and leg impact protection, and changes of interior handles and knobs have been dropped." The news article marked one of the few times that the automotive standards-making process had broken the surface of bureaucratic anonymity and made its way into the public eye. The message was simple: the National Highway Safety Bureau had been defeated.

~

This defeat had deep ramifications for the future of American automotive safety regulations. As other historians have shown, in the following years the federal auto safety agency increasingly shied away from setting ambitious, far-reaching regulations, in part because the country's legal environment discouraged such behavior.[63] But as the case of Standard 201 shows, the agency was, from the beginning, weak and unable to push safety technologies forward in any significant fashion. It never lived up to the hopes of safety advocates such as Haddon or Nader.

PART THREE

POLLUTION

CHAPTER SIX

Discovering and (Not) Controlling Automotive Air Pollution

In the early 1950s the chemist Arie Haagen-Smit, best known at the time for having isolated the chemical makeup of pineapple flavor, turned his techniques and instruments on a different topic: the chemical constituents of the smog that had been terrorizing Los Angeles, California, for more than a decade. Through a series of experiments, Haagen-Smit made a striking discovery and developed an argument for a new way of looking at urban air pollution. Automobiles, Haagen-Smith asserted, were the primary cause of smog. As we have seen, from the 1890s through the 1940s, the public saw accidents as the greatest problem associated with automobiles. People had long known that cars produced potentially deadly carbon monoxide, and early auto manuals warned users not to run vehicles in enclosed spaces. But the idea that aggregated exhaust and other effluents from cars could create pollution in the ambient air hardly received any attention at all before the 1950s.

Haagen-Smit's discovery initiated a prolonged struggle that eventually culminated in federal regulation. That struggle involved a drawn-out process of redefinition. First, Haagen-Smit and other advocates had to redefine the problem and redirect regulatory efforts from focusing on industrial sources, the traditional locus of pollution control, to focusing on cars. Second, environmentalists, legislators, and others came to argue that automotive and other forms of air pollution were best addressed through federal regulation rather than through rules set at the local or state level. This chapter examines this twin process of redefinition and refocus and the many battles that shaped it. The federal government did regulate automobiles in the mid-1960s, but these first regulations were seen as such a massive failure by environmentalists and legislators alike that they created an opening for the much tougher regulations examined in later chapters.

Early Air Pollution Control Efforts in Los Angeles

The early history of smog and automotive pollution control is fundamentally a story about Los Angeles, California. Sitting in a topographic bowl, with the Pacific Ocean on one side and mountains on three others, Los Angeles has always had problems with air pollution. Indeed, when the Spanish explorer Juan Rodríguez Cabrillo first saw the place in 1542, he named San Pedro Bay "La Bahía de los Fumos," or the Bay of Smokes.[1] Smoke from the fires of resident Native Americans filled the skies above. The problem, known as inversion, arises from the city's weather patterns: cool breezes blow into the city from the ocean, while warm air comes down onto this cooler air from over the mountains, trapping the cool air and any man-made smoke, fumes, or gases in the city.

Not until the mid-twentieth century, however, did the area's propensity to trap air pollution become a major public problem. Los Angeles County experienced enormous growth in terms of population, industrial production, and automobile use during World War II.[2] In 1941 the city experienced the first of what would later become known as "smog attacks." *Smog*, a portmanteau word derived from combining *smoke* and *fog*, was later found to be a misnomer, having little to do with either smoke or fog, but this fact was unknown in the 1940s. In urban legend, in the midst of war, the public widely mistook the first appearances of smog for Japanese gas attacks.[3] Incidents of smog only increased over the next few years.

The early efforts to remediate smog in Los Angeles fit within the longer history of attempts to control smoke in American cities. In the late nineteenth and early twentieth centuries, Pittsburgh, Chicago, St. Louis, Boston, and some other cities worked to reduce the urban smoke that blackened skies and choked citizens. St. Louis was the site of the most famous and important pollution control efforts in the first half of the twentieth century.[4] Raymond Tucker, a professor of mechanical engineering at Washington University, led the successful abatement efforts there, which focused on forcing users to switch to higher-quality fuels. Smoke in the city decreased greatly when high-quality coal shipped from Arkansas arrived in local markets, and Tucker's efforts were seen as a massive success.

As with those earlier pollution control efforts, leaders in Los Angeles initially focused on industrial pollution sources, including factories, oil refineries, and chemical plants. This focus threatened powerful entrenched interests in Los Angeles County and the surrounding locales. Oil and the industries that grew up around it had been responsible for a good deal of the region's economic development.[5] But the focus on industrial sources was unsuccessful. Sometimes after shutting down a factory or plant, the smog would seem to dissipate. Celebrations were held. The cause had been found! Then the smog returned.

The public fretted. Smog was uncomfortable, burning eyes and throats, but perhaps more important, it conflicted with the region's image as a natural paradise, an image that the real estate, advertising, and other industries were deeply invested in selling to the rest of the nation.[6] Citizens began predicting doom in op-eds and other venues. Smog would force people out of town, they said, especially those who had moved to Los Angeles for the Southern California climate's supposed health benefits. Norman Chandler, the publisher of the *Los Angeles Times* and a local landowner, made his newspaper the chief advocate for doing something about the smog issue, initiating a series of articles on the topic in late 1946. The series made little pretense about being objective or merely descriptive; it was meant to get city leaders moving.

As part of the series, the *Times* brought in Raymond Tucker from St. Louis to write a report about the Los Angeles smog problem.[7] Tucker's report was published on the front page of the newspaper in January 1947, and it focused almost completely on the kinds of causes that he knew so well from St. Louis—namely industrial gases and backyard fire burning.[8] His analysis focused especially on the role of sulfur dioxide, the egg-smelling gas earlier found to be one of the major problems in St. Louis and other cities. Others had suggested previously that automobiles might be one of the causes of California smog, but Tucker set aside such concerns, arguing that cars at best "contribute to the nuisance." Everything in Tucker's report, therefore, suggested that Los Angeles should follow the model set in St. Louis and other major cities. His report ended by recommending that the state of California create an air pollution control district in Los Angeles County, since the problem went beyond city lines and had to be addressed at the regional level.

Responding to Tucker's report and a number of other pressures, the California state legislature soon passed the Air Pollution Control Act of 1947. California governor Earl Warren signed it into law, creating the Los Angeles County Air Pollution Control District (LACAPCD).[9] Warren appointed as agency head Louis McCabe, an engineer who had conducted fuel studies with the Bureau of Mines for decades. Under McCabe's leadership, the pollution control district more or less followed the model laid out by Tucker: it targeted individual industrial operations and threatened to close them down if they did not clean up their ways.

Such an approach required surveillance. McCabe's people taught recruits at the Los Angeles Police Department Academy how to use the Ringelmann smoke chart.[10] Observers held the chart, which consisted of five boxes ranging from white through darkening grays to black, at arm's length while looking at a pollution source, such as the smoke pouring out of a smokestack.[11] Experts knew that the chart was of limited usefulness and that many problematic gases were not visible; yet the chart's seeming objectivity guaranteed its continuation as a tool for remediating public air pollution problems. Ringelmann chart–equipped police officers stalked Los Angeles streets looking for offending businesses, and, at least in urban mythology, McCabe also had a volunteer force of pilots, the "Flying Vigilantes," who flew low over the county, attempting to spot violating industrial sources.[12] Throughout the 1940s, then, Los Angeles smog control efforts followed the same pattern: officials and staff members of the APCD would identify some new source of pollution, "control" it, and then celebrate victory, only to have the choking, tear-inducing pollution come back, often worse than it was before.

Industry often pushed back against these efforts. A trade group, the Western Oil and Gas Association, began funding studies by the Stanford Research Institute (SRI), an independent think tank not formally connected to Stanford University, to discredit ideas that oil refineries and associated industries caused smog. SRI researchers conducted a wide variety of studies and published a series of reports on their findings. As the causes and even the makeup of smog were poorly understood, scientists lacked the ability to determine what was inducing the problems so often associated with it, such as eye irritation.

The SRI researchers set as their goal, therefore, to translate subjective impressions into objective measures: human sensations became the metaphorical litmus test. The researchers put individuals into a "smog chamber," a small room into which aerosolized gases from different industrial substances were pumped, including formaldehyde, sulfur, and hydrogen peroxide.[13] The SRI researchers found that none of the substances they investigated fully captured the effects of smog, and that even their attempt to make "synthetic smog" by mixing various substances did not produce the same sensation as the ambient air in Los Angeles. "This provides strong

Figure 6.1. The Ringelmann smoke chart was a state-of-the-art pollution control instrument when smog became a problem in Los Angeles in the 1940s. The charts covered a spectrum from a nearly white page to this dark grid, which was the "equivalent of 80 percent black." Trained observers held these charts up against the sky to determine the darkness of smoke coming, say, out of an industrial chimney. The charts were believed to offer an objective measure of air quality. US Bureau of Mines, Ringelmann Smoke Chart, Information Circular 8333, May 1967, http://www.cdc.gov /niosh/mining/userfiles/works/pdfs/ic8333.pdf.

evidence," they wrote, "that the eye irritation of the smog is not a product of any one substance but is an enhancing effect."[14] Yet the SRI reports insisted that smog problem was "highly complex" and could not be pinned on the petroleum industry.

In December 1949 President Harry Truman called the first federal conference on air pollution, largely in response to the Public Health Service's report on the 1948 air pollution event in Donora, Pennsylvania, which killed at least twenty in a day and perhaps as many as fifty more in the following month. Louis McCabe chaired the United States Technical Conference on Air Pollution, as the meeting was called. It was organized by a newly formed Interdepartmental Committee on Air Pollution, made up of members from different cabinet-level departments (including Interior and Agriculture) and several branches and divisions of the US military. At the conference, there were palpable worries and uncertainties about what had gone wrong in Donora. As the published conference proceedings later announced, "The purpose of this volume is to bring together . . . an authentic body of information on the nature and control of air pollution," and the same could be said of the conference itself.[15] As we have seen, the federal government had been using conferences since the 1920s to shape the relationship between industry and society, including meetings focused on auto safety, but this conference was the first to address air pollution.[16]

The aspiration announced in the conference proceedings—that it would bring together "an authentic body of information"—went to the heart of the matter: in the late 1940s, no body of knowledge concerning air pollution stood as credible and based in consensus. Moreover, much of the knowledge that had been collected about air pollution rested with far-flung municipal bodies and in obscure professional journals that not all those interested in the topic of air pollution would read. (The 1951 founding of the special interest journal *Air Repair*, later renamed the *Journal of the Air Pollution Control Association*, somewhat mitigated this latter issue.) The conference, then, pooled preexisting but disconnected knowledge. The fact that it was a "technical conference" was also meaningful. Nearly all of the conference panels were dedicated to technical and scientific topics—including agriculture, health, meteorology, research equipment and instrumentation, analytical methods, and chemical properties—*not* regulation or control of the problem. The sole exception was a panel on legislation, encompassing 9 papers out of the conference's total of 97. Most of those 9 were dedicated to pollution control regimes in various US cities and states (and one in Great Britain). Only one paper dealt with the federal government's role in pollution control, and it suggested timidly that the prospect of strengthening this role should be further explored. During the discussion period after the paper, one attendee said, "I dread the day when the Federal Government will enter control work."[17] Nearly everyone assumed that air pollution was a local problem best solved through local solutions—perhaps with some objective technical knowledge handed down from higher levels of government.

Fittingly then, research on Los Angeles County smog continued, and it did not focus solely on its causes or its health effects. Parts of the county were famous for agriculture. Farmers and botanists had long suspected that smog was damaging the crops and forests around Los Angeles, and researchers at Caltech used their laboratories to study how smog affected plant health.[18] The Los Angeles Chamber of Commerce was also concerned about smog's deleterious effects on agriculture and other parts of the city's economy. In 1946 the Chamber of Commerce had made the chemist and scientific instrument manufacturer Arnold O. Beckman a member of its scientific advisory council.[19] In large part due to his smog work, Beckman would go on to become the president of the chamber in 1956. In the late 1940s, however, he began to work closely with McCabe on smog. (Beckman also later saw economic opportunity in manufacturing scientific instruments for the study of smog and air pollution. Much like the companies that earlier had produced brake- and headlight-testing instruments, Beckman's firm made mobile smog-sensing devices and automotive tailpipe emissions measurement systems and capitalized on the markets opening as a result of public concerns and, eventually, regulation.)

McCabe, like Tucker before him, became obsessed with sulfur dioxide, which had been shown to be the chief cause of air pollution in other cities.[20] McCabe proposed that the air pollution control district should buy closed synthetic ammonia plants and use sulfur dioxide from pollution to create fertilizers, which would then be sold for profit. The proposal worried Beckman. Not only would such plants be extremely expensive—and be run using taxpayer dollars—but Beckman also was not convinced that sulfur dioxide was the main problem in Los Angeles. LA smog lacked the egg smell so characteristic of such pollution. Beckman knew that a sulfur dioxide control program was under way in Gary, Indiana, at the US Steel Corporation works, the world's largest steel plant. He flew to Gary and found that, though the sulfur dioxide there was already mostly under control, the place still smelled of eggs. He flew back to Los Angeles and convinced McCabe that sulfur dioxide was not the problem. They would need to do more research.

Oct. 7, 1952   A. O. BECKMAN ET AL   2,613,126
RECORDING APPARATUS FOR RECORDING GAS
CONCENTRATIONS IN THE ATMOSPHERE
Filed Feb. 18, 1946   3 Sheets-Sheet 3

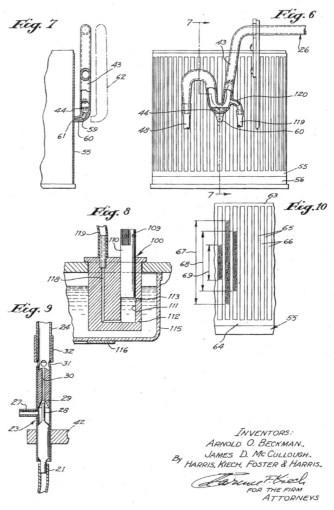

Figure 6.2. One of the scientific instruments Arnold Beckman invented to measure air pollution. Beckman helped foster the science of smog in Los Angeles. Also foreseeing the market potential in pollution regulation, he created testing and measuring instruments that were eventually sold to governments around the nation. Beckman patented this device to determine "minute concentrations of poison gases or other contaminating substances in the atmosphere." The figures include no. 6, the main operating tube used to draw in treated air and no. 10, the record member used to register concentrations of contaminants. From patent application 2,613,126, "Recording Apparatus for Recording Gas Concentrations in the Atmosphere," filed 1946, granted 1952, A. O. Beckman, et al.

Discovering the Causes of Smog

Beckman turned to his Caltech colleague and fellow chemist the Dutchman Arie Haagen-Smit. Haagen-Smit was an expert in flavor chemistry. In the 1940s he and his Caltech associates did work for the Pineapple Research Institute, a Hawaii-based organization that shipped six thousand pounds of pineapples to Haagen-Smit's lab.[21] Through various methods of fractionation, Haagen-Smit processed the tons of pineapples down to a tiny amount of condensed fruit essence and was able to uncover the chemical constituents of pineapple's smell and flavor and to reproduce them synthetically.

Haagen-Smit rejected the most widely accepted explanations for what was causing the Los Angeles smog. Arguments for problematic industrial emissions, such as sulfur dioxide, simply did not add up. Nor did these well-known pollutants create the kinds of damage that individuals were witnessing around the county. For instance, John T. Middleton, a professor at University of California Riverside's Citrus Experiment Station, had been researching the effects of smog on agriculture throughout the region and had found that smog created a unique form of plant damage that gave the underside of spinach and other plant leaves a silvery coating.[22] Middleton found, however, that this plant damage was not consistent with that known to result from sulfur dioxide pollution.[23] Similarly, researchers and others had realized that car tires and other rubber products wore out faster in Los Angeles than they did elsewhere. Something seemed to be damaging them too.

Beckman, McCabe, Haagen-Smit, and others spent months collecting samples of Los Angeles air and processing it to produce a condensate of the air pollution, a small amount of "smelly brown stuff."[24] Haagen-Smit then used the same processes that he had used to break pineapple down into its constituent parts to examine the chemical makeup of smog. He found that the air pollution slurry contained a complex mixture of chemicals, but he hypothesized that, out of all of them, it was ozone that was creating both this "peculiar damage to plants" and the effects on rubber.[25] Scientists had known since at least the 1930s that ozone caused cracking in rubber.[26] Indeed, they used "low concentrations of ozone" in materials testing to judge the resiliency of different forms of rubber.[27] One of Haagen-Smit's insights was that standardized rubber (of a known and dependable composition and consistency) could, conversely, be used to measure ozone concentration.[28] Standardized rubber thus became a scientific instrument. He found that whereas under normal conditions of ambient air, rubber began cracking in 45 minutes, under smog conditions, the time dropped to 6 minutes. Through laboratory experimentation he found that this difference amounted to 0.03 parts per million of ozone in ambient air versus 0.22 parts per million in heavy smog.

Moreover, no previously tested substances created the kinds of plant damage seen around Los Angeles. But when spinach, sugar beet, and endive plants were fumigated with ozone for a few hours, they developed the silvery condition that was "indistinguishable from that noticed on plants exposed to smog."[29] The plants developed this damage at the same concentration of ozone that had been shown to damage rubber, which was also found to create eye irritation similar to that experienced during smog attacks.

The question then became what was creating the ozone? Haagen-Smit's crucial realization was that "the cause of 'smog'" involved "a photochemical reaction between sunlight, hydrocarbon, and nitrogen oxide."[30] That is, hydrocarbons and nitrogen oxide were reacting with each other in the presence of sunlight, a process that Haagen-Smit was able to re-create in an outdoor Plexiglas room exposed to the sun. This finding explained why others, such as the Stanford Research Institute, had not been able to make smog simply by mixing various substances. The missing ingredient in these earlier experiments was sunlight. While Haagen-Smit's theory about smog would be refined over the following decades, his basic insight about its causes held true, and his work started air pollution control along a wholly different track. Haagen-Smith unveiled his findings and reasoning through a series of publications culminating in a 1952 article that outlined his overall theory of smog formation.

Haagen-Smit's discovery had two important implications. First, the main sources of nitrogen oxide and hydrocarbons—which was really just unburned gasoline—were automobiles, followed distantly by petroleum refineries, *not* all of the other industrial sources on which the Los Angeles County Air Pollution Control District had focused. Of course, these control efforts had without doubt reduced pollutants that harmed human health, but they had not affected the formation of *smog*. Second, smog's true causes rendered the APCD's strategies beside the point. The question became how to reduce the emissions coming out of the millions of cars driving around Los Angeles County, but this quest would require a set of regulations entirely different from the ones McCabe and his colleagues had developed. The path forward was unclear.

The automobile industry resisted the thrust of Haagen-Smit's findings. In 1953 the automakers formed a technical committee on air pollution within the Automobile Manufacturers Association, and in early 1954 ten engineers from all of the major carmakers visited Los Angeles to learn more about what was going on there.[31] After the visit the companies pledged to spend $1 million dollars a year on a research program for reducing automotive emissions. Moreover, the firms entered a cooperative research agreement, first informally in 1954 and then via a formal cross-licensing agreement a year later. This cross-company research venture would become highly controversial in the late 1960s when environmentalists and then the federal government argued that the automakers were using the agreement to avoid, rather than foster, technological change. In the 1950s, critics argued that $1 million dollars a year was far too little commitment from an industry that made many times that amount in profits. In general, however, the auto industry asserted that Haagen-Smit's findings were overblown and that their products were not the core cause of urban air pollution problems. In 1953 Ford Motor Company wrote to Kenneth Hahn, a member of the Los Angeles County Board of Supervisors, claiming that automobile exhaust gases "dissipated in the atmosphere quickly and do not present an air-pollution problem."[32]

Ironically, while Haagen-Smit argued that automobiles emitted most of the gases that caused photochemical smog, it was the petroleum industry that targeted his findings most directly and vociferously. Its attack also had the largest repercussion for the future of air pollution control, as it transformed Arie Haagen-Smit, the interested scientist, into a fierce crusader.[33] SRI, while working under contract for a petroleum industry trade association, had replicated Haagen-Smit's experiments in which ozone was created out of hydrocarbons and nitrogen oxide in the presence of sunlight.[34] But SRI researchers did not believe that ozone created eye irritation and the other discomforts and problems associated with smog. In an academic talk at Caltech, an SRI researcher told those gathered that it was "unfortunate that a chemist of Haagen-Smit's caliber could be misled" to believe that ozone was the chief problem in smog.[35] Haagen-Smit had never been attacked in public before. He was livid. He spent the next several months working day and night to vindicate his smog research, and in a sense, he never left smog science after that day. He would go on to become the first chairman of the California Air Resources Board in 1968, then a new smog control agency that would come to greatly influence American air pollution and clean energy politics.

The development that did the most to settle the controversy surrounding Haagen-Smit's findings was the formation, led by Beckman and other powerful actors, of the California-based Air Pollution Foundation. Like the Stanford Research Institute, the Air Pollution Foundation was funded by donations from industry, including petroleum and auto companies, but in both rhetoric and practice, the foundation prided itself on its independence and objectivity. By the mid-1950s, scientists working with the foundation had corroborated Haagen-Smit's findings. While his work required minor alterations and improvements, the core of Haagen-Smit's theory held true: photochemical smog formed in the way he described. And if California and other states and localities were to control smog, which had begun appearing in cities throughout the nation, they would have to change their regulatory strategies. But no one was sure how to do that.

Should the Federal Government Regulate Air Pollution?

Beginning in the early 1950s, a string of US senators and representatives from California, both Democrat and Republican, including Representatives Morris C. Poulson and Gordon L. McDonough and Senator Thomas H. Kuchel, began pushing legislation that would have created federal executive branch programs focused on air pollution.[36] The air pollution problem was tremendously complex, they reasoned, and solving it was simply too expensive to be undertaken with local and state funds alone. Most other politicians believed that air pollution was a local problem, however. For instance, Kenneth Roberts, the conservative Democrat from Alabama, was, as we have seen, one of the earliest proponents of federal regulation to promote automobile safety, just as he had earlier done with refrigerator locks. Yet Roberts did not apply this same logic to automotive emission control, which he believed the affected cities should address.[37]

Kuchel and the other California politicians introduced a series of air pollution bills every year throughout the early to mid-1950s, but most of them died in committee and none stood a chance of becoming law. After yet another of these bills was defeated in 1954, Kuchel persuaded President Dwight Eisenhower to look into the air pollution problem. Eisenhower formed the Ad Hoc Interdepartmental Committee on Community Air Pollution, led by the US surgeon general, at that time a position within the Department of Health, Education, and Welfare.[38] The findings of this committee supported increased federal involvement in air pollution research, and in January 1955 Eisenhower gave a speech backing the initiative.

The law that eventually passed in 1955 was tamely named the Act to Provide Research and Technical Assistance Related to Air Pollution Control, often referred to as the Air Pollution Control Act. It furnished research funds to the Public Health Service (PHS), which, two years earlier, had become part of the then-new Department of Health, Education, and Welfare. This research was to provide technical knowledge to local and state governments but granted no regulatory authority to the federal government. Whereas the US Technical Conference on Air Pollution involved federal authorities pulling together existing scientific and engineering knowledge from sources outside government, the Air Pollution Control Act enabled federal staff members to generate new knowledge that local and state governments, for want of resources, were unable to discover for themselves. Moreover, since the science of air pollution in California had often devolved into a morass of claims, counterclaims, and even name-calling, lawmakers hoped that the Public Health Service would be a source of credible, authoritative findings.

Federal work on automotive air pollution was centered at the Robert A. Taft Sanitary Engineering Center in Cincinnati, Ohio.[39] This early research was focused on "characterizing" vehicle emissions, that is, describing what chemical compounds cars emitted and in what amounts. On a concrete level, this meant that researchers built a smog chamber at the center to study photochemical smog created from car exhaust under controlled conditions.[40] This research marked one of the first major studies of vehicle emissions, and the Public Health Service reported it in a series of publications. The science of smog was in such an immature state that much of the center's work focused on developing pollution measurement procedures, both for motor vehicle tailpipe emissions and for the ambient air, methods that would later become standard practice for research and regulation.

The full import of Haagen-Smit's findings led to a kind of catatonic stasis in California air pollution control. While individuals proposed wild technological solutions, from blowing away inversions above Los Angeles with huge ground-based fans or a fleet of helicopters, to neutralizing smog in the air by seeding it with chemicals, no one was sure how to proceed.[41] With the California state government seemingly paralyzed, the Los Angeles County Air Pollution Control District announced its intention to promulgate automotive emissions standards. But in January 1959 the California Department of Motor Vehicles struck down this program, which would have required access to the state's vehicle registration system.[42] Others worried that the APCD's policies might lead to a confusing set of requirements whereby automobiles within the county would have had to be manufactured or altered in the aftermarket to be different from automobile's in the rest of the country. In the worst-case scenario for automakers, if such developments continued indefinitely on a county-by-county or city-by-city basis, manufacturers would have had to make different products for every locality. Both the control district and state legislators looked to the federal government for guidance.

Perhaps the most vocal proponent of the federal government's moving beyond a merely supportive role came from within the federal government itself. By 1959, Secretary of the Health, Education, and Welfare Arthur S. Flemming, a liberal advocate of social security and later head of the US Commission on Civil Rights, had gained experience implementing the Water Pollution Control Act of 1948.[43] He came to believe that powers similar to those exercised in cleaning up water, including the ability to promulgate national pollution standards and to create pollution management programs, should be applied to air pollution. But the surgeon general, the head of the Public Health Service, a division of Flemming's own department, resisted and criticized this idea, arguing that even holding hearings and making recommendations—that is, going one step beyond providing technical assistance—would lead to resentment from state and local governments.

Pressure had also begun to mount within Congress. In a hearing on automobile safety in 1956, a witness tied safety problems to automotive emissions: both were the result of bad design and inaction on the part of automakers.[44] A year later, Congressman Paul Schenck introduced a bill that would have required the surgeon general to establish a national standard for safe levels of auto emissions and would have banned the interstate sale of cars that did not meet this standard. Both the auto industry and the Department of Health, Education, and Welfare opposed this form of the bill, the latter because no one had defined what level of emissions was "safe." Eisenhower signed a much-weakened version of the bill, known as the Schenck Act, into law in 1959. It directed the surgeon general to study the health effects of automotive exhaust.

The resulting study, *Motor Vehicles, Air Pollution, and Health*, was published in 1962.[45] Its findings were not a major step forward for the science of smog but rather a compendium of existing knowledge, some of which the Public Health Service had created. Still, the volume testified to how quickly scientific, technical, and medical knowledge about air pollution was growing. A section of the book on health effects included 548 citations, most of which had been published in the previous ten years, and the lion's share of *those* in the previous five. By this measure it was clear that the science of smog and automotive air pollution had come a long way since the first smog attacks had hit Los Angeles in the early 1940s. Yet, in the late 1950s, no one—neither experts in Los Angeles County, nor politicians in the California State Assembly, nor bureaucrats in the US government—knew how to get urban air pollution under control.

The Beginnings of Automotive Air Pollution Control Regulations

In 1959 Los Angeles County had attempted but failed to pass a law regulating automotive emissions.[46] A year later a version of that same measure was introduced in the California legislature, though in a weakened form. The law, the California Motor Vehicle Pollution Control Act of 1960, created a Motor Vehicle Pollution Control Board within the state's Department of Public Health. Among the law's significant weaknesses was its assumption that effective pollution controls were in the offing. The pollution control board was charged with evaluating control devices as they became available, and the law would not go into effect until at least two such devices were found, supposedly to prevent monopoly and encourage economic competition between device manufacturers. Yet the board had no standardized test by which to judge the effectiveness of pollution controls, and its members had little, if any, technical expertise. Indeed, the county of Los Angeles, which had been working on the issue since the 1940s, had more experts than the state did.

Nevertheless, the pollution control board certified four emission control devices made by independent producers, including a direct-flame exhaust afterburner, which ignited unburned hydrocarbons and other problematic effluents.[47] Faced with the prospect of having to put other people's inventions in their cars, the automakers suddenly announced that they had just created workable emission control technologies. The automakers' unexpected "discovery" reflected two long-standing, interconnected dynamics: the so-called not-invented-here syndrome, according to which organizations prefer to rely on resources developed internally, and corporations' reticence to rely on external intellectual property.[48] General Motors began producing a new pollution control technology known as positive crankcase ventilation. As part of the combustion process in engines, unburned gasoline, or hydrocarbons, would inevitably leak past the piston rings into the crankcase. Traditionally, these emissions were simply vented into the ambient air, a significant source of hydrocarbons that played a role in smog formation. Positive crankcase ventilation recycled these unburned hydrocarbons into the combustion process, using instead of venting them. General Motors voluntarily added the systems to all of its vehicles and made them available, royalty free, to other automakers. By 1964 nearly all auto manufacturers were using the systems on cars sold across the country.

The California air pollution control programs were not highly effective. Automotive air pollution remained a real problem, and it seemed the state could do little about it. Yet, by creating the first automotive air pollution regulations in the United States, California began a trend eventually enshrined in laws whereby the state set tougher standards than the rest of the nation, a trend that, with a few important exceptions, continues to this day. Put bluntly, in air pollution control, California led; the federal government followed.

Indeed, the US government moved into automotive regulation slowly, with much reluctance on the part of leaders in the legislative and executive branches. The Public Health Service continued to conduct scientific studies on air pollution, mostly focused on emissions characterization. In 1963 the PHS moved its automotive air pollution efforts from Cincinnati to a new laboratory in Fairfax, Ohio, still relatively distant from Auto City, Detroit, Michigan. It also began testing automotive emission control technologies created by independent inventors, usually in response to congressional pressure or at the request of other, higher-level officials.[49] These tests of independent inventions constituted an extension of activities pursued earlier by the National Bureau of Standards, which was required to the same kind of work, and the federal government has continued to do such tests ever since. Nearly all of these inventions, both aftermarket add-on devices and various fuel dopes, made insignificant reductions in auto emissions or increases in fuel economy.

Also in 1963 the US Congress passed the first Clean Air Act, a landmark, though flawed, law. The Clean Air Act was the result of at least a decade of advocacy, particularly by federal legislators from California and New York. As Senator Edmund Muskie later put it in his memoirs, "The early 1960s were the first time that many people, certainly legislators, began to think of air pollution as controllable."[50] The politics of the fight to reduce air pollution during the 1950s and early 1960s involved redefining the issue as one of federal concern. This process was spurred both by new individuals entering Congress with ideas different from their predecessors' and by established individuals *changing their minds.* For instance, Kenneth Roberts, who had long believed pollution control was a local or state issue, eventually came to a different position. Alluding to continued problems that went unabated by local efforts and to a recent smog attack in London that had killed hundreds, Roberts declared, "Consistency is a hobgoblin of little minds. Someone said, 'The wise man changes his mind and the fool never does.'"[51]

The Clean Air Act gave the National Air Pollution Control Administration (NAPCA), formed within the Public Health Service, authority to set maximum levels for pollutants in ambient air, known as "air quality criteria," and limited abatement powers, but the process by which the agency set these rules was poorly structured, and the resulting rules were often quite weak.[52] The Clean Air Act also encouraged the auto industry to begin developing and implementing automotive emission controls, but such efforts were strictly voluntary.

Soon, however, legislators came to believe that the voluntary approach was not working.[53] This realization eventually led Congress to eschew the voluntary approach and to pass, in October 1965, the Motor Vehicle Pollution Control Act. This law was the first to authorize the government to set national automotive pollution control standards and, therefore, marked a milestone in the history of governance in the United States. Even so, it was weak in several ways. First, the Motor Vehicle Pollution Control Act limited initial federal standards to levels already set in the 1960 California law. Moreover, although legislators had proposed that the standards be specified within the law itself, in the end other forces prevailed, ensuring that the Public Health Service would set the standards through administrative procedures.[54]

Automotive regulations differed from most federal air pollution control laws, which depended on cooperative agreements between federal, state, and local governments and were actually administered at the local level. In contrast, automotive regulations centered on the point of production; they controlled companies.[55] The automakers originally resisted federal regulation, saying, via the Automobile Manufacturers Association, that they preferred to work with each state to meet its needs.[56] But they changed their tune quickly when Pennsylvania and New York passed automotive pollution control laws, with New York's standards being tougher than California's. The automakers' ploy to work with the states had been little more than an attempt to stave off federal action. In reality, the industry's greatest fear was a patchwork of state and local laws, each requiring a different kind of vehicle.

The act's most glaring weakness was that it required the Public Health Service to prove the "feasibility" of technology necessary to meet standards. Before making a rule, the Public Health Service had to prove that an adequate pollution control technology was available—these are sometimes called off-the-shelf" technologies. For this reason, the agency started putting increasing amounts of funding into research and development efforts: that is, the federal government built an R&D lab for emission controls. The most important technology to come out of this research was exhaust gas recirculation, which the agency developed with the Esso Corporation and which was a key emission control technology before the advent of the catalytic converter.[57] The law did not give the agency any power to force automakers to create controls of their own. All of the research burden was on the federal government, and the agency had seriously limited resources, in terms of both money and available expertise.

For a few years the Public Health Service unit dedicated to air pollution also lacked a leader. In 1967 John T. Middleton was appointed to head the National Air Pollution Control Administration. A botanist by training, Middleton had been studying smog's negative impacts on plants since the 1940s. He joined a large California air pollution research community, including Arie Haagen-Smit and others, and before coming to the Public Health Service, he headed the California Motor Vehicle Control Board.[58] By most accounts, Middleton was a kind and charming man.[59] Professorial in his tastes, favoring tweed suits and wearing a handlebar mustache, Middleton was known for inviting colleagues to a glass of sherry in his office at the end of the day.

With equal unanimity, however, people viewed Middleton as an ineffective leader. After the success of *Unsafe at Any Speed*, Ralph Nader began gathering young people, often known as Nader's Raiders, to conduct "study groups" investigating other societal problems. One Nader study group looked into air pollution, and it criticized Middleton for not fighting harder to garner more attention for his program within the Public Health Service or asking for more money from Congress. As the study group pointed out, Congressman Paul Rogers, a Democrat from Florida, asked Middleton in a hearing why he was not more forthcoming with budget requests, pointing out that Rogers and other legislators wanted to help him in any way they could.[60] Similarly, a political scientist interviewed agency staff during this period and found that Middleton was "generally acknowledged and praised in interviews for his knowledge in the field, but rather uniformly criticized for his lack of administrative and political savvy."[61] The agency lacked aggressive leadership.

It also lacked resources and expertise. The agency had 407 employees in 1963, "a figure roughly equal to the number employed in air pollution control by Los Angeles County at that time."[62] Its annual budget increased more than ten times after 1963—from less than $5 million to about $50 million—and the staff grew to number more than a thousand. But it still lacked the capacity to deal with the manifold issues in air pollution control.[63] Resources were even more limited for automotive pollution control. The Nader study report noted damningly, "Whether fraudulent or farcical, there is no doubt that NAPCA's automotive emissions program is a smashing failure. . . . Though [the automobile] accounts for 60 percent of the total problem, the program for control of automotive emissions has never accounted for more than 5 percent of NAPCA's budget."[64] Even to the degree that the agency had capable experts, few of their efforts were dedicated to the car, and the agency's budget paled in comparison to the resources and manpower of the automakers themselves.

Yet critics were off base to the extent they suggested the agency's work was fruitless. A good deal of institutional learning was going on, learning that was eventually folded into the Environmental Protection Agency. Perhaps the most important outcome of all was the creation of the Federal Test Procedure, the standard automotive tailpipe emission control test that, with some refining, remains the basis of pollution standards to this very day.[65] Unlike the National Highway Safety Bureau, which used numerous tests to judge the safety of automotive systems, federal automotive emissions standards have always been based on one central test. The Federal Test Procedure has two basic components: a procedure for collecting and measuring tailpipe emissions and a "driving schedule," or a series of accelerations, static driving speeds, decelerations, and idle times. The test is run in a laboratory on a chassis dynamometer, a device that allows the car's wheels to spin during the test in order to simulate road driving.

Like much of the federal emission control program, the Federal Test Procedure can be traced back to earlier work in California, which established a tailpipe emissions test in the early 1960s.[66] When Congress passed the Motor Vehicle Pollution Control Act in 1965, the National Air Pollution Control Administration adopted the California test almost completely. These federal standards took effect with the 1968 model year. That year the NAPCA announced that it was tightening standards that would take effect in the 1970 model year. For the first time, the agency expressed the standards in mass units (grams per mile) instead of in concentration (parts per million or a percentage), which made it more difficult for automakers to influence the tests by increasing airflow through the exhaust system.

NAPCA staff made further refinements to the test, many of which were controversial with automakers because they effectively made the test harder to pass. One point of contention was which driving schedule should be used in the test. Just as psychologists in the 1920s and 1930s had created profiles of the average driver, and impact biomechanics researchers had created measures for how much force the average adult body could withstand, automotive air pollution regulators attempted to create models of the average driving pattern. The idea was that regulatory tests should be based on limiting the amount of pollutants the average American put out during his or her daily driving routine.

From 1966 to 1968 the National Air Pollution Control Administration experimented with several different schedules based on average driving patterns in Cincinnati, Ohio, and Ypsilanti, Michigan, but ultimately the agency settled on a modified version of a test representing the average Los Angeles commute, known as the Urban Dynamometer Driving Schedule.[67] Many, including the automakers, questioned whether this California test really represented the average American driver, but a 1971 survey conducted by the Coordinating Research Council found the test to be broadly representative.[68] Since then, scientists and engineers both within and outside the EPA have repeatedly studied the Federal Test Procedure, taking stock of its strengths and weaknesses.

⌒

Since the 1940s, the fight over smog had helped redefine urban air pollution problems as stemming from automobiles, and a variety of factors had led legislators and others to view the problem as a federal issue, rather than only a state or local one. By the late 1960s, environmentalists and consumer advocates were fed up with the lack of progress on automotive pollution control, and legislators came increasingly to see federal pollution control efforts as a failure. The pieces were in place for environmental and public health advocates and lawmakers to argue effectively for a significant shift. In 1968 the National Air Pollution Control Administration moved its automotive branch office from Fairfax, Ohio, closer to Detroit, originally renting an empty hanger at the Willow Run Airport in Ypsilanti, Michigan, to house both a makeshift office and a lab for certification and other tests. In many ways, the move was symbolic: the federal government's automotive pollution control efforts had focused on research and development, but soon the NAPCA would begin conducting surveillance of the automakers' own efforts to find a solution to the pollution problem. With the Clean Air Act Amendments of 1970 on the horizon and the new laboratory in Michigan, the administration, which would soon become a part of the Environmental Protection Agency, was ready to make a change.

CHAPTER SEVEN

Command and Control

The year 1970 marked a milestone in the history of American environmentalism and regulation. Legislators had passed major new environmental and antipollution laws, and President Richard Nixon was feeling pressure to act on the environment from those inside and outside his party, including Senator Edmund Muskie, a potential presidential contender.[1] Nixon felt additional pressure after the first Earth Day was held on April 22, 1970, in which nearly 20 million people participated. By midyear his administration had decided to centralize all of the federal government's environmental and pollution control activities into a single, independent agency. The Public Health Service's pollution programs, including the National Air Pollution Control Administration, moved into the new Environmental Protection Agency (EPA). Nixon appointed William Ruckelshaus, then a young assistant attorney general, as the agency's first administrator.

Congress also passed the Clean Air Act Amendments of 1970, which created some of the toughest regulations in US history, mandating that automakers reduce selected emissions by 90 percent within five years. In the 1980s conservative critics and some economists began calling these laws "command and control." The phrase *command and control* first sprang into use in military contexts during the 1960s and was only later applied to regulations that featured direct, forceful restraint of an industry. Command and control regulation was contrasted with free market solutions to environmental problems, such as trading pollution rights. As commanders were to the military rank and file, so regulators were to the workings of markets, or so critics argued. For critics, command and control was a dysphemism, a derogatory term used to attack rules and regulations that, they believed, should be opposed.

But viewed from another angle, the term *command and control* raises an important question: How do regulatory agencies attain the authority and develop the organizational capabilities and everyday routines to oversee and influence regulated firms effectively? This chapter tells a twin history of intertwining legal and organizational changes that helped automotive pollution control efforts succeed. The legal history of automotive air pollution control laws is relatively well known: through the efforts of advocates such as Ralph Nader and legislators such as Senator Edmund Muskie, the weak federal regulations of the 1960s were plowed over by the Clean Air Act Amendments of 1970, which had a larger influence on automotive technology than any other single set of statutes. Yet strong laws are not enough; regulatory agencies also require leadership and insightful organizational strategies to meet their goals.

Less well known are the organizational changes within the executive branch that—sometimes following, sometimes leading—accompanied the legal ones. Yet these organizational changes within what would become the EPA were just as important for the creation of command and control in automotive pollution regulations. Since the early twentieth century, government regulation had always dealt with technology-related issues and concerns by enlisting experts in problem solving. As federal administrators built a bureaucracy around combatting automotive air pollution, they hired experts, primarily scientists and engineers, onto their staffs. But the question arose, What would these experts do all day? How administrators answered that question changed over time, in part because they sensed opportunities in the evolving legal environment, and these changes had important ramifications for the relative success of automotive air pollution regulations in the United States.

The Clean Air Act Amendments of 1970

By the late 1960s, environmentalists, consumer advocates, legislators, and ordinary citizens had grown sick of the automobile industry's lack of progress on air pollution. The individual who did the most to change this situation at the federal level was Senator Edmund Muskie. Muskie, a Democrat, had been elected to the US Senate in 1958, after serving for four years as the governor of Maine. As governor, Muskie had worked on water pollution issues, including pushing for building more water treatment plants in the state.[2] In 1959 he became a member of the Senate's Public Works Committee, which traditionally oversaw environmental issues.[3] Four years later, Senator Pat McNamara

(D-Mich.), chair of the Public Works Committee, put Muskie in charge of the newly created Special Subcommittee on Air and Water Pollution. Over the coming years, Muskie used this position to make a name for himself as an environmental crusader, eventually earning the moniker Mr. Clean.

Initially, Muskie focused on water pollution, but increasingly he came to believe that air pollution, far more than water, "provided much of the momentum for our fight against all other forms of pollution."[4] Most Americans did not live within sight of water, Muskie reasoned. "You can escape water pollution . . . at least for a time. You cannot escape air pollution" (ellipsis in original). Muskie also believed that air pollution "provided its share of villains for environmentalists and the general public to attack," attacks that eventually mounted, building energy for more significant laws and regulations. Originally, environmentalists had focused on factories, power plants, refineries, and other smoke emitters, but Muskie saw the target change over the course of the 1960s. "The automobile has become the symbol of our air pollution crisis," he wrote, "and with good reason."

That automobiles had become the core symbol of air pollution heightened the sense that their manufacturers were stalling in order to avoid creating meaningful emission controls. One of the most salient markers of this procrastination was the industry's failure to generate new knowledge for understanding and decreasing automotive emissions. As we have seen, Colonel John Stapp's Car Crash Conferences were coalescing researchers working on auto safety into a single community by 1955, but no equivalent community had emerged within automotive pollution control even by the mid-1960s. Indeed, before 1965 few papers on automotive emission characterization or control technologies appeared in the various publications of the Society of Automotive Engineers. Publications increased briefly in 1962–1963, when California regulations pushed positive crankcase ventilation systems into cars and then automakers put them on all cars sold in the United States; but thereafter publications decreased again. This inactivity began to change somewhat with the passage of the Motor Vehicle Pollution Control Act of 1965. Researchers, including some automaker employees, published a trickle of papers over the next few years, but they made little headway. Moreover, because the pollution control act put the burden on government to prove that control devices were available, the auto companies had little incentive to invest in pollution research. Indeed, the law actually pushed the companies' incentives in the other direction. It was in their interest to make sure that devices, which would inevitably raise production costs, did not develop.

Furthermore, there was no consensus about which technologies would best solve automotive air pollution. Some hoped to move automobiles away from the internal combustion engine. Others had more modest aspirations. In November 1968 the National Air Pollution Control Administration and the Franklin Institute of Philadelphia cosponsored the First National Symposium on Heterogeneous Catalysis for Control of Air Pollution. It was the first attempt to form a research community around the technology that would become known as the catalytic converter, which used chemical processes to transform tailpipe emissions into harmless compounds.

The catalytic converter was originally conceived by Eugene Houdry, a mechanical engineer who spent most of his career working in the petroleum industry.[5] During the 1930s Houdry had introduced a catalytic "cracking" process in petroleum refining plants to more efficiently produce high-quality gasoline. By the late 1940s he had become concerned about air pollution from automobiles. He believed—probably incorrectly—that automotive air pollution was a major source of lung cancer in the United States. Applying the same kind of thinking that he had used in the petroleum industry to the problem of tailpipe emissions, in 1954 Houdry filed a patent for a "catalytic muffler."[6] The patent came through in 1962, the year Houdry died. Although Houdry was unable to develop the technology further, by the mid-1960s some individuals had come to believe that technology was the most likely near-term solution for automotive air pollution, and thus the first national symposium was organized. Representatives from chemical, petroleum, and materials firms, along with researchers from institutes and universities, dominated the conference. Many of these individuals and groups saw economic opportunity in the development of emission controls, while others, especially those from the petroleum industry, saw risks. Only a few engineers and scientists from the automakers attended. Of these, most were from General Motors. Chrysler sent no one.

The catalysis conference marked an ongoing transformation within the auto industry that involved the increasing importance of chemists in automotive research and development. J. M. Campbell, the retired assistant vice-president of General Motors' Research Laboratories, gave one of the major addresses at the event. He recalled that, when he began working at General Motors in the mid-1920s, the company had only three other chemists, including T. A. Boyd, the codiscoverer of tetraethyl lead. For a primitive initial survey they undertook of exhaust gases in GM cars, the research group designed a catalytic burner for the exhaust gases not unlike the catalytic converter.[7] Despite this early interest, over time the issue of automotive emissions receded in importance for GM's research groups.

Only Haagen-Smit's discovery of the role of automobiles in photochemical smog raised the topic again, and, as Campbell emphasized, individuals working outside the automobile industry made the important breakthroughs and discoveries. Yet Campbell was extremely proud of his profession, and he highlighted repeatedly that it was chemists who made these strides. The chemists he referred to included Eugene Houdry, Haagen-Smit, and Gordon P. Larson, the former director of the Los Angeles County Air Pollution Control District who, now directing the Environmental Resources Center at the Franklin Institute, had organized the conference. And the league of chemists encompassed all of the individuals who had created the scientific instruments, such as nondispersive infrared devices and flame ionization detectors, that sped up research results from weeks to seconds and enabled, Campbell exclaimed, "quantitative measurements of individual hydrocarbons in a sample of only a few milliliters in terms of parts per billion!"[8]

Yet most of this chemical knowledge lay outside the auto industry, and the manufacturers were slow to hire chemists. The papers presented at the conference included almost no references to automotive-centric publications, such as the various journals of the Society of Automotive Engineers. In this way, the conference featured the application to an automotive problem knowledge hard won in other industries. And the automakers had almost no incentive to put this knowledge into practice.

Muskie gave the banquet address at the catalysis conference. He emphasized the kinds of technological transformations over the past century that had fundamentally altered life for the human species. "That changed condition has accelerated since World War II in a series of quantum jumps which amounted to a technological revolution."[9] But the United States had not seen anything like this revolution in terms of pollution control. Indeed, industrialization had only increased pollution. "We still require major breakthroughs," Muskie declared. The issue was one of priority setting—care for the environment had to become a corporate imperative. As Muskie put it, "A nation that built a better beer can, surely has the technological capacity to dispose of that beer can."[10] The nation needed corporations to apply ingenuity to the public problems that resulted from their products and that they were ignoring.

Just how little incentive the automakers had to improve emission control technology became evident one year later, in 1969. In 1954, American automobile manufacturers had entered into an unofficial research collaboration centered on emission control technologies, which they formalized as a cross-licensing agreement in 1955.[11] Observers had long criticized this arrangement. Ralph Nader's study group on air pollution recalled Adam Smith's dictum, "People of the same trade seldom meet together but that the conversation ends in a conspiracy against the public."[12] In 1964 the head of the Los Angeles County Air Pollution Control Board, S. Smith Griswold, argued in a speech that the cross-licensing agreement "has served to guarantee that no manufacturer would break ranks and bring into this field of air pollution control the same kind of competitive stimulus that spokesmen for the industry frequently pay homage to as the force that has made them what they are today."[13] Lawyers soon pointed out that Griswold was describing a classic restraint of trade. In 1965 the Los Angeles County Board of Supervisors resolved to investigate anticompetitive practices on the part of the auto industry, and by 1966 the board was urging the US attorney general to do the same.[14] The US Justice Department took some time to act, but in January 1969 it brought a suit against the automakers under the Sherman Antitrust Act.

The Nixon administration did not have the appetite for pursuing the suit very far, however. Lloyd N. Cutler, the influential D.C. lawyer who worked for the Automobile Manufacturers Association and had earlier played a role in weakening auto safety standards, took part in secret discussions with the Justice Department, which decided to settle the matter via a consent decree. The consent decree forbade the automakers from conducting collaborative research on automotive pollution controls, in hopes that the companies would begin competing with one another on the introduction of control devices. Ralph Nader (who had called the suit "the antitrust case of the century"), leaders in California, and other advocates were extremely disappointed by the settlement.

This disappointment added to the sense that the automakers would put off pollution control for as long as they could. Environmentalists and consumer advocates grew frustrated not only with the automakers but also with Muskie and legislators, who they felt were too focused on compromise with industry and not enough on passing strong laws. In 1970 Nader and his "raiders" published the results of their air pollution study group in a book titled *Vanishing Air*. Muskie, who prided himself on his "Mr. Clean" moniker and saw his environmental work as one path to the presidency, was surprised to find himself attacked in the volume. *Vanishing Air* took aim at many dimensions of the 1967 Air Quality Act, which Muskie had shepherded through Congress, as well as Muskie's inability or unwillingness to go after the automobile industry. The book recounted the poor performance of John Middleton, head of automotive emission control within the National Air Pollution Control Administration, before a House subcommittee exploring environmental issues. Middleton repeatedly avoided directly answering questions, *Vanishing Air* asserted, "artfully dodging those areas which might underscore his agency's poor performance or which might embarrass the architect of the legislation and Middleton's best friend on Capitol Hill, Senator Muskie."[15] The book made Middleton sound like Muskie's pet, and a poorly performing pet at that.

Muskie took these criticisms seriously enough to hold a press conference the day after *Vanishing Air* was published. He defended his record as an environmentalist and stood up for the soundness of the Air Quality Act of 1967. He also questioned the Nader group's assumption that legislators had to take an antagonistic stance toward industry and could not work via compromise. "We don't think it is necessary to be noisy" to pass meaningful environmental laws, he told the reporters gathered at the press conference. "We don't think it is necessary to be nasty to do this."[16]

At the same time, however, Muskie began feeling around for a legislative solution for addressing the automotive air pollution issue once and for all. He decided to introduce a bill that built large reductions in emissions into the law itself, rather than relying on administrative procedures to set them. But where should the reduction levels be set? By how much did automotive emissions need to be reduced to clean the air? The answer came in a paper that Delbert S. Barth, the director of the National Air Pollution Control Administration's Bureau of Criteria and Standards, presented at the annual meeting of Air Pollution Control Association in 1970.[17] Barth proposed reducing automotive-generated carbon monoxide (CO), nitrogen oxide (NO_x), and hydrocarbons (HC) by 92.5 percent, 93.6 percent, and 99.0 percent, respectively. These numbers would entail stringent, even draconian, standards.

Barth's paper has a peculiar place in historical memory. It is unclear whether he arrived at the 90 percent reduction on his own, or his paper justified numbers that had been arrived at through political means, perhaps orchestrated by Edmund Muskie's mercurial chief of staff, Leon Billings.[18] Either way, the reduction level in the standards was not set by anyone with deep knowledge of the auto industry or its organization. This fact formed one basis for the automakers' criticism of the law for years afterward. As a *New*

York Times journalist summarized in 1973, "The industry has maintained that the standards were drawn up by people who had little [experience] with the technology and the production process to meet the 1975 levels."[19] Yet the fact that outsiders wrote the law may have been essential in pushing the industry beyond what many of its members thought possible.

In June 1970 the US House of Representatives passed the Clean Air Act Amendments by a vote of 374-1. The House version was based on a bill put forward by the Nixon administration that lacked the stringent deadlines that would come to define the law's reputation. In September the Senate passed the tougher, Muskie version of the bill by a vote of 73-0. Passage of the Clean Air Act Amendments had gotten a boost, one news article noted, when "massive air inversions had blanketed the East Coast, including Washington, with hot, sticky, dirty air several times during the summer."[20] As Senator Eugene McCarthy claimed, "It's better than motherhood. At least a few people are against motherhood. Nobody can be against clean air."[21]

In contrast to the earlier Motor Vehicle Pollution Control Act, Muskie got his wish with the Clean Air Act Amendments: he built the emission standards into the law itself. The law required that automakers reduce carbon monoxide and hydrocarbon emissions by 90 percent by the 1975 model year and nitrogen oxide by the same amount for the 1976 model year. Legislators and regulators knew that the automakers currently lacked available technology necessary to meet these standards. In this way, the law was "technology forcing": it was meant to induce technological change in the auto industry, most likely through the industry developing or enhancing research and development programs. The law shifted the research burden to industry's shoulders. The Clean Air Act Amendments did contain a safety valve, however. If automakers believed that they could not meet the standard, they could petition the EPA to suspend the standards for one year.[22] Still, the Clean Air Act Amendments constituted an extremely tough law. It was perhaps the most stringent regulation of technological hazards in the history of the United States. In an interview with the journal *Automotive Industries*, Eric Stork, who would later helm automotive regulatory efforts at the EPA, said: "The fine art of regulatory management involves the ability to identify, get hold of, and hang onto the short hair. And the Clean Air Act provided EPA the best grip on the short hair of the industry that any government agency has ever had."[23] The law, one journalist noted, might induce the automakers to "recall Dr. Johnson's comment about the man who is told he will be hanged in a fortnight: it concentrates the mind wonderfully."[24]

Technical and Organizational Change at the NAPCA and EPA

In January 1970 President Richard Nixon signed the National Environmental Policy Act, which provided a broad framework for federal environmental policy, established the President's Council on Environmental Quality, and required federal agencies to conduct environmental assessments and environmental impact statements for proposed actions. The law displayed Nixon's credentials as a political pragmatist, willing to appease the public concerning issues such as protecting the environment.[25] Yet even this action did not appease the growing environmental movement, and Nixon faced constant pressure from Edmund Muskie, his presumed rival in the 1972 election. At the first Earth Day on April 22, 1970, crowds had cheered louder for Muskie than they had for Phil Ochs and other heroes of youth culture.[26] One of the student leaders of Earth Day questioned Nixon's seriousness when talking to journalists: "President Nixon seems to think that the environment issue is a good thing to quiet down the campuses and patch up the country."[27]

One strategy whereby Nixon might further prove his environmental bona fides would build on an initiative he had started a year earlier. In April 1969 Nixon had appointed the President's Advisory Council on Executive Reorganization to find ways to reduce the bureaucratic inefficiencies and duplication of effort in which the president believed the federal government was mired.[28] In the late 1960s more than eighty federal agencies worked on pollution control in one way or another. Among other recommendations, the Ash Council (as the advisory council was known, after its head, former Litton Industries CEO Roy Ash) proposed the creation of an Environmental Protection Agency to centralize all pollution control efforts in one body.[29] Creating the EPA would also score Nixon further points with environmentalists, or so the president and his staff hoped. Nixon announced the EPA and other new agencies as part of an executive branch reorganization message in July 1970, and the plan experienced no resistance from Congress.

After considering a number of candidates to be the first EPA administrator, the Nixon administration settled on William Ruckelshaus, then a thirty-eight-year-old deputy attorney general in Indiana. Ruckelshaus had spent his undergraduate years at Princeton before going on to receive a Harvard law degree. He had been involved in Republican politics at the state level, unsuccessfully running for office more than once before working on environmental issues in the Indiana Attorney General's Office. Ruckelshaus had developed a reputation of a principled and consummate professional who fiercely enforced state environmental laws and went after individual businesses when necessary. He would develop a similar reputation as the first head of the EPA—and then later as acting director of the Federal Bureau of Investigation, as assistant attorney general (from which position he would resign in the midst of the "Saturday Night Massacre"), and as EPA administrator once again under Ronald Reagan—though, of course, like anyone who would lead the EPA, Ruckelshaus was sometimes criticized by environmentalists.

These organizational changes in the upper echelons of the executive branch would forever change American environmental politics, but just as important for the history of automotive pollution control regulation were changes taking place in the bowels of the federal bureaucracy. Organizational transformations that would come to define federal automotive pollution control efforts throughout the 1970s began a short time before the EPA was formed. By 1970 it was clear that the automotive section of the National Air Pollution Control Administration was in need of repair. (Sometimes known as the Office of Mobile Source Air Pollution Control, this office within the NAPCA, and eventually within the EPA, has had a variety of names throughout its long history. I refer to it as the Office of Mobile Sources throughout this book.) The office was disorganized and lacked a clear institutional vision. John T. Middleton, the agency's head, had taken a beating from Congress when he admitted that around 85 percent of cars exceeded the emission standards promulgated under the Motor Vehicle Pollution Control Act by 15 to 20 percent.[30]

Moreover, aware that the Clean Air Act Amendments were on the horizon, agency leaders were looking for talent. An agency official recruited one of his former staff members, Eric Stork. Stork was a career bureaucrat. From 1962 to 1966 he had been an executive officer of the Federal Aviation Agency's Airports Service, after which he became a deputy director of the Food and Drug Administration's Bureau of Regulatory Compliance. Thus, Stork had worked in regulation for nearly a decade when he came to the NAPCA. He was an expert regulator. Unlike the early heads of the National Highway Safety Bureau, Stork understood and embraced the adversarial relationship in regulation.

Stork set about reforming the agency in various ways. Perhaps his most consequential decision was to shift the agency's focus from research and development to technology assessment. He was convinced, he later explained, that "we would never be able to get the huge resources required to do the research and development to demonstrate the feasibility of meeting the emission standards. . . . [E]ven if we were to succeed in such a task the NIH (not invented here) factor would mean that the industry would focus its efforts on proving that we were wrong rather than on adopting and improving upon our work." This dynamic was precisely what had developed when the National Highway Safety Bureau had tried to promulgate its first wave of standards. The automakers and the Automobile Manufacturers Association had poured resources into knocking the agency down rather than improving automotive technologies. Stork concluded, "I had very limited technical resources and I wanted to focus those resources on analysis and rulemaking that could force the industry to develop the required technology."[31]

Stork surely saw the formation of the EPA as an opening for overhauling the government's automotive emission programs. These large structural shifts allowed for changes that could not be made in an established agency, or at least would be much more difficult to accomplish. He also saw potential for a shift in regulatory strategy in the Clean Air Act Amendments of 1970. The National Air Pollution Control Administration had developed its own R&D capabilities to prove that technologies were feasible, as required by the Motor Vehicle Pollution Control Act, but the 1970 amendments no longer required such proof from government. Stork realized that staff members could move from R&D to begin spending their time studying where the automakers were with *their* efforts. Federal staff members would thus assess the promises of technologies as industry developed them. Such assessments would enable the federal government to object whenever the companies argued, as they inevitably would, that viable control technologies were too far out of reach. To frame his idea, Stork drew on the notion of *technology assessment*, which had become popular in Washington, D.C., after a seminar in the late 1960s at George Washington University introduced it to the area; it was also the subject of congressional hearings in 1969 and 1970.[32] Most discussion of technology assessment focused on having technical experts, including engineers and scientists, explain the promises and perils of technologies to non-experts, including members of Congress and executive branch officials. The Office of Technology Assessment, formed in 1972 within the US Congress, was the outcome of such thinking. But Stork slightly modified the idea for the EPA's purposes: his staff members would assess the industry's pollution control technologies to inform his decisions and those of the EPA administrator. Their efforts would ensure that such decisions were based on pertinent, reliable information, and it would keep the industry from feigning technological impossibility to disguise its lack of will and action.

In a November 1970 memorandum, Stork outlined his belief in the priority of technology assessment over R&D and defended his decision to move the agency in this direction.[33] The decision came as a disappointment to some younger staffers who were excited by the reformist spirit of the emission control efforts and had come to the NAPCA believing that they were basically signing up for an R&D project. Knowing this, Stork always trod a careful line between maintaining what he saw as the agency's primary goal and alienating his staff. He sometimes let projects closer to pure research through the budgetary process to keep staff members happy. But the primary product of the Office of Mobile Sources became a series of semiannual technology assessment reports.[34] Two processes converged in the reports' compilation: what economists sometimes call "search," and various in-house studies and experiments. In the first process, the EPA staff familiarized themselves with all available knowledge on emission control, including knowledge emerging in such venues as journals and professional societies. In the second, staff carried out various studies of their own to ground the reports.

Stork took a strong, antagonistic stance against the auto industry. He later recalled that he had his "hippie" daughter cross-stitch a tapestry that hung above his desk at work.[35] (Years later it was prominently featured in his home office.) In Latin, it read, "Orchides eorum trahite cordes et mente venient"— which, roughly translated, means "If you have them by the testicles, their hearts and minds will follow." Hearts and flowers surrounded the words. Stork cultivated a regulatory style that mirrored this adversarial understanding of his role. When an ambiguity in the Clean Air Act arose or, more often, if the automakers postulated such an ambiguity to increase their strategic position, Stork would issue so-called advisory circulars, papers that explained how the agency was interpreting the law.[36] The advisory circulars did not merely announce the agency's opinion about what the statute meant: they effectively *determined* the law. In this instance, within limits, meaning followed power. Stork's use of advisory circulars partly relied on the structure of the Clean Air Act Amendments of 1970. By contrast, because of the way the traffic safety act was written, the National Highway Safety Bureau had created the protocol for commencing rule-making procedures to set its own standards. Congress did not set these standards for the agency. Yet with this rule-making habit firmly ingrained, the safety bureau also began relying on the process to handle any disagreements about the law, working out interpretive differences through "democratic" processes of eliciting opinions and even holding hearings. During the early 1970s, leaders at the Environmental Protection Agency entertained no such considerations. The Clean Air Act Amendments set the standards for them, and the staff did not form the habit of commencing rule-making proceedings, even informal ones. When questions arose, the agency announced how the law was going to be interpreted. Stork took it upon himself to issue the advisory circulars, though he no doubt discussed the issues at hand with EPA administrator William Ruckelshaus and other leaders at the agency. Stork signed all of the circulars, however. In 1978, after Stork left the EPA, the agency's Office of General Counsel ordered that advisory circulars should be discontinued because of their questionable legality.[37]

The Arrival of Expertise

Recruiting experts is absolutely essential for any organization. But this is one of the most daunting challenges for government entities, especially regulatory agencies. Industry can consistently outbid the government in terms of salary and benefits for the employee. The Office of Mobile Sources had no trouble recruiting experts. On the contrary, the agency won a great deal of young, hot talent—exceptionally bright men and women who were dedicated to the cause of curbing automotive emissions. Stork later recalled that, aside from a few of his division heads, "all the others were just young kids."[38] These young experts began arriving at the agency before Stork did. One of agency's appealing features was that, because the National Air Pollution Control Administration was part of the Public Health Service, they could enter the agency as officers of the Commissioned Corps, thereby avoiding military service in the Vietnam war. Inevitably over time, some of these recruits drifted off to other opportunities, but a surprising number remained personally committed to the Environmental Protection Agency.

Indeed, several of them remained in automotive emission control at the agency. One such employee, Charles Gray, came to the National Air Pollution Control Administration in 1970 after earning a bachelor's in mechanical engineering from the University of Mississippi. Although Gray did not have an advanced degree, as many of his peers did, he had a talented mind for technical issues. In 1976 Gray was appointed to head the regulations branch of the Office of Mobile Sources, and played a central role in passing a number of regulations during his tenure there. He had an abiding interest in research, especially in alternative fuels and energy efficiency. In the early

1980s when the Reagan administration made passing new regulations virtually impossible, Gray turned seriously to studying the technological possibilities of methanol. He eventually coauthored a book on the topic.[39] By the 1990s, Gray was one of the chief researchers in the so-called Supercar program, a collaborative project between industry and the EPA.

Joseph Somers was another staff member who was originally hired through the PHS Commissioned Corps. He also came to the National Air Pollution Control Administration in 1970, at the age of twenty-five, having just completed a PhD in chemistry at Duke University.[40] Somers quickly began working on emerging scientific issues, including characterizing exhaust emissions and unregulated (but dangerous) emissions at the EPA's Ann Arbor office (work he continued until his recent retirement). He was one of the central researchers involved in the study of sulfate emissions, discussed later in this chapter. Over time, Somers's work moved him increasingly into the field of health effects research.

As Stork turned the agency's focus toward technology assessment, two men took the helm of these efforts—Thomas C. Austin and Karl Hellman. They began publishing annual reports outlining the current state of emission control technologies, and in automotive emission control circles these publications came to be known as the Austin-Hellman Reports, leading some to think that Austin-Hellman was one person. Austin had earned his bachelor's degree in automotive engineering from the University of Michigan in 1969. From 1969 to 1971 he worked as a mechanical engineer for the US Army Tank-Automotive Command (TACOM) at its headquarters in Warren, Michigan. There, Austin worked on automotive projects that would define the rest of his career. The projects focused on the energy efficiency of military vehicles, including a jeep modified with a stratified-charge engine. He also helped develop emission controls for jeeps, research that involved his taking the systems to the EPA's labs to have them tested. Eventually the EPA staff asked him to join the agency. Austin came to the EPA in 1971 and worked as project manager on technology assessment. In 1975 he left the EPA and went to work for the California Air Resources Board, first as a deputy executive officer and then, from 1978 to 1981, as the executive officer. In 1981, Austin cofounded a consulting firm, Sierra Research, where he continues to work.

Karl Hellman grew up in Gary, Indiana. He received a bachelor's in engineering sciences at Purdue University before entering Northwestern University, where he completed his PhD in mechanical engineering in 1970.[41] At that point, he came to the National Air Pollution Control Administration and stayed on when it became part of the EPA. Hellman devoted most of his time at the agency to technology assessment. But the Office of Mobile Sources was an incredibly dynamic place during the 1970s, and staff members could be called upon to work on whatever project was hottest at the moment. Austin and Hellman became deeply involved in fuel economy measurement during the early to mid-1970s, and Hellman continued to develop expertise in that area after Austin's departure. Hellman later became the branch chief in charge of characterization and technology assessment, where he continued to lead teams that produced annual assessments. He remained at the EPA until he retired in the mid-2000s.

The agency also required experts in his Washington, D.C., office. Stork recruited Joseph Merenda to become his technical assistant there. Merenda finished a master's in chemical engineering at Stanford before coming to the Public Health Service in 1970.[42] Stork often relied on Merenda when he needed quick advice on technical issues. Merenda also wrote a number of policy issue papers on various topics for Stork. In 1977 Merenda moved to the EPA's new Office of Pesticides and Toxic Substances (created as a result of the Toxic Substances Control Act of 1976).

The dynamic environment of federal air pollution control created excitement at the agency. A corollary to this vitality was that staff members worked long hours, including many weekends. Stork himself recalled, "I worked six days a week and half a day on Sunday. I worked from seven to seven and then came home and passed out in my armchair, much to the consternation of my wife and children."[43] One former longtime employee suggested that these long work hours may have led to the divorce of one of his peers. By their own account, the men and women of the Office of Mobile Sources worked passionately. Their passion went largely unpaid—none of them earned overtime— but it yielded personal rewards, a palpable organizational vitality, and the satisfaction of eventually breathing cleaner air.

Organizational Responsiveness at the Early EPA

Within a few years of forming, the Environmental Protection Agency's Office of Mobile Sources had become a formidable regulatory agency. Whereas, earlier federal automotive pollution control efforts had focused on R&D, the Office of Mobile Sources moved decisively toward the production of knowledge through technology assessment. With this information in hand, the agency and its administrators could do battle with automakers in the regulatory environment. The full import of this move toward technology assessment became clear only in the 1970s as the agency's constant pressure on the industry took the form of a series of public hearings (the subject of the following chapter) in which agency staff members argued that they *knew* the automakers were much closer to having viable emission controls than the companies claimed. Yet the expertise and organizational capabilities that developed in the Office of Mobile Sources was valuable in other ways, too, especially a flexibility that enabled the agency in its early days to respond quickly to any problems that arose. "Back in those days, we just did things," one staff member later recalled.[44] As the agency was breaking a path toward its goals through a jungle of uncertainty, many unexpected obstacles inevitably arose. Its flexibility enabled it to address these issues rather than remaining stuck in predetermined routines. This organizational reality conflicted with the Environmental Protection Agency's later reputation—whether accurate or not—as a monstrous bureaucracy, known for its plodding and tone-deaf rigidity.

The Office of Mobile Sources faced a number of tests in its early years. Some of these were direct threats to the integrity of the Clean Air Act Amendments, while others endangered the agency's reputation, the image it projected of organizational competence. One of the earliest tests came in May 1972 when staff members from Ford Motor Company's environmental group appeared at the Environmental Protection Agency's Ann Arbor office demanding a meeting.[45] The Ford staff announced that the company's emission control testing lab had been falsifying certification data. The Car Product Development Group, a subdivision of Ford's Engine and Foundry Division, was responsible for testing, and it reported its findings to Ford's Automotive Emissions Office, which was organizationally separate from that division and worked closely with the Environmental Protection Agency on certification paperwork.[46] The Car Product Development Group had effectively kept two sets of books on its tests—one for itself and the other for sending to the Automotive Emissions Office. The latter set of data was passed onto the EPA, and it was these data that failed to record that Ford employees had completed hundreds of acts of maintenance on test vehicles. In other words, the Engine and Foundry Division was cheating the Clean Air Act. It slipped up by keeping records in its computer system of all the maintenance it failed to report to the EPA. For some time, the division had been complaining that it received "inadequate support from Ford's Central Computer Group."[47] The computer group finally responded and decided to make a report of the division's computer records. When it printed the data, however, the group realized that there were two sets of records and that they diverged wildly. It reported the discrepancy to a company vice-president. For five days, Ford officials debated how to handle the situation and eventually decided to report it to the Environmental Protection Agency and to withdraw its certification applications.

The Clean Air Act Amendments stipulated that certification cars could receive a significant tune-up only after being driven 25,000 miles. If, say, a part on the car broke unexpectedly, the automakers had to request permission from the EPA to carry out "unscheduled maintenance." The agency usually granted permission because, after all, the law did not require *all* of the automobile's systems to last for 25,000 miles. If a normal malfunction occurred, it should be fixed. The distinction between a normal and an abnormal malfunction was important. For example, a particular engine might foul the spark plugs and render them ineffective, which would greatly increase emissions. Although the EPA took such problems seriously, it would allow a repair if a broken system or part not directly related to emission control, such as an axle, was holding up a test. The Ford employees, however, had been jiggering with the engines during the entire testing process, changing spark plugs, filters, and other parts. While such maintenance may seem minor, it made cars run much more smoothly and, thus, produce fewer emissions. Over the course of the succeeding weeks, Ford admitted to an increasing number of such repairs. As the General Accounting Office ultimately reported, "Ford identified 442 instances of unauthorized maintenance on 26 test vehicles."[48]

Ford's cheating threatened the Environmental Protection Agency's reputation in at least two ways. First, the agency faced potential criticism for not having earlier discovered the double booking on its own. Second, the EPA seemed legally required to force Ford to shut down production until it ran new certification tests. But such tests would take months, and in the meantime, Ford would have to lay off thousands of its workers, a dramatic development that might lead to public retaliation against the agency. For several reasons, Administrator William Ruckelshaus sought an alternative solution.[49] He believed that regulators should be merciful when companies self-reported errors or wrongdoing, because harshly punishing firms after such confessions would discourage openness and foster conspiracies. He also believed—contrary to some interpretations of Clean Air Act Amendments— that it was important to take economic impact into account, or at least not to disregard the effects of the law's enforcement cavalierly. The Environmental Protection Agency and Ford came up with a complex arrangement that enabled the company to continue production. The arrangement required the agency to redirect significant resources, both time and money, to conduct new tests. Ford's retests entailed more than 1,300 hours of unscheduled overtime at the agency, a huge cost for a cash-strapped organization, and something that it could pull off only because it was flexible.[50] The Justice Department eventually fined Ford $7 million. A US attorney claimed that the size of the fine helped "demonstrate that the Government means business in enforcing the provisions for the Clean Air Act."[51] Inevitably, however, environmentalists and the popular press interpreted the settlement as a sign of the EPA's feebleness. As a subtitle of a *Wall Street Journal* article read, "Ford Imbroglio Shows Clean Air Enforcers How Weak They Are."[52] The article claimed that, although automakers cast Washington as a "growing regulatory monster," the testing scandal demonstrated that "the monster is in truth a toothless tiger."

The specter of companies cheating emissions tests haunted the Environmental Protection Agency throughout its early years, and it continues to do so today. In mid-1972, technicians from the American Motors Corporation (AMC) brought two cars to the EPA's Ann Arbor labs to be certified. The cars were pulled into the testing room, put on dynamometers, and had hoses that collected the vehicles' emissions connected to the tailpipes. The company's cars failed the test. They produced nitrogen oxide levels above the federal limit. The AMC technicians were displeased and stood scratching their heads. "It must have been a bad sensor," one of the AMC techs suggested.[53] "What bad sensor?" asked an EPA staffer. The answer to that question created a controversy that hung over the agency for half a year.

Staff members from AMC soon explained what had gone wrong. Many emission control devices created or experienced intense heat that could destroy them. The AMC cars were equipped with exhaust gas recirculation systems, a technology first introduced in 1973 model year vehicles to address nitrogen oxide emissions.[54] The systems recycled inert exhaust emissions into the combustion process, lowering the combustion temperature, thereby reducing the amount of nitrogen oxide produced. The technology had some negative side-effects, including lowering the engine's power output and thus its performance. AMC put compensating valves on its vehicles that used heat sensors to turn off the recirculation systems under extreme heat conditions. Apparently, company staff members had put the wrong valves on the test vehicle, and these improper compensators turned off the system at a lower temperature than intended, leading the car to fail the test. AMC vehicles with the right compensating valves met emission standards.

Yet this development opened up a larger concern for EPA staff members. What if other automakers were putting similar systems on their cars but were using them not to protect control devices but to cheat emissions tests? Staff in the Office of Mobile Sources simply did not know to what extent automakers were installing systems that shut down under selected circumstances and engine conditions.[55] In a July 1972 memo, the head of the office's certification unit, Ed Brune, described these conditional systems as "defeat devices." While Brune recognized some appropriate purposes for such technologies, such as saving emission controls from destruction, he noted that they could also be used "to reduce control system effectiveness under ambient or operational conditions which are not represented on the Federal test in order to improve engine economy and/or performance."[56] That is, designers could use the systems to cheat federal emission controls by "hacking" the test. Federal tests mandated laboratory-like controlled conditions, including ambient air between 68 and 86 degrees Fahrenheit. All that auto companies would have to do to undercut the law is add devices that shut down systems when temperatures fell outside this range. Brune argued that earlier federal tests were so weak in this regard, they virtually guaranteed cheating.[57]

This legal and regulatory structure left much room for corporations to interpret the rules, however.[58] Automakers cast its heat-sensing and other systems as necessary components that guaranteed the safety and functionality of the car, while technicians in the Office of Mobile Sources claimed that they broke the law. In this way, the agency's coinage and dissemination of the term *defeat device* was highly rhetorical. In press releases and follow-up interviews, EPA spokespeople argued that the devices "defeated" pollution control efforts, breaking the spirit and letter of the Clean Air Act Amendments. The agency first banned the devices through an advisory circular, but only a few days later, EPA administrator William Ruckelshaus stepped forward and gave the automakers ninety days to get the potentially fraudulent systems off their cars.[59] An auto executive later reported that he knew the automakers had been beaten when he read the words "defeat device" in the newspaper.[60] Just as in the case of the Ford cheating scandal, handling the defeat device issue required organizational resources, especially staff time. Organizational flexibility enabled the early EPA to address many unforeseen circumstances.

The Office of Mobile Sources had to respond not only to problems that arose from the automakers but also to ones that came from other parts of the federal government, including even other branches of the Environmental Protection Agency. In late 1972 the Dow Chemical Company, which was working as a contractor for the EPA, found a surprising fact: particulate emissions were higher on vehicles equipped with catalytic converters than on ones without them.[61] The Ford Motor Company had provided the catalysts to the contractor, and after hearing about the particulates issue, the company asked to see the filters in order to do some testing of its own. Finding sulfuric acid on the filters, Ford warned the EPA about the presence of the toxin.[62] The greatest concern was that the catalytic converter, which then looked like the most promising technology for addressing the automotive air pollution problem, was going to create an even worse hazard. Within the Office of Mobile Sources, Joseph Somers, a chemist, was assigned to work on the sulfuric acid problem. He would spend most of his time in the following three years on this project and would thereby become a resident expert on the health effects of air pollution.

Unfortunately for the Office of Mobile Sources, another part of the EPA—the Office of Research and Monitoring, based in Research Triangle Park, North Carolina—also took up the sulfate issue.[63] Some of the Office of Research and Monitoring's most important work centered on the Community Health and Environmental Surveillance System, or CHESS, an epidemiological study of air pollution in eight cities. Sulfur oxides were among the air pollutants that the office studied, so the news that automobiles might be increasing the amount of sulfur oxides in the ambient air alarmed its staff.[64] In September 1973 the office announced the results of its research: automobiles had typically emitted the sulfur in gasoline as SO_2, a basic precursor of sulfuric acid.[65] This chemical compound would, in turn, combine with water, creating a mist of particulate matter that had deleterious effects on human lungs (and other organic matter as well—sulfuric acid was a precursor to what would later be known as acid rain). Therefore, catalytic converters were potentially harmful to public health. Sulfate was not the only chemical byproduct of the catalytic converter that worried regulators. At one point, someone suggested that catalytic converters would also produce hydrogen cyanide. But EPA staff member Joe Merenda did a "back-of-the-envelope equation" and discovered that carbon monoxide would kill a person many times over before hydrogen cyanide did.[66]

The Office of Research and Monitoring's announcement started a bureaucratic face-off within the EPA. The Office of Mobile Sources was deeply invested politically in the catalytic converter, and its credibility was on the line. That office's job was to make sure that the automakers met the emission standards, and its staff believed that, in the short run, the catalytic converter was the only way to achieve those standards. Meanwhile, Office of Research and Monitoring staff believed that the EPA should delay any standard that would force automakers to deploy potentially harmful catalytic converters. "It's like watching two armies form on a battlefield," one EPA staff member said.[67] The battle soon went public, with spokespeople for both offices speaking to the press. The head of the Office of Research and Monitoring, Stanley Greenfield, and one of his chief research scientists, John Moran, became vocal critics of the catalytic converter. [68] Moran went so far as to say that he would not buy a catalytic converter–equipped car because of the technology's potential health risks.

Eric Stork cast the dispute as merely a bureaucratic issue. He believed that the Office of Research and Monitoring was trying to get a piece of mobile sources, which at the time was the most prominent issue on which the EPA was working.[69] The Office of Research and Monitoring had seen its budget slashed for the 1974 fiscal year from $881,000 to $150,000.[70] Moran told some congressional staffers about the sulfuric oxide issue while they were visiting Research Triangle Park. Congressman Paul G. Rogers (D-Fla.) then pressured the EPA's interim administrator (between Ruckelshaus and Russell Train), and Moran was given $2 million to study the relationship between catalytic converters and sulfuric oxides. Stork described Moran as "a researcher who is using every possible means to get research money for his own interests."[71] Moran responded, "I hope it works out that we don't have a problem. But I'll be damned if I'll apologize for sticking my neck out and insisting this program be funded the way it is now being done." Meanwhile, Robert S. Leventhal, the executive vice president of Engelhard Minerals and Chemicals Corporation, the company that was supplying catalysts to both General Motors and Ford, accused Moran of fabricating a "controversy" in order to win funds. As Leventhal claimed, "I have no knowledge of what Mr. Moran's study is. I don't think he even has a study."

Over time, however, the new EPA administrator, Russell Train, who took office during this sulfate flap, moved to the side of Mobile Sources. In testimony before Congress, he defended catalytic converters. The agency as a whole had staked much of its credibility on regulating the automobile, and the catalytic converter appeared to be the key to its effectiveness. If the EPA had to set aside this technology, it would have been at a loss as to how to proceed. Thus, Train's early moves on the matter may have been based more on his sense of the personal and organizational stakes than on good science. The controversy surrounding catalytic converters more or less forced Train to suspend the 1977 emission control standards on March 5, 1975.[72] In mid-1975, Moran was scheduled to appear at hearings of the House Subcommittee on Environment and the Atmosphere. Indeed, the subcommittee had specifically requested Moran to appear. But at the last second the EPA substituted Stork for Moran.[73] Moran appeared as a "backup witness," but it was Stork who spoke, and he did so in defense of catalytic converters. Stork, not Moran, represented the EPA's policy stance. Train thus muzzled Moran.

Luckily for the Office of Mobile Sources and the Office of the Administrator, another powerful group was also deeply invested in catalytic converters—General Motors. The controversy surrounding catalytic converters boiled down to a single issue: automobiles contributed less than 1 percent to overall levels of sulfur oxides in the ambient air.[74] The real question was what effect auto-generated sulfur oxides would have in places with dense traffic. From the beginning, General Motors challenged the Office of Research and Monitoring's findings on catalytic converters.[75]

General Motors settled the sulfuric acid issue through a road test at its proving grounds. On June 11, 1975, the Office of Mobile Sources met with representatives from GM in Ann Arbor to discuss the possibility of carrying out a real-world study of sulfate emissions.[76] The Office of Research and Monitoring studies had all been based on dynamometer tests that were then extrapolated using models. In October 1975 General Motors attempted to create real conditions: to simulate rush-hour traffic, it drove a large fleet of cars around and around its proving ground in Milford, Michigan. As one publication later put it, "In this study, GM operated a fleet of 352 catalyst equipped cars on their 10 km north-south straightaway. The cars ran for 2 hours at a speed of 80 km/h (50 mph), which is equivalent to a traffic density of approximately 5500 vehicles/hour on a four lane highway."[77] With GM, the EPA, and other groups and automakers collecting data on the ambient air, the study found that sulfates from catalysts were insignificant. The sulfate flap was over.[78]

Greenfield, Moran, and other Office of Research and Monitoring staff left the EPA between 1974 and 1976. The controversy over the catalytic converter may not have been the central reason for their departures, but Russ Train's decision in favor of the catalytic converter left them in an untenable position at the agency. Stork later recalled that Office of Research and Monitoring staff once verged on threatening Train in a meeting, claiming that, if the administrator continued down the road with catalysts, they would not support him.[79] After the GM experiment, these staff members lost credibility with both Train and the public. The EPA eventually regulated sulfur oxides, using Section 211 of the Clean Air Act to force their removal from gasoline. But the agency promulgated the sulfur oxides regulations not because of their health effects but because, like lead, they "poison" catalytic converters. This decision had real health benefits, as sulfur oxides contributed significantly to the automobile's overall output of particulate matter.

⌐⌐

Federal automotive air pollution control regulation underwent massive legal and organizational transformations during the 1960s and 1970s. Federal auto pollution laws began rather weak, with the government needing to prove that control technologies were available before it could promulgate new standards. Consequently, the National Air Pollution Control Administration built a federal research and development program around emission control devices, though the program, at best, met moderate success. The Clean Air Act Amendments of 1970 created a tough new regulatory regime, however, that incorporated stringent automotive air pollution standards in the law itself, rather than requiring a federal agency to set them later. Leaders at the new Environmental Protection Agency took advantage of this law to transform the work undertaken by federal staff members. Instead of doing research and development, staff members began conducting technology assessment, which involved studying the cutting-edge and state-of-the-art pollution controls used within the auto industry, which was now responsible for conducting the R&D. The move toward technology assessment fit a more aggressive image of regulation in the new EPA: it would *force* the auto industry to address its pollution problem, and it would keep a close eye on the companies as they conducted this R&D to ensure that they would not falsely claim an inability to reach the standards. The outcomes of this new dynamic are best seen through a series of hearings that the EPA held throughout the early 1970s, a story to which we now turn.

CHAPTER EIGHT

Establishing the State of the Art

The Clean Air Act Amendments of 1970 required automakers to meet the most stringent emissions levels in United States history by 1975. But the law also contained a safety valve in case the automakers could not attain the standards. If the manufacturers believed they would fail to meet the deadline, they could petition the Environmental Protection Agency to suspend the standards for one year. The automakers would then have to demonstrate that they had made "good faith" efforts but that the technology was not yet "feasible." The difficulty in defining these two terms set up a battle over what was technologically possible in the world of emission control. Ultimately, the EPA administrator had to decide whether to delay implementation of the regulations. Everyone at the EPA knew that the petitions were coming. The question was not if but when the automakers would protest that they could not meet the standards.

The stakes proved much higher than the suspension decision, however. The automakers criticized the Clean Air Act Amendments of 1970 for having been written by individuals who knew little to nothing about automotive technology, and insisted that they could therefore not meet the standards.[1] The suspension hearings provided automobile manufacturers an opportunity to attack the law in public and potentially to undermine its perceived legitimacy. As the deadline for suspension applications approached, the EPA began planning how it would run the hearings that would determine the administrator's decision. It held the first hearings in April 1972. On May 12, 1972, William Ruckelshaus, the first EPA administrator, announced that he would not grant a one-year suspension of the standards. In response, several of the automakers sued the EPA, claiming that

Ruckelshaus had not provided sufficient reason to deny the applications. On February 10, 1973, Judge Harold Leventhal ruled, in *International Harvester v. Ruckelshaus*, that the auto companies had a point: Ruckelshaus and his assistants at the EPA had not given the automakers, the courts, or the public adequate justification for the EPA's rejection of the applications. The EPA then held a second round of hearings, and, this time, Ruckelshaus suspended regulations until the following year. Although the story of automotive emission control was not yet over at this point, it was settled in several key senses.[2]

This chapter recounts and examines the process by which the EPA held hearings over whether to suspend the automotive emission standards and ultimately decided to do so. In many ways, this chapter is a meditation on how regulation produces knowledge. It argues that regulation, rather than impeding "innovation," as some contend, works as an engine of knowledge production of all sorts.[3] These forms of knowledge production include the research that regulations force companies to do; interpersonal sharing of knowledge between regulators and the regulated; making meaning of the law and the balance between public and private power (often through court decisions); and understanding the current technological state of the art.

This last point was particularly evident in the Clean Air Act regulations. In 1969 the auto companies had signed a consent decree that forbade them to share research on emission control technology. For this reason, no one truly knew where emission control technology sat. Technology-forcing regulation contains an intractable uncertainty, namely what will happen in the future. But in creating and enforcing standards that force technology, regulators face another kind of uncertainty—knowing fully the present state of a technology. To predict where a technology can be in the future, one has to have some sense of where a technology is currently. The hearings created an environment in which automakers reported openly on what they knew and how they knew it. Thus, regulation produced a great deal of information useful to government, antipollution advocates, *and* industry. The dynamic by which regulation created and aggregated knowledge was so clear in the EPA hearings of the early 1970s that multiple historical actors referred to them as "the continuing seminar on automotive engineering."[4]

This chapter proceeds chronologically through four sections. It begins with the members of the EPA deciding how they would organize hearings arising from the automakers' petitions for suspension. Next, it turns to the first set of hearings and Ruckelshaus's ensuing decision. Several automakers challenged this decision in court, so the third section examines the District of Columbia circuit court's decision to remand the suspension to the EPA and Ruckelshaus for further consideration. Finally, the chapter turns to the second round of hearings and the EPA's creation of a statistical methodology by which to settle the issue of what *feasibility* meant. Before embarking on this narrative, however, we must examine the technological systems that were under debate during the period, particularly the catalytic converter and the thermal reactor. In many ways, the limits of these systems constituted the limits of regulation.

Technical Considerations

The Clean Air Act Amendments of 1970 set a high bar for automotive emission control. It required automakers to reduce carbon monoxide (CO) and unburned hydrocarbons (HC) by 90 percent of their 1970 levels by 1975. Likewise, nitrogen oxide (NO_x) emissions had to be cut the same amount by 1976. This translated into 0.41 grams per vehicle mile of HC, 3.4 grams of CO, and 3.1 grams of NO_x in 1975. In 1976, NO_x would be reduced to 0.4 grams. To understand how dramatic this prospective change was, we should consider that, according to one estimate, uncontrolled 1968 vehicle emissions were around 17 grams per mile of HC, 125 grams per mile of CO, and 6 grams per mile of NO_x.[5]

The automakers had met the standards set by the Motor Vehicle Pollution Control Act of 1965 through a number of engine modifications, including "adjustments to the carburetor to provide leaner fuel-air mixtures and improved mixture uniformity, controlled heating of the intake air, increased idle speed, retarded spark timing, reduced engine compression ratio, and improved cylinder head design."[6] Different manufacturers used a variety of other systems as well, including air pumps to burn up more HC and CO in the exhaust manifold.[7] Yet major technological breakthroughs were not forthcoming. Indeed, as the previous chapter shows, in 1969 the Justice Department sued the four largest domestic automakers, accusing them of conspiring to hold up the development of emission control technologies.[8] The manufacturers, it seemed, were unwilling to deploy any emission control technologies that they did not all have. The suit was eventually settled by a consent decree in which the companies agreed to stop sharing research about emission control technologies.

New technologies would be necessary to meet the standards set by the Clean Air Act Amendments of 1970. Engine modifications alone would not suffice. In passing the amendments, Congress knew that the technologies were not available to the automakers and that their research divisions would have to develop them. In this sense, the law was "technology forcing": it would push the auto firms to develop new technologies. The automakers, both domestic and foreign, experimented with a number of different technologies to meet the standards, including alternative engine types (electric, turbine, etc.) and thermal reactors. The Big Three automakers quickly came to focus on the promises of the catalytic converter. During the early 1970s, automakers experimented with catalytic converters that had two different physical configurations—beads and monoliths (the latter resembling honeycombs), and two basic kinds of materials—noble or base metals. In each case, the catalyst was applied to the material, either beads or a monolith, and this material was in turn placed in a "can" installed in the automobile's tailpipe.

Automakers preferred base metals simply because they cost less. The most commonly used noble metal in catalytic converters was platinum, one of the most expensive precious metals. The bulk of platinum—nearly 1.5 million troy ounces yearly—came out of the Soviet Union and South Africa, two countries considered politically unstable by the day's standards. One GM employee estimated that moving the entire US auto fleet to platinum catalysts would require between 1.3 and 1.4 million troy ounces a year.[9] This would nearly double global production, requiring major changes in the platinum industry, greatly increasing the number of miners needed, and probably, vastly increasing the price of platinum. Others, however, including employees at the catalyst-producing companies Matthey Bishop and Engelhard, believed that producing enough platinum would not be a problem.[10] This dynamic of automakers making pessimistic predictions and suppliers making optimistic ones formed a pattern throughout the era.

During this period, catalytic converters also had two basic flaws constraining their successful development—physical durability and chemical "poisoning."[11] A number of factors could cause the catalytic materials to break down physically, but the most common were intense heat and simple wear from vibration during operation. The converters had to work under extremely hot conditions, typically at least 1,200 degrees Fahrenheit. At GM, beaded catalysts sometimes became so hot that they burned through the grid plate that held them, though they were retained by the external casing.[12] In at least one case, after researchers tested a converter at high speeds, the floor mats began to "smolder."[13] In addition, the threat of sparking a fire on asphalt or, especially, in leaves or grass was a constant concern.[14] In GM tests, converters did occasionally "char" grass that the cars were parked over. For these reasons, converters had to be insulated on both top and bottom, which increased unit costs. The beaded catalysts were especially susceptible to breaking down from vibrations. Often, beaded catalysts would slowly be turned into dust and would blow away through the exhaust system. The automakers were concerned about the possibility of extremely hot catalysts breaking down physically, "blowing out" the exhaust system, and injuring someone standing behind the vehicle.[15]

Poisoning or adulteration of the catalysts was a more complex issue. Under different conditions, chemicals would bond to the catalytic materials, reducing the effective, catalyzing surface area and stopping the catalysis. The chemical industry had long used catalysts but under strictly controlled conditions. Conditions in cars varied widely and chaotically, depending on such mundane factors as weather and the user's driving style. Lead was the primary culprit, but sulfur and phosphorous also damaged the catalysts.[16] To deal with the lead problem, the automakers pushed for creating unleaded gasoline.[17] General Motors had once owned a hefty share of Ethyl Corp, the primary manufacturer of the gasoline additive tetraethyl lead, so it was significant that General Motors was advocating the removal of lead from gasoline. Unsurprisingly, petroleum companies pushed back. GM president Ed Cole famously announced at a California Air Resources Board meeting that if the government would ban lead in fuel, he would deliver a functional catalytic converter.[18]

Organizing the Hearings: Rationalities of Regulation

In 1972, as expected, the applications for suspension of the Clean Air Act standards began rolling in. Volvo sent the first one, which arrived on March 13. Chrysler, Ford, General Motors, and International Harvester sent in their respective applications soon thereafter.[19] The Clean Air Act Amendments specified that whenever manufactures submitted a petition, the EPA must hold hearings in order to discern the verity of the manufacturers' claims. The law presented the EPA with an uneven field of dictates. Some requirements, such as the mandated emission levels, were spelled out in great detail in the law. Others, including the administrative regulatory procedures, barely contained any direction at all. The suspension hearings belonged to the latter category. Other laws, especially the Administrative Procedure Act (see chapter 5), set protocols for rule making and other government actions but were silent on how these hearings should proceed. In such cases, bureaucrats often turn to examples from other agencies, but Ruckelshaus and his advisors believed that they had no models to follow.[20] Without such legal direction or a model to work from, employees of the EPA became responsible for structuring these sui generis hearings.

In creating the procedures for the suspension hearings and decision, the EPA had two primary goals. First, it had to gather necessary information. Second, its public actions must be seen as rational.[21] Between the concerns of the automakers and those of environmentalists and consumer advocates, the EPA's activities would be closely watched. Controversy could not be avoided. Thus, the EPA had to build procedures that would make the decision seem eminently well reasoned and, therefore, unassailable when it was handed down.

The information request for the suspension hearings built upon the Office of Mobile Sources' technology assessment program (see chapter 7). In August or September 1971, the Office of Mobile Sources sent out a technical questionnaire to the automakers, and by October or November the agency started receiving replies.[22] Having secured a great deal of information through the suspension applications, the EPA still had to decide how to structure the hearings. What form should they take? Ruckelshaus and George Allen, the head of the EPA's Office of General Enforcement, were both lawyers. Stork was not, but his father had been one. All were suspicious of modeling the suspension hearings on court proceedings. They believed that industry lawyers, if given the chance, would bog down the hearings through administrative requests and cross-examination. Given the strict timeline EPA faced, Ruckelshaus and company did not think they could afford follow such procedures. Moreover, Ruckelshaus did not want to put his scientists, engineers, and technicians on the stand. If industry lawyers cross-examined EPA employees, they might undermine the EPA's already-delicate credibility, and Ruckelshaus always paid attention to maintaining the agency's trustworthiness. As Allen put it, "We didn't want the hearing to turn into a process in which counsel would interrogate witnesses in the usual way that counsel do, trying to elicit peripheral matters that seem highly prejudicial or dramatic and yet never groping with the entire mass of data that we thought had to be analyzed."[23]

This assertion went to the heart of the disagreement between the EPA and the automakers: What truth would the hearings produce? As evidenced by industry testimony early on in the hearings, the firms sought to put the Clean Air Act and the EPA's procedures on trial. For industry, the question was whether the law made sense and whether the government was acting rationally and justly. Ruckelshaus did not want to place the law or his staff members in doubt. For the EPA, the question was whether the industry proved that it could not meet the emissions standards.

As important, Ruckelshaus did not believe that the automakers' economic interests entitled them to any special treatment in the hearings. He did not agree that "the automakers' interests rises above the interests of the general public." Furthermore, he asserted,

> even if we consider only the kinds of economic and property interests which the [Ford Motor Company argued for], we have witnesses here who represent equally strong economic interests; the interest of suppliers in the outcome of this proceeding is as strong as the interests of the manufacturers of the basic vehicle. . . . The interests of business men who do business in urban areas that are threatened by unhealthful air pollution caused by automobile emissions is an economic interest which is in many cases as strong as the economic interests of the manufacturers.[24]

And so the hearings went as follows: the witnesses would show up in the morning, and EPA employees would begin grilling them about their emission control programs. In at least one way, the EPA took a casual attitude toward structuring the hearings. The agency did not think much about judicial review when creating the hearings, though it did when writing the decision.[25] This attitude and the agency's whole approach probably helped foster the set of circumstances that led the automakers to sue the agency and eventually persuaded Judge Harold Leventhal to rule against the EPA in *International Harvester v. Ruckelshaus*.

The First Hearings: Questioning the Companies, Trading Analogies

The hearings began on Monday, April 10, 1972, in an auditorium at the Department of Commerce in Washington, D.C. Ruckelshaus started the hearings off with powerful words: "I fully share Congress' conviction that the automobile cannot be permitted to continue to menace the health and welfare of large numbers of the people of this nation."[26] In accord with his objectives of limiting legal jousting and focusing primarily on technical issues in the hearings, he announced, "For the purpose of making the decision, the wisdom of that standard is not debatable, nor is the need for prompt development of more effective automobile control systems debatable."[27] He broke the decision about the suspension down into four factors. First, it had to be the case "that suspension is essential to the public interest or the public health and welfare of the United States."[28] In defining the public interest, Ruckelshaus included both the automakers' ability to produce enough cars to meet consumer demand and the avoidance of massive layoffs or economic damage.[29] Second, the automobile manufacturers had to show that they had made "good faith efforts," though there was no agreement on how they would do this. Third, the companies had to demonstrate that there were no feasible emission control technologies available, which required them more or less to prove a negative. They could also show, in case an effective technology *was* available, that they did not have sufficient lead time to meet the standards. Fourth, Ruckelshaus would take into account information from other sources, including the National Academy of Sciences.

Although only five automobile companies applied for suspension, the EPA subpoenaed the other major vehicle manufacturers, both foreign and domestic, that sold cars in the United States. Although these companies could not meet the standards, either, they had not yet filed petitions; the EPA subpoenaed them so that everyone would be at the table at once. The agency also subpoenaed suppliers of catalysts and other devices and materials that could potentially be used in emission control. Ruckelshaus announced in his opening statement that the EPA had invited these suppliers because they would likely contradict the auto companies that had applied for suspension.[30] The agency had invited public interest groups for the same reason. If the manufacturers were going to get a suspension, they were going to have to fight for it.

Crucial to learning and teaching in the suspension hearings were analogies. On the one hand, the auto industry used them to teach regulators about industry structure and the traditional division of intellectual labor between the automakers and their suppliers.[31] On the other hand, the regulators put forward analogies to question the automakers' practices and to encourage increased research and development. In this way, the hearings proceeded by individuals trading analogies.

The people who testified at the first hearings can be roughly broken down into four groups: the Big Three (General Motors, Chrysler, and Ford) and other domestic manufacturers, foreign automakers, catalyst companies and other suppliers, and public interest and governmental groups. Each of these groups generally towed a party line that corresponded to their respective interests in emission control. The domestic auto companies all sounded the same note: the standards could not be met. The Big Three were the major voices here. American Motors Company, for instance, depended almost wholly on General Motors for research because it did not have the scale to carry out major research projects in its own and because it had been cash-strapped and barely profitable for years.[32] The catalyst makers all claimed that the standards could be attained, and, of course, if the automakers adopted catalytic converters, they would require hundreds of tons of catalyst daily, opening a huge new market. The catalyst makers' potential profits could not have been tidier. (Indeed, there was so much at stake for the catalyst makers that they formed a lobby organization, the Manufacturers of Emission Controls Association, which worked in Washington, D.C., to ensure that everyone knew that control technologies *were* within reach.[33]) Environmentalists and consumer advocates invariably believed that the standards fell within automakers' grasp and argued that the companies were simply dragging their feet by applying for suspension.[34]

Foreign manufacturers were, by far, the most divided group. Interestingly, the main divide fell along East-West lines. European automakers were pessimistic about meeting the regulations' requirements. They had also invested far fewer resources into research and development on emission controls. Asian companies were much more optimistic about reducing emissions. To some degree, they had a jump on domestic automakers on many of the factors that influenced emission controls. Their cars were smaller, and they had put more into alternative power sources, including the Wankel engine and Honda's CVCC engine.

Automakers were centrally concerned about three issues, uncertainties that they believed the EPA could alleviate. First, no one had a clear picture about what the EPA's eventual policy on vehicle maintenance would be. None of the automakers were making it anywhere close to the 50,000-mile durability requirements of the Clean Air Act Amendments. By that point, catalytic converters failed and thermal reactors often cracked or melted. Later, the EPA allowed for the automakers to swap out the catalytic converters during the 50,000 miles but only after 25,000 miles. This decision was controversial both with environmentalists and with the courts, since it was unclear what incentives consumers would have to get their emission controls serviced.[35] Second, and along similar lines, the manufacturers were unsure where the final lead levels in "unleaded" gasoline would be set. Everyone knew that petroleum companies would have to reduce lead levels for catalytic converters, but no consensus existed about what "unleaded" gasoline was. The manufacturers used widely different lead levels in their durability tests; each had its own notion of de-leaded gas. Third, the EPA had not officially announced how or whether it would average the automakers' emissions test results to judge whether the automakers had met the standards. Couched in this larger issue was the question of whether the EPA would carry out tests on the manufacturers' assembly lines or would rely on some other system to ensure compliance. With these three questions hanging in the balance, automakers lacked knowledge and were quite uncertain about the meaning of the law and the structure of their regulatory environment. While the EPA was sensitive to the automakers' uncertainty, it sought to keep the hearings focused on its own priorities.

Early on in the hearings, Allen often handled the questions. His goal was to address the legal issues as quickly as possible before handing the baton to Stork and EPA's technicians, who would question the witnesses about technical issues. Ruckelshaus did not want "legal fencing." Instead, Stork and his staff staged a good, adversarial joust over engineering details. As Stork recalled, "I suppose [the debate over engineering] made a better and more credible show in terms of how deeply EPA was really digging into what the heart of the issue was."[36] The Office of Mobile Sources found an occasion in the hearings to build credibility by showing off the fruits of its technology assessment program. Still, for all of their work setting up hearings that would be focused on technological feasibility, EPA members felt as though the early discussions were dominated by industry lawyers and executives who knew little about technical details. Stork characterized the presence of lawyers as "a dismal kind of thing."[37]

Examining the testimony of two auto firms, General Motors and Chrysler, offers a broader perspective on the issues involved. On the one hand, General Motors was the world's largest carmaker at the time, and its fortunes had large ramifications for the auto industry and, albeit less so, for the whole US economy. On the other hand, one of Chrysler's emission control test cars, Car 333, which first came to public light during the hearings, became an important symbol and subject of debate throughout the period. To understand the disagreement over the meaning of Car 333 is largely to understand Chrysler's and the EPA's diverging views on innovation. In another sense, if the two fundamental questions in the 1970 Clean Air Act Amendments were what "feasibility" and "good faith" meant, then the EPA's debate with Chrysler focused on feasibility, while its exchange with GM centered on the question of the company's good-faith efforts.

General Motors: Trading Analogies of Good Faith

On Monday, April 17, 1972, the EPA held the first hearing in which it questioned GM. Although the EPA had already carried out six days of hearings, GM was the first of the Big Three to appear. Expectations were high as people wondered how the United States' domestic automakers would make a showing. The testimony of GM's employees and the questions that the EPA asked exemplified the tension between the two parties. The EPA wondered why GM had not put more effort into developing catalyst systems; GM implicitly questioned how much authority the EPA had to force the auto companies to change the industry's traditional industrial organization.

As with most companies, EPA's initial questioning of GM dealt primarily with spelling out the firm's current emission control programs, including its relationship with external suppliers, and its most promising control system to date. GM had begun preparing for the ramifications of the 1970 Clean Air Act Amendments even before Nixon signed it into law on December 31, 1970. Indeed, GM president Ed Cole viewed the catalytic converter as a patentable and potentially profitable innovation, and he started a laboratory focusing on catalysts in 1968. In the summer of 1970, General Motors surveyed the potential of six companies as catalyst suppliers—W. R. Grace, Engelhard, American Cyanamid, Air Products, Monsanto, and OXY (Occidental Petroleum).[38] Several months later, UOP (Universal Oil Products) was added to the slate. General Motors estimated that it would require a hundred tons of catalyst a day once manufacturing operations were up and running.[39] This potential market would be a huge prize for whichever company won the contract to supply GM with catalysts. GM also had an extensive and intensive internal testing program. It ran durability tests on the catalyst substrate using eighteen vehicles with catalysts from different suppliers.[40] The cars were assembled in late December 1971 and January and February 1972.[41] GM's AC Delco Division had 128 people working on the catalysts.[42]

When the questioning moved beyond outlining these basic issues, the EPA essentially began to quiz GM about its "good faith" efforts—that is, whether the company was really taking emission control seriously. EPA officials were unimpressed by GM's early showing and became clearly frustrated when lawyers and financiers dominated the discussion on GM's side. "We were pissed," Eric Stork recalled.[43] People with real knowledge of the company's technical systems hovered in the background (only later did they become the primary respondents to EPA's questions). This dynamic led EPA to demand that GM make a second appearance at the hearings.

General Motors depended almost entirely on outside suppliers for catalyst development, but the way it relied on the catalyst companies without aggressively supporting their research "disturbed" Ruckelshaus. He questioned whether General Motors would leave any other part that was important to its bottom line and that might give the firm "competitive advantage," such as an automatic transmission or power steering, to outside suppliers to devise on their own.[44] Starkman pointed out that the auto companies often relied on external organizations to develop components that were essential to auto manufacturing. For instance, the auto companies relied on steel companies to make the alloys that formed the car's body. This was not to say that the company did no R&D. By the time of the hearings, GM had produced fifty-eight catalysts.[45] Catalyst manufacture was a highly specialized field, however, and the catalyst suppliers had more expertise in "chemistry and chemical engineering." One employee claimed that GM's "area is more so engineering systems, putting the components together, if you will."[46]

Ruckelshaus did not relent: "Supposing you were trying to achieve—let's go back twenty years and you wanted—it was a question of competitive advantage over an automatic transmission. How was that developed?" In response, a GM technician characterized automatic transmission as a "good example" of what the GM employees had been describing to the panel.[47] The mechanical aspects of automatic transmission, including the gears and the casting, had been relatively easy to figure out. The transmission fluid, which he called "the slip factor," was much more difficult. When GM first put out an Oldsmobile with an automatic transmission—"one of the first fully automatics on the market"—no fluid or oil could tolerate the heat generated. GM was not the company that developed the fluid that finally made automatic transmissions viable. Petroleum companies created it for GM, though GM "did the testing."[48] Thus, the development of automatic transmission had a great deal in common with the way automakers were developing catalysts. The automakers simply did not have the in-house expertise to address every problem that arose in the automobile's complex technical systems. Ruckelshaus misread the companies when he assumed that they did.

Yet Ruckelshaus believed the issue hinged on the capital needed to conduct the research and testing. "Probably with the oil companies," he said, "there wasn't a question of adequate resources as there might be here."[49] If the auto companies poured more money into catalyst research, perhaps the endeavor could reach a larger scale and scope. The catalyst companies had expressed some doubts about whether they had enough funding to continue R&D, and the auto companies were not supporting the research. The GM employees claimed, however, that "a very significant part of this development is the cost to test."[50] Therefore, by carrying out the tests, GM was sharing R&D costs. Even so, Ruckelshaus wondered whether the auto companies had enough incentive to really push for innovation in catalysts, including supporting the research. He questioned whether the incentive was as strong "as it might be if you were to gain significant competitive advantage [through the research] as opposed to a statutory deadline."[51] GM's representatives insisted that the catalyst companies were not cash-strapped and that they had plenty of incentive because becoming GM's sole provider of catalysts would be a great boon.[52] GM held that it was providing value to the catalyst manufactures by carrying out the testing and that the information these tests generated was critical to the work the catalyst companies were doing. Yet an EPA staff member pointed out that employees of Air Products testified that they had sent catalysts to GM in January 1972 but had never received word from GM about how the tests went.[53]

For years, General Motors had run a general test on all of its vehicle makes. One technician describe it as a "matrix or statistical testing arrangement, whereby the engine transmissions, car bodies, size, are matched against the various parameters, such as emission performance, fuel economy, the various durability problems, and then on a selective basis, we can cover the high percentage or high-volume applications, so we are assured that we have a proper number."[54] GM used this methodology to ensure that it could produce enough vehicles at a low enough cost to sell vehicles profitably. So far, no catalytic converter–equipped cars had passed the matrix. Members of the EPA were skeptical that this test ever limited the company's activity where competitive advantage was at issue.

The EPA asked the panel when the test had ever become an issue during a moment of competition, as opposed to regulation.[55] A GM employee pointed to the case of sports cars.[56] He recounted that Volkswagen had brought out sports cars that started doing well with US consumers, and then Ford began offering the "boss Mustang, a really sporty type vehicle."[57] The horsepower wars had taken a new turn. GM began competing for this emerging consumer demand with the Oldsmobile 442. The 442 sported an open "cold air hood" that gave it an aggressive, masculine image. These hoods posed several potential problems, however, including "hood blow up, water entry into the engine, noise, exhaust noise involved with that system, and the durability of that hood."[58] GM submitted this system to its testing matrix. Since Oldsmobile planned only to release some two thousand cars equipped with the hood, the firm's engineers were content when five cars passed its tests.

The EPA pointed out that other manufacturers said that they would go forward with production if they could get five cars to pass the federal test procedures for emission control. The GM employee replied that his company would not be so easily satisfied. Five cars passing the test was fine for a small-scale item like a muscle car, but emission controls would be put on hundreds of thousands, and eventually millions, of vehicles. Moreover, emission control was considerably more complex than a cold-air hood. Although the hood required engineers to check its effects on other systems of the 442, emission control would almost certainly require multiple new components, including new carburetors and the catalytic converter. Each new system introduced new complexities and multiplied the difficulties of others. For this reason, GM had taken to component testing: some cars had new carburetors, while others had catalytic converters.

General Motors returned for a second round of testimony on Wednesday, April 26, 1972. From the beginning of the session, George Allen pushed GM harder on why it had not poured resources into the catalyst companies in order to hasten development of emission controls.[59] The GM employees expressed surprise at even being asked about this, as the catalyst companies had never come to GM seeking resources. One GM spokesperson said, "We supplied considerable aid, we think a disproportionate amount of aid to the catalyst manufacturers and their development programs by doing the testing for them." In the company's eyes, the support was "disproportionate with respect to other parts of our product, the extent to which we will take samples of materials in various ways, paints, fabrics, plastics, metals, and test them and then say, well, they are good, bad, or indifferent I think is much less than the case of what we did in the catalyst area."[60]

In furnishing this analogy, the GM staff member hoped to show that the company was making a serious effort to develop emission controls by taking a more active role in testing converters. Allen was still not convinced. "Let me make . . . let me try to put the record clear on this point: Do you contend that it would have constituted a good faith effort by General Motors to meet these standards if you had simply taken the position that you are willing to purchase from a catalyst supplier any proven devices that he can prove?" No, that wasn't it, the GM employees insisted. All of their examples were aimed

at showing that they weren't sitting around waiting for external developments but were actively participating in the testing process. Finally, Allen
put the situation in stark terms, "Well, I have a note here that . . . advising
me that the total catalyst market is about a hundred million [dollars] per year.
The automobile market is fifty billion [dollars] per year."[61] But the employees insisted that the first time GM had considered funding research at the
catalyst companies was "last Monday," and that they had found the very idea
surprising. Robert Fri, the EPA's deputy administrator, suggested that GM
would at least want to consider using capital infusions to break specific, targeted bottlenecks in catalyst research.[62] One GM staff member replied, "I
guess what the question really comes down to is after having focused on
those resources did we recognize that there was an area which might have
been enhanced? I can only say in all due honesty obviously we didn't, and I
think also obviously neither did the catalyst suppliers."[63]

By furnishing analogies drawn from the corporation's history—analogies
meant to explain how the industry "worked"—GM employees sought to preserve the industry's traditional organization. They were also *educating* the
EPA. The EPA, on the other hand, was trying to determine whether GM had
made good-faith efforts by asking whether good faith in this instance involved
changing traditional practices, including supporting outside research.

Chrysler: Trading Analogies about Technical Change and "Feasibility"

The EPA and automakers clearly had differing models of technical change,
differences that stemmed partly from the relationship between prescriptive
knowledge about technology (knowing how to make something) and propositional knowledge about it (knowing why something works).[64] One EPA
staffer later compared the agency's understanding to Edison's work on the
lightbulb: the automakers should take system components that worked and
move forward with them. Propositional knowledge about why a particular
converter was successful was not nearly so important as carrying out systematic experiments on models that did work. The automakers, on the other
hand, were focused on creating systems that worked over a wide spectrum
of their products, typically studied via statistical tests like the one GM used,
and they believed that they had to understand why the systems worked in
order to ensure their robustness. At the heart of these differing visions lay a
Chrysler automobile known as Car 333 in regulatory parlance after the number it was given in the company's emission control tests. The vehicle had a

manifold catalytic converter that had melted during testing, a common problem at that time. Normally, when the catalysts melted, the test was effectively over; the device was not going to make the cut. Yet Car 333 was different. Although the catalyst had melted, it continued to function.

Stork's technicians at EPA discovered the prospects of Car 333 as they sifted the suspension submissions. In one version of the EPA's story of Car 333, Engelhard, the company that made the catalyst installed in the car, had tipped off the EPA that Chrysler was going to bring the melted converter to the hearings as an object lesson in the company's inability to produce a viable emission control system.[65] Once alerted, Stork's technical staff examined the data on Car 333 and realized that, even though the catalyst had melted, the converter was relatively successful. If a Chrysler employee were to hold up Car 333's melted, ugly converter during the hearings, EPA staffers planned to ambush him with their findings that it met the standards. Whether it was tipped off or not, the EPA questioned Chrysler employees about Car 333's performance.

Sydney "Syd" L. Terry, the company's vice president of environmental and safety relations, took the first questions after giving an opening statement. Unlike representatives of the other automakers, Terry began by saying that the company had made great headway on the catalytic converter and that it had some converters that performed above the 1970 Clean Air Act Amendment's requirement for tailpipe emissions. But none of the company's efforts had come close to meeting the law's 50,000-mile durability requirements. Terry testified that Chrysler was asking for suspension for three reasons: uncertainty about the standard's requirements (especially about the availability of lead-free gasoline, the permissibility of carrying out maintenance within the 50,000 miles the act required, and whether the EPA would do assembly line and field testing); inability to produce a system that met the 50,000-mile requirement; and concerns about "lead time," that is, the amount of time required for design and planning before production could begin.[66] "Once [solutions to] these [problems] are established," he assured the agency, "we could immediately finalize engineering programs to determine and develop an optimum system."[67]

Soon after the EPA asked Terry to clarify some of his points, the conversation moved to Car 333. The company outfitted the vehicle with a converter filled with Engelhard PTX, a platinum monolithic catalyst. Chrysler employees explained that while Car 333 had fared well compared to other vehicles equipped with catalysts, it did so only because the engineers and test drivers had effectively "babied it." They took great care not to overheat the catalytic converter, limiting it to "1400 degrees, with the occasional excursion to 1500." They equipped the car with a "direct reading pyrometer" that allowed the driver to see the engine's heat as he was driving.[68] If the temperature spiked, he would back off the accelerator, thereby lowering the temperature. Because of this treatment, when the hearings began, Car 333 was still running.

Terry and his fellow employees were surprised at the EPA's interest in the car. "I think maybe we missed the point here somewhere," he said. "I tried to explain that the purpose of this test was to determine whether or not catalysts would live for say 40,000 or 50,000 miles, we hope, if we could keep the conditions of the catalyst the way it [the catalyst] wanted it [i.e., in ideal conditions]. So, we ran the test in such a way that it did not fail. We were trying to make it live."[69] Later, Charles Heinen, Chrysler's executive engineer of materials, claimed that until it "became a cause celebre, we did not realize that Car 333 was all that much fun. When you fellows got into Car 333, it became the most prominent car we had ever had."[70]

Beyond the issue of Chrysler's having tested Car 333 gingerly to guarantee the durability of its catalytic converter, the auto company also carried out frequent maintenance on the vehicle, which became a rub between the automakers and the agency.[71] Chrysler changed the spark plugs on Car 333 every 5,000 miles, regardless of whether they had failed, to prevent the overheating that would have ruined the catalyst. Chrysler argued that this was too much maintenance to expect from consumers. The EPA didn't think so. Indeed, the agency expected that automakers would need to communicate to the customer how the car should be used, including how it should be maintained. EPA's belief that regular, even frequent maintenance was OK ran counter to long-standing traditions in the American automobile industry, according to which cars should require as little maintenance as possible and their components—the tow hitch, for example—should meet as many people's demands as possible. The disagreement about maintenance and vehicle design went beyond mere quibbling over technicalities to the core of the federal automotive emission standards. If regular maintenance was expected, the automakers would be better able to meet the standards, since the catalyst would be less likely to overheat or otherwise fail.

Although little eventually came of Chrysler's Car 333, at the time of the first hearing, it represented a difference in understanding between the EPA and the auto industry—Chrysler in particular—about how "innovative" activity worked. On the one hand, the EPA held up Car 333 as comparable to Edison's first successful lightbulb. No one had any propositional knowledge about why the converter continued working even after it had melted, but in the eyes of EPA staff, Chrysler failed to follow up on the converter in Car 333 because the company's engineers did not view it as an example on which to build. Chrysler, on the other hand, asserted that nothing could be made of one marginally successful case. Many more successful converters would be necessary—and converters that "succeeded" much more clearly—before the company could consider itself as making any headway. In time, the EPA came around to viewing things as the industry did, but only after the automakers took the agency to court.

International Harvester v. Ruckelshaus

On May 12, 1972, Ruckelshaus denied the suspension applications of all five automakers—Volvo, Ford, Chrysler, General Motors, and International Harvester.[72] In defending his decision, he argued that the automakers had simply not made a convincing case that they could not produce a car that met the standard. Ruckelshaus observed that the "most effective systems typically include: improved carburetion; a fast-release choke; a device for promoting fuel vaporization during warm-up; more consistent and durable ignition systems; exhaust gas recirculation; and a system for injecting air into the engine exhaust manifold to cause further combustion of unburned gases and to create an oxidizing atmosphere for the catalyst."[73] The administrator characterized systems that contained all of these subsystems as utilizing the best "available technology," and he believed such a system could meet the standards. Soon, however, Ruckelshaus learned that he would not have the last word.

International Harvester, General Motors, Chrysler, and Ford quickly responded to Ruckelshaus by suing the EPA, claiming that he had not adequately supported his decision. The companies argued their case before the US Circuit Court for the District of Columbia on December 18, 1972. Two months later the court ruled in the companies' favor.

Judge Harold Leventhal wrote the court majority's opinion in *International Harvester v. Ruckelshaus*. Prior to becoming a federal judge, Leventhal had spent considerable time thinking about business and industrial policy. He had worked as chief of litigation for the Bituminous Coal Division of the Department of the Interior during the New Deal (1939–1940) and spent two stints at the Office of Price Administration during World War II (1940–1943,

1946) and one at the Office of Price Stabilization during the Korean War (1951–1952). After the Korean War, Leventhal went into private practice for almost fifteen years. In 1965 President Johnson appointed him to replace Wilbur K. Miller, a long-standing judge in the D.C. Circuit of the US Court of Appeals. Leventhal was confirmed by the Senate and began his tenure on April 7, 1965.

In many ways, Leventhal was a leader in the court; he set many of the doctrines that would come to dominate judicial practice. Leventhal staged what one judge called the "debate of the decade" with Judge David Bazelon, the D.C. court's senior judge.[74] The two disagreed over how deeply judges should look into agency practices, including the science that "supported" the agencies' decisions. Bazelon argued that judges should only ensure that agency practices would lead to "truth-finding." Leventhal, on the other hand, contended that judges needed to examine the details of agency practices to see how fully bureaucrats were fulfilling their mandates. Leventhal's position came to be known as the "hard look" approach, and it became the court's de facto philosophy after the US Supreme Court, in *Vermont Yankee v. NRC*, struck down Bazelon's view in 1978.[75] Although Leventhal advocated looking hard at federal agency activities, even he felt extremely restricted by his inability to master all of the technical details under consideration. Nowhere was this truer than in dealing with environmental regulation, which often required extensive knowledge of science and technology.

Leventhal's reservations about Ruckelshaus's first decision were based in part on a National Academy of Sciences report that seemed to contradict the reasoning behind Ruckelshaus's decision, especially on the issue of feasibility.[76] The Clean Air Act Amendments of 1970 directed the EPA to contract with the National Academy of Sciences (NAS) to complete a study on the technological feasibility of emission control systems. The contract was meant partly as a check on the EPA's internal judgment; an external "non-interested" scientific group would also determine whether the automakers could meet the standards, granting, it was hoped, greater credibility to the process. The NAS formed the impressive Committee on Motor Vehicle Emissions, including A. J. Haagen-Smit, the father of the science of air pollution and codiscoverer of the causes of California smog, and Edward L. Gintzon, an eminent electrical engineer, who was the committee's chair. The committee first met on June 16, 1971, and came together monthly thereafter.[77] Members made their recommendations on the basis of advice from eight panels of experts. Seven of the panels began work in 1971, but the catalyst panel was not assembled until "early in 1972," when the committee realized how complex and controversial the subject was.[78]

The committee focused almost solely on feasibility, with subordinate and related interests in long-term performance and cost. Unlike the automakers, the committee was especially keen to "open" the standard gasoline engine and consider alternative systems; indeed, its members feared that, if pushed too aggressively, federal regulations might lock in a suboptimal system instead of allowing more ideal systems to emerge. This was a popular fear at the time. As one news article asked, "Did the Clean Air Act's strict timetable push the auto industry down the catalyst road to the detriment of more promising antipollution technology, as Ford and Chrysler have suggested?"[79] Stork doubted such was the case: "I think there is a far greater likelihood that the automobile industry now will seek alternatives to the catalyst than it would if they didn't have to face up to the catalyst."[80] He added that the industry was more likely to develop alternative engines if they were better than the already-existing technology, namely, the traditional internal combustion engine with a catalytic converter installed in the exhaust system.[81]

Stork had no inkling of how locked-in the catalyst would become over time. Perhaps the NAS committee members did. "Four types of systems will meet the prescribed emissions standards during certification testing," they wrote. "These are: the modified conventional engine equipped with an oxidation catalyst, the carbureted stratified-charge engine, the Wankel engine equipped with an exhaust thermal reactor, and the diesel engine."[82] The committee went on to claim that automakers must be allowed one catalyst change during the 50,000-mile durability tests for catalytic converter systems and that the EPA had to create a rigorous national inspection program because no one knew how the converters would perform under daily stress. More important, the committee stated that automakers *could* produce these systems in adequate quantity to meet demand for the 1975 model year at a per vehicle cost increase of about $160 over that of 1973 cars.[83] Yet the NAS ultimately suggested creating interim standards greater than the 1974 but less than the 1975 legislatively mandated standards.

With the NAS's report in hand and his own concerns in mind, Leventhal remanded the decision to the EPA for reconsideration. He placed no restraints on Ruckelshaus other than to remind him that, as the NAS had suggested, interim standards were a possibility within the mandate of the 1970 Clean Air Act Amendments and were a probable short-term solution to the disagreement between the EPA and automakers.

Second Decision: Trust in Numbers, Trust in Others

Once the DC circuit court remanded the decision back to the EPA, Ruckelshaus and his staff began planning a second round of hearings. The format would be roughly the same as in the first round. The majority opinion had ruled against Bazelon's proposal to incorporate cross-examination into the proceedings. Yet the EPA made at least one crucial change: coming out of the Leventhal decision, the EPA realized it had to create a more rigorous methodology for statistically averaging the companies' test results. The methodology that the EPA created in many ways settled the meaning of "feasibility." The methodology came to function as a mediator between the EPA and the industry—an agreed-upon, dependable standard that both parties understood.[84] As Stork later pointed out, "We were not sued on the second decision."[85] Over time, EPA staff members and automaker employees came to know and trust each other, but in this early period, the methodology played an important role as arbiter. Yet, although Ruckelshaus had earlier agonized how to appear trustworthy in the public's eyes, and although the EPA's adoption of the methodology—and Ruckelshaus's second decision, which emerged from this choice—built trust with the auto industry, it alienated other parties, including consumer advocates and some members of Congress.

The agency announced its preliminary methodology, which employees had developed in-house, on March 9, 1973.[86] Then, on March 17, during the second hearing, the EPA held an informal, off-the-record meeting with the automakers to hear comments and to hash out a compromise on the methodology's details.[87] Stork characterized the meeting as a "bull session." The EPA participants had debated internally whether the meeting should be recorded and transcribed and finally decided against it, probably because they believed that the industry employees would not be candid if they knew they were being recorded or, more likely, that the industry lawyers would limit what the engineers said if a tape recorder were running.[88] Clarence M. Ditlow III, the well-known consumer advocate and critic of the automobile industry, said of the decision to hold the meeting off the record, "That was one of our greatest public interest objections."[89]

The Office of Mobile Sources and the automakers met in Ann Arbor during a massive snowstorm that added a sense of urgency to the proceedings. Stork recalled the meeting as "very, very informal, noisy, and useful." During the week following the meeting, the automakers sent in criticisms of the methodology.[90] Those criticisms that the EPA found to be justified were incorporated. The finalized methodology was based on Monte Carlo statistics. First developed to solve intractable problems with nuclear weapons design, Monte Carlo simulation came to be used in modeling the random, or stochastic, processes inherent in any real-world procedure.[91] Whereas Monte Carlo simulations earlier modeled bomb tests, in emission control, they represented other realities full of variability—namely, variations in the performance of emission control devices and the vagaries of the shop floor. Although automakers were at this time implementing increased mechanization and automatic machine control, tolerances (and materials) were—and still are—wide enough that one car would perform differently from another of the same automotive model. As observed earlier, catalytic converters also performed over a broad spectrum, partly because of their inherent properties and partly because of the wide variability in the cars in which they were installed. The EPA's Monte Carlo methodology became a way to model this wide range of performance. Some cars would perform better, others worse. But once the statistically average vehicle exceeded the mandated standard, both the EPA and the automobile industry would agree that the technology was "feasible." The method captured well the distance both the EPA and the automakers had moved since the EPA's staff members fixated on Chrysler's Car 333 during the first hearings. No single car provided substantial enough ground on which to base a policy. By developing the methodology, the EPA had created a mode of analysis that the industry could not gainsay as a rational basis for rule making and enforcement. But in the process—and due in large part to Leventhal's decision—the agency had moved much closer to the auto industry's vision of innovation. The mathematization of the EPA's decision making was also a harbinger of things to come. Increasingly in the late 1970s and, even more so in the 1980s and afterward, agencies were required to carry out cost-benefit analyses and other formal decision-analytic tools before implementing any consequential rule.

On March 12, 1973, the EPA commenced over two weeks of hearings in answer both to the remanded applications for suspension (by Volvo, Chrysler, Ford, General Motors, and International Harvester) and to American Motors Corporation, which had filed an application on March 2, 1973. Things moved quickly. The court had announced its decision on February 10; the hearings began a month later; and on April 11 Ruckelshaus announced his second decision: he would grant the suspensions and set interim standards.

Five days later, Ruckelshaus, Allen, Stork, and Sansom appeared before Senator Edmund Muskie's Subcommittee on Air and Water Pollution, the body that had birthed the Clean Air Act Amendments of 1970. The first three men testified for three days. Muskie was not pleased. He began the hearings by noting that Congress had passed the Clean Air Act Amendments in 1970 after more than fifteen years of public research on air pollution. Yet, now, the automobile companies said they could not meet the mandated standards. "I want to know why not," he demanded. "I want to know what the industry has done in the past 3 years. I want a public explanation from the industry for the path they have chosen, a course that has not been altered since 1969. I want to know what the industry is going to do in the coming year to overcome past failures. I want to know what commitment the industry is willing to make to the American people. And, I intend to challenge the assumptions on which the industry's failures have been based."[92] Of course, Ruckelshaus could not answer for the industry per se, but he defended his decision to suspend the regulations for a year and to impose interim standards.

This chapter explores the varying ways regulation produces knowledge both on the part of regulators and on the part of the regulated, and also for the public, including advocates of various causes. Both the regulators and the regulated have to learn about the regulatory environment as defined by the law and the courts while simultaneously trying to shape that environment to fit their respective visions. The opposing bodies also learn about each other. The most widely useful knowledge of all is gained about the underlying science and technology.

The 1969 consent decree banning collaborative research between the automakers had created something of a knowledge vacuum as to the current state of automotive emission control technologies. The auto manufacturers regularly petitioned the government to drop the consent decree and its ban on sharing information. In the first suspension hearings, the EPA's George Allen surreptitiously maneuvered the automakers into a place where they had to admit that the consent decree and not sharing research was the best arrangement. The automakers themselves required catalyst makers to carry on independent research that avoided sharing research with one another. Allen asked one of GM's employees about this arrangement: "Do you think this is desirable, this independent research by these catalyst companies?" "Yes," the employee answered, "I think it is very desirable. I think because of this approach it has furthered the state of the art, and I think that is why there has been this great improvement over the last two years." Allen responded, "That is the only trick question I asked you guys today, and I am glad to hear you say that because about every month Mr. Hilder [a lawyer who acted as GM's assistant general counsel] shows up and tries to get me to go to the Justice Department Anti-trust Division to get permission for you guys to exchange information with your competitors."[93]

The consent decree would stand, but the automakers were free to share information when testifying before the EPA. After the auto companies realized that they would not face a suit from the Justice Department, they began to share information more openly at the hearings, which became an effective, if decidedly unofficial, means for circumventing the consent decree. The companies still could not join one another in R&D projects, but they learned, in a basic way, how their competitors' technologies were developing.[94] One could understand, then, why Stork called the hearings "the continuing seminar on automotive engineering."[95] Allen characterized them this way: "The decision process itself, in that sense, undid the disabilities created by the consent decree; that is, it became a clearing house for information, and perhaps it is better that it worked that way than that the exchange of information occurred privately prior to the hearing."[96] It was better, perhaps, because the exchange happened in an open, on-the-record environment that freed the process from suspicion. In developing a mechanism whereby the EPA produced the truth it needed to make its decision, the hearings also aided the automobile firms in securing valuable markers by which they could guide their research. However unintentionally, the state had produced knowledge for which firms otherwise often pay large fees to consultants and analysts.

Following the second decision, the debate over the potential of catalytic converters changed considerably. As described in chapter 4, by late 1973, the primary question was whether catalytic converters created sulfate particles as a by-product, material that would be more harmful than the chemicals the converters were supposed to control. As a lead line in *Business Week* read: "Warning: Catalytic converters may be harmful to your health."[97] As these new issues came to the fore, the struggles over what constituted "feasibility" and where the state of the art lay faded into the background. Discussions about how automakers would meet emission standards, especially the one for nitrogen oxide, continued throughout the 1970s. But, in many ways, the first suspension hearings accomplished their goal; they produced vital knowledge and, ultimately, established the state of the art.

PART FOUR

BUREAUCRACY

CHAPTER NINE

The Bureaucratic Struggle over Fuel Economy

From 1965 to 1970, Congress passed laws that brought automobile production under unprecedented levels of federal regulation. These laws focused specifically on safety and automotive air pollution. In the early 1970s, however, legislators and policymakers considered regulating fuel economy. Concerns about automotive fuel economy were nearly as old as the industry itself. The first car buyers were perhaps rich enough that they did not need to think about fuel cost. After the Ford Motor Company began building the Model T, it encouraged auto dealers to highlight how much less fuel the T used because its vanadium steel frame made it lighter than other cars on the market (see chapter 1). Worries about gasoline shortages after World War I engendered research programs at General Motors, the National Bureau of Standards, and other organizations, efforts that eventually gave rise to the inclusion of tetraethyl lead as a gasoline additive (see chapter 2). Yet, for most of the twentieth century, consumers in the United States enjoyed cheap fuel, partly because of government energy subsidies. With the oil embargo mounted by the Organization of the Petroleum Exporting Countries (OPEC) and the energy crisis of 1973, however, policymakers considered using the same regulatory tools to push automakers to improve fuel economy as they had with safety and air pollution—namely, performance standards.

Regulating fuel economy was different in some ways from previous auto regulations, however. First and foremost, rules regarding fuel economy emerged within a federal government that had already created bureaucracies for administering automotive regulation. Fuel economy created tensions within the federal government itself; it pitted members of the executive branch against one another. After he left the Environmental Protection

Agency, Eric Stork hung a poster-sized drawing in his home in Arlington, Virginia. As discussed in the previous two chapters, Stork had headed the EPA's automotive pollution control efforts from 1970 to 1978, and he received this drawing from his staff when the Carter administration forced him out. At the center of the drawing stands a caricature of Stork, characteristically puffing on his pipe and dressed in plate and chainmail armor. He looks like a medieval knight. His foot rests on a dragon he's slain, a sword still sticking out of its side. The dragon resembles a car. The caption below the central image reads, "Eric 'The Iron Duke' Stork." Other caricatures surround the central image, each accompanied by one of Stork's oft-repeated phrases.

One of those images and phrases stands out for the purposes of this chapter. It pictures Stork waist-deep in water. Still puffing on his pipe, he's using a stick to beat back alligators, each of which has the name, or acronym, of a federal institution on its back—Congress, the Office of Management and Budget, the Department of Energy, the Department of Transportation, and two other offices within the EPA, the Office of Planning and Evaluation and the Mobile Source Enforcement Division. Below the image is the caption "It's hard to remember you're supposed to drain the swamp"—part of a longer saying that concludes, "when you're up to your ass in alligators." In Stork's context, this meant, "It's hard to remember I'm supposed to be decreasing auto emissions, when I'm fighting off all of these other federal organizations." Clearly, bureaucratic infighting and Beltway battles had an important place in the work of regulation.

During this period, scholars studying government discussed something they called bureaucratic autonomy, the ability of an agency to demonstrate that it provided value and to find reliable, stable resources. As a classic text on the topic put it, "By *autonomy* we refer to the extent to which an organization possesses a distinctive area of competence, a clearly demarcated clientele or membership, and undisputed jurisdiction over a function, service, goal, issue, or cause."[1] As later chapters show, conservative critics of government in the 1970s and 1980s increasingly lambasted the autonomy, power, and reach of federal agencies. This chapter examines how EPA's Office of Mobile Sources (formally, the Office of Mobile Source Air Pollution Control) endeavored to retain bureaucratic autonomy over a single issue— measuring and reporting on the fuel efficiency of automobiles.

In many ways, this chapter is a study in bureaucratic struggles with internal and external forces. It moves through four sections. First, it tells the story of how the Office of Mobile Sources, seeking to protect itself from external criticism, built the capabilities to measure automotive fuel economy. A central worry was that citizen-consumers would lose faith in federal clean-air programs. Initially, then, these measurements were part of an effort to preserve the credibility and reputation both of the agency and of the air pollution laws it enforced.[2] The second section describes how the EPA came to publish data about automotive fuel economy for public use. In this case, the agency sought to aid consumers in making wise, thrifty choices at car dealerships. The third section shows how the Office of Mobile Sources then struggled to retain the capabilities to measure and report on fuel economy when other federal agencies sought to take them for their own. In other words, the office fought to preserve its turf. Bureaucratic organizations that grow new capacities must sometimes strive to sustain them in the face of threats, even threats that come from other organizations within the state.[3] Finally, the fourth section examines the emergence of federal fuel economy standards, sometimes called CAFE standards. The unique multiagency structure of these standards has largely resulted from the EPA's success in retaining fuel economy measurement as part of its territory.

The First Fuel Economy Study: Capacity Building as Self-Defense
Throughout the early 1970s, the automakers and some members of the public increasingly attacked federal automotive emission standards for increasing fuel use or, put another way, for decreasing fuel efficiency. The trend was due partly to the auto industry's simple misallocation of blame. Yet many early (i.e., pre–catalytic converter) engineering solutions to automotive emission control, including altering the air-fuel mix and retarding the spark timing, definitely increased fuel usage. The Office of Mobile Sources encountered consumer dissatisfaction with fuel efficiency mostly through letters that citizens sent to their representatives in Congress, who then passed the comments on to the EPA administrator before they worked their way down to lower organization levels.

In late 1970 through 1971, the EPA received a sustained rash of letters from people concerned about the effect of emission controls on fuel efficiency. Stork asked his staff members to find some way to figure out to what degree emission controls were affecting fuel efficiency.[4] The problem was that there was no accepted, "standard" method for measuring fuel economy, but the staff members quickly found a solution.[5] Tom Austin, Karl Hellman, and the other persons in charge of the agency's technology assessment program hit upon a preexisting method for measuring fuel economy. Created at General Motors, the method used hydrocarbon tailpipe emissions from dynamometer tests—that is, the same kind of procedure used in emission control testing—to estimate the amount of fuel consumed. Although none of the Office of Mobile Sources' publications from the period cite the methodology's origin, it was probably Donald L. Stivender's 1971 paper, "Development of a Fuel-Based Mass Emission Measurement Procedure."[6] For the Office of Mobile Sources, the real value in developing Stivender's procedures lay in its wider application to the agency's needs, and the EPA's technology assessment program itself grew into an important source of generating new knowledge.[7]

The EPA's fuel economy procedures used a "carbon-balancing method": by knowing the amount of carbon in the fuel and then measuring how much carbon the vehicle emitted (in different forms—CO, CO_2, and hydrocarbons) during the test cycle, testers could calculate how much fuel the vehicles had burned. For the Office of Mobile Sources, the advantage of this method over others was that the only information the test required was carbon emissions—data the office already had in spades. Indeed, the agency was the largest repository of such cross-firm information in the world: it had emissions data on all of the manufacturers who sold vehicles in the United States going back to the mid-1960s when the original Motor Vehicle Pollution Control Act took effect. The key innovation on the part of agency staff was employing these historical data to carry out a number of studies on vehicle fuel efficiency. They could draw on previous emissions certification tests and other in-house databases in examining historical trends in fuel efficiency. Using the pre-1968 data as the baseline, the Office of Mobile Sources could determine with certainty the percentage of lost efficiency due to emission controls.

In November 1972 Austin and Hellman internally published the agency's first fuel economy study.[8] They found that although emission controls had reduced fuel economy, they were not the primary factor in fuel efficiency. The primary determinant was (and remains) vehicle weight. Emission controls came in third place after air conditioning on the list of factors affecting fuel efficiency. The report proved extremely useful for answering public inquiries, as well as those from legislators and executive branch agencies.[9] Thus, the EPA created the first measures of automotive fuel economy not because they were looking to increase efficiency or because they particularly cared about it, but for defensive reasons—because automakers and other groups were criticizing emission controls for increasing fuel use. The timing of this development, however, was propitious given the looming OPEC oil embargo and the first "energy crisis."

The Office of Mobile Sources also carried out other studies as a form of self-defense. Public dissatisfaction with automotive emission control found its most extreme expression in attempts to remove emission controls from cars.[10] Automotive repair garages began advertising emission control removal services in local newspapers around the nation, while consumers both sought out such services and attempted to disconnect the controls on their own. While much of this consumer behavior could be attributed to simple self-interest—the desire to lower personal gasoline costs—the removal of automotive emission controls was also an early sign of resistance to federal regulations. Future presidential candidate Ronald Reagan began to stoke this resistance in his weekly radio addresses in the mid-1970s, and it would culminate in the tax revolt and deregulation movements of the late 1970s and early 1980s.

The Office of Mobile Sources reacted to this problem creatively. Building on Austin and Hellman's earlier fuel economy study and other ongoing fuel efficiency studies, the office decided to look into how tampering with emission controls affected fuel economy. To do this, it tested cars before and after taking them to automotive garages where mechanics removed the emission controls. The office hoped that its study would convince drivers to stop taking their cars to mechanics for the purpose of removing their emission controls, but the real motivation for the study was much more direct: less than a month after the beginning of the OPEC Oil Embargo, GM president Ed Cole told the media that, as far as cars were concerned, the best way to respond to the "energy crisis" was to remove emission controls. He even proposed that Congress repeal—or at least relax—the Clean Air Act. Firing a shot over the Office of Mobile Sources' bow, Cole told an Associated Press reporter, "We've got some people in Washington who are acquainting people on the Public Works Committee with this proposition. It's being put into perspective."[11] He claimed that the country could save "5 billion gallons of gas a year" if it removed emission controls from "40 million late-model cars" that had been sold since emission controls came into effect in 1968. In the same article, Eric Stork fired back: "All they have to do, to save fuel and have clean air, is stop building those huge behemoths on which they make the most profit, and build small cars." He added, "Emission controls have adversely affected only the fuel consumption of heavy cars, of the type Mr. Cole builds." Stork concluded, "The auto industry is just using the opportunity of the energy shortage as yet another attack on the Clean Air Act."[12]

While Stork publicly battled with Cole, he asked his Ann Arbor staff to begin the study on the fuel economy gains from removing emission controls. That the Office of Mobile Sources could so quickly switch to a new project is a testament to the organization's flexibility in its early days. The scientists, engineers, and technicians at the EPA's Ann Arbor lab quickly responded to the agency's needs.

The study involved ten cars that represented the "full range of typical vehicle weights encountered in the existing vehicle population."[13] (Recall that the Office of Mobile Sources had already found that vehicle weight was the most important factor in determining fuel economy.) Staff members then contacted a "number of garages," asking that "they do whatever they could do" to improve fuel economy. Surprisingly perhaps, only a quarter of the garages contacted agreed to tamper with the emission controls. The EPA was not sure whether the mechanics declined "for the reason that they thought such work was illegal or because they did not want to contribute to deterioration of air quality," but eventually the staff members found the requisite ten garages needed for the study. The researchers made sure that the garages "represented a cross-section of the automobile service industry, and included corner gas stations, commercial tune-up centers, as well as a garage that is widely advertised in the Detroit metropolitan area as a specialist in the removal of emission control devices." Some of the garage proprietors claimed boldly that they "could improve both performance and fuel economy with little impact on emissions," while others asked customers to get back to them on how they thought the performance had changed. The staff members did not tell any of the garages that they worked for the EPA, though two of the garages eventually figured out.

In January 1974, two months after Cole's comments appeared in the press, the Office of Mobile Sources finished its report, "A Study of Fuel Economy Changes Resulting from Tampering with Emission Control Devices." At first, the report was distributed only internally. It showed that the EPA modifications could improve fuel economy by about 9 percent on average by "tuning up the 'as received' rental vehicles to manufacturers' specifications."[14] But the majority of auto garages actually reduced the cars' fuel economy by 3.5 percent by tampering with emission controls. The report thus demonstrated that having mechanics remove emission controls, as Cole had proposed, would decrease fuel efficiency while greatly increasing automotive emissions, to levels above the nitrogen oxide standard and nearly three times the standards for hydrocarbons and carbon monoxide.[15] The best way to increase fuel economy, therefore, was to keep the car tuned up. The report also remarked that tampering would probably hasten engine deterioration. Stork had pointed out one reason for this deterioration in his earlier jousting with Cole: many of the changes that automakers had made to engines to decrease emissions were so basic to the engine's design that they "cannot readily be varied independently."[16] As the EPA staff wrote, "This would indicate that application of a hot rodder's knowledge does not insure achievement of improved fuel economy."[17] This last point was the theme that Stork took up as he talked to journalists.

"The application of the hot-rodder's skills to fuel economy just doesn't work," Stork told *Detroit Free Press* reporter Douglas Williams. "It's preposterous to think that you can redesign 40 or 50 million cars in the garages of America and get any fuel economy gain."[18] In an earlier article, Stork had explained that he had developed the tampering study, which he described as a "little vest pocket program," to defend his office. "We're not exactly disappointed by the results" of the study, Stork said.[19] In the end, the tampering study demonstrated that Stork and his staff were willing to use the agency's capabilities in flexible ways in order to generate knowledge useful to the agency and to disseminate that knowledge to the media in order to fight off critics.

Building the EPA's Fuel Efficiency Statistics

The advent of the "energy crisis"—really, if anything, more a series of interlocking "crises" than a unified event—was a long time coming. Best remembered is the OPEC Oil Embargo, which set in overnight on October 20, 1973, when the organization announced a ban on sale of member countries' oil to the United States (and soon after, other industrialized nations) for its support of Israel in the Yom Kippur War.[20] But the United States had already faced shortages of natural gas during the winter of 1972–1973, and gasoline prices were already high in the spring and summer of 1973. The country had become a net importer of petroleum—that is, its imports exceeded its exports—in 1970. The population's use of energy, an important pillar of the "American way of life," was outpacing both domestic production and reliable foreign exports.

On April 18, 1973, President Richard Nixon delivered a special energy message to Congress.[21] Nothing in Nixon's message was revolutionary. Writing in the July issue of the *New York Review of Books*, the historian Emma Rothschild described Nixon's speech: "The presidential message achieved a tone that mixed the calm and shrill, major and minor, multinational and nationalistic, Eastern and Texan voices of different energy interests."[22] The president's "major projects," she thought, "followed most of the recommendations of the American Petroleum Institute, the National Petroleum Council, the American Gas Association, and the other institutes and associations that had described America's energy crisis." In other words, Nixon toed industry's line. But this could not be said of Nixon's relationship with the auto industry. His earlier policies remained a burr under Detroit's saddle.

In the days before Nixon's speech, his staff members looked for policy ideas to include in the address. Since the federal government had no centralized energy agency (the Federal Energy Office, which was another recommendation made in the speech, would not be created for several months), staff members polled a number of agencies for recommendations. The EPA was one such agency, and the Office of Mobile Sources was contacted. The office suggested that the EPA start publishing its fuel economy numbers as a consumerist measure.[23] In his energy address, Nixon directed the EPA to begin publishing fuel economy data and creating an energy-labeling program for automobiles.[24]

The EPA published the first fuel economy statistics without ceremony or ornament in the *Federal Register* two weeks after Nixon's energy message, on May 2, 1973.[25] It instantly created a hubbub in the press and a furor among the automakers. It was the first publication to compare comprehensively the fuel economy of all makes and models of cars. The Office of Mobile Sources always struggled to communicate the real meaning of fuel economy statistics (and continues to do so to this day). The first *Federal Register* notice itself warned, "EPA must caution against attempting to compare these published fuel economy values with other values obtained under different conditions or by different techniques." Fuel usage was determined by "a wide range of factors including the manner in which the vehicle is driven, type of route and terrain traveled, speeds at which the vehicle is driven, frequency of cold-starts, use of power-absorbing accessories [for instance, air conditioning], vehicle weight, axle ratio, ambient conditions, and many others."[26]

After publishing the fuel economy data, the Office of Mobile Sources began to work on the "voluntary" automobile-labeling program. All of the automakers agreed to the program except American Motors Corporation, the maker of Jeeps. As Nixon had directed, the EPA was supposed to work with the Department of Commerce and Council on Environmental Quality, but little came from these interactions, and the EPA ended up developing the automotive labeling program on its own.[27] On August 27, 1973, the EPA announced the details of the voluntary labeling program, including a sample of the standard format of the label that would be posted on all new cars sold in the United States.[28]

The EPA handled its second publication of fuel economy data completely differently. Instead of simply publishing the results in the *Federal Register*, Russell Train, the new EPA administrator, unveiled the data on the 1974 model year vehicles at the National Press Club on September 18, 1973. The room was packed with journalists, and many major news media carried the story the following day.[29] It was some of the biggest positive news the agency had ever generated, and the publication curried great favor for the Office of Mobile Sources with the powers within the EPA, including Russell Train, who basked in the spotlight at the Press Club only four days into his new job.

The automakers were not happy, but no company was as furious as Mazda, whose vehicles scored poorly in the EPA tests.[30] The debate between the agency and the company went on for months. To respond to Mazda's criticisms that its vehicles did better on the highway than did other companies' cars, the EPA developed the highway test cycle, which became a standard part of fuel economy labeling. EPA staff created the "highway" and "city" fuel economy numbers that continue to this day. Yet, even after running the combined urban and highway test cycles, Mazda's rotary engine–equipped cars fared very poorly, getting about the same fuel economy as the Ford Torino, which weighed nearly 2,000 pounds more than Mazda's cars.[31] Subsequent studies found that Mazda's cars performed even worse than indicated in the initial estimate. These new tests emboldened the EPA to create a fact sheet on the Mazda rotary engine, which the agency then sent out as an official statement to any interested party, including inquiring journalists.[32]

The EPA made other changes to its testing and labeling program, for reasons other than automakers' complaints. One change, however, arose from pressure from journalists and the public. The Associated Press reporter Stanley Benjamin called the Office of Mobile Sources, complaining that the EPA's fuel economy publications were too complex.[33] It was too hard to figure out who the winners and losers were, and that was the story that mattered to the journalists—the narrative of gas-guzzlers. And so the agency began ranking cars by fuel efficiency. The EPA made other changes as well. For instance, in the September 1973 fuel economy report, the vehicles had been divided into weight classes, but a subsequent consumer survey that EPA contracted found that these classes held no meaning for consumers.[34] Few people knew how much their car weighed. So the EPA broke the cars down into broader, easier-to-understand categories.

By 1974 the EPA also wanted to broaden the reach and quality of its publications, but the Office of Mobile Sources had no resources to spend on publishing. Stork contacted Robert Hemphill, deputy assistant administrator of the new Federal Energy Agency, and made him an offer.[35] If Hemphill would foot the bill for printing, Stork proposed, the fuel economy statistics would be released as a joint publication between the two agencies. The energy agency would get a piece of the publicity touting automotive fuel economy without needing to do any work. Hemphill gladly signed on. On September 20, 1974—once again at the National Press Club—EPA administrator Russell Train and John Sawhill, the administrator of the Federal Energy Agency, announced the 1975 model year fuel economy data to great fanfare. Later, the Federal Energy Agency's successor entity, the Department of Energy, would publish the federal fuel economy statistics. The agency's buy-in to this program was also the first moment when federal responsibility for fuel economy—in this case, the responsibility for publishing data—became divided. Ever thereafter, the federal fuel economy program would become a Hydra with several different agencies handling different aspects.

Fighting to Retain Fuel Economy

In Washington, D.C., when a topic is hot, agencies compete to get a piece of the action. The OPEC Oil Embargo made fuel economy and other energy issues piping hot. Energy prices spiked, and the federal government scrambled for any and all means of providing the country with relief. Energy issues, including fuel economy, suddenly became desirable commodities for federal agencies. Seemingly every office, agency, and bureau had its iron in the fire of energy policy. In such an environment, several organizations made a play for the EPA's fuel economy program. Stork instinctively knew that systems become stronger when they gain associations with and the loyalties of powerful actors.[36] He had already won over the Federal Energy Agency. Soon he set about doing the same with the Federal Trade Commission and the Society of Automotive Engineers, each of which he "co-opted," as he put it, to the EPA's cause.[37]

At the same time, the EPA also found a powerful adversary in the auto industry. The agency's fuel economy publications had infuriated the automakers. The Society of Automotive Engineers was also displeased that it had been left out of the standards-making process. The EPA had introduced—by fiat when its data were picked up by the Nixon administration—what was quickly becoming the standard fuel economy testing procedure. The engineering society's biggest complaint was that a dynamometer test was at the center of the EPA's federal test procedure for emission control, instead of an on-road test, which the society preferred.[38] Stork largely dispelled the SAE's concerns and co-opted the organization by asking it to help the EPA revise the agency's fuel economy test and by flattering the organization in the press.[39]

Other influential members of the auto industry, most visibly and vocally Chrysler CEO Lee Iacocca, pressured the executive branch to take the measurement of fuel economy out of the EPA's hands and put it in the hands of a more industry-friendly agency. Iacocca was close to Roy Ash, the director of the Office of Management and Budget under the Nixon and Ford administrations. He visited Ash's office in the first week of March and told Ash that he would like to see the publishing of fuel economy data moved to the National Bureau of Standards (NBS), housed at the Commerce Department.[40] Ash promised his friend he would do so. In April 1973 he told Betsy Ancker-Johnson, a scientist who was then the Department of Commerce's assistant secretary for science and technology, to take control of the fuel efficiency program and move it to Commerce.

Stork first learned of Ash's promise to Iacocca from a contact at Ford, and a contact at the Office of Management and Budget subsequently confirmed the fact.[41] Stork saw it as an attempt to make the fuel efficiency information "of the industry, by the industry, for the industry."[42] "The real issue," he told a reporter at the time, "is that the auto industry would like to get those mothers at the EPA out of fuel economy testing. All real influence depends on the ability to obtain, assimilate, and process information."[43] Stork began pushing journalists to question Ash about the matter. On April 25 the *Los Angeles Times* wrote a story about the issue.[44] It quoted Ash as saying that the issue was under consideration. Stork claimed that having any other agency conduct the tests would be inefficient because the EPA could base the data on tests it conducted anyway. The next day, the *Detroit Free Press* published an article that said more or less the same thing. Ash said that the decision regarding fuel economy would be made "in a few days." Yet the article also paraphrased and quoted a "high level civil servant close to the White House [who] said . . . the decision to move fuel economy testing to Commerce had already been made at [the Office of Management and Budget]. He termed it 'a dumb idea done for purely political reasons and one that had been hard fought within the administration.'"[45]

That same day, April 26, Stork dashed off a one-page memo to EPA administrator Russell Train outlining his understanding of how the Office of Management and Budget was trying to take the fuel economy program out of the EPA's hands.[46] As was typical for important, sensitive memos of the day, Stork made only three copies—one for Train, one for Train's chief of staff, and one for his own personal files. On Sunday morning, May 4, Stork rose from his bed and went downstairs. During breakfast his wife pushed the newspaper over to him. There, in Rowland Evans and Robert Novak's column in the Sunday *Washington Post*, Stork saw his memo liberally quoted. He was dumb struck. He had no idea how the columnists had gotten his memo, but he didn't like it. It put him in a delicate position.

Even the conservative Evans and Novak explicitly framed Ash's move as a pro-industry, anti-EPA policy. "If you took a poll in Detroit, no agency in Washington would be more unpopular than EPA," one "top-level Ash aide" told the columnists.[47] Evans and Novak argued that "behind the White House effort to strip testing from the consumer-oriented EPA is a more dangerous game which looks suspiciously like a clandestine administration effort to gut its own auto-pollution controls program." Following the Nixon administration's "well-established rule of cottoning big business and ignoring the consumer," the columnists noted, Ash was thinking of moving fuel economy either to the Commerce Department or the Department of Transportation, "both safely industry-oriented." The editorial then quoted Stork's April 26 memo: "What we see as really involved in the proposed move of fuel economy testing out of EPA is an effort to emasculate EPA in any future fight over auto emission standards [a fight in which] . . . all real influence depends on the ability to obtain, assimilate, and use information."[48]

Heading into work on Monday, Stork was nervous and on edge. Not long after he arrived at the office, Stork's secretary called: Russ Train was on his car phone. Stork had her put Train through, and greeted him. Stork later recalled that Train said, "Eric, it's about the Evans-Novak column." "Yes, sir," Stork replied. "I hope it didn't embarrass you," Train continued. "You see, Evans is an old friend, and I had him over for dinner. While I stepped out to take a phone call, I must have left your memo out on the coffee table, where Evans must have looked at it."[49] Stork's memo had been leaked from above. Train was protecting the EPA's and Stork's position by tapping his high-level media contacts.

At this point, Ash backed off from moving the fuel economy program to the National Bureau of Standards or the Department of Transportation.

Several factors motivated this shift. The press gave the issue a fair amount of coverage, particularly after the Evans and Novak column. Meanwhile, the Nixon administration was already deeply immersed in the public fallout from the Watergate scandal. In April 1974 the administration released its first Oval Office transcripts; in May, Democrats would open impeachment hearings against him; and in July, the House would vote in favor of impeachment.[50] To put it mildly, things were not going well for the president, and his administration may have been avoiding further controversy. Transferring fuel economy tests from the EPA to the NBS would certainly have resulted in an outcry from environmentalists and consumer advocates. The longtime auto safety activist and critic Clarence M. Ditlow III, who in 1974 worked for Ralph Nader's Public Interest Research Group, strongly supported fuel economy remaining in EPA hands.[51]

On May 17, 1974, members of the Office of Management and Budget and the EPA appeared before the Senate Subcommittee on Science, Technology, and Commerce, which was a part of the Committee on Commerce. The hearing focused on a bill, the National Fuel Economy Testing Act of 1974, that never became law, though its intention would be captured in the Energy Policy and Conservation Act one year later. Ash claimed that he was still considering moving fuel economy to the National Bureau of Standards, but that for the time being he would leave it in the EPA's hands. When a senator confronted him with one of the bureau's pamphlets on air conditioners, Ash agreed that it was totally incomprehensible to the average consumer and confessed that the NBS had not always done the best job at communicating energy conservation information. Even so, Ash and the other OMB staff members insisted that the National Bureau of Standards should play *the* central role in developing fuel economy standards because it was the key agency with expertise in developing standards. That's what the agency did. To this kind of talk, Stork and the EPA consistently replied, as Stork put it in his seminar: "Because we had the only comprehensive data base, we had the best data base."[52] By the time Russell Train and Stork testified that day, the majority of the senators were in the EPA's corner.

The following day, the *Washington Post* ran an article titled "EPA Wins Its Fight to Keep Car-Mileage Tests in Agency."[53] That wasn't the message that Ash wanted to send, but that was what the public read. A week later, the *Automotive News* claimed that "EPA still holds the reins in mileage labeling."[54] The issue died down over the summer. In August the General Accounting Office published a report on the issue of fuel economy programs within the federal government, in response to an April request for an investigation from Congressman Henry S. Reuss (D-Wisc.), the chair of the House Subcommittee on Conservation and Natural Resources.[55] While the accounting office report purposely did not take sides on the issue of which agency should run the fuel economy program, it did show the EPA's program in a rather positive light. The report highlighted the EPA's use of preexisting capabilities (the federal emissions tests) to create the fuel economy data, the new and expensive capabilities the Department of Commerce would need to develop if it wanted such a program, and the hard work the EPA had done over the past few years to improve its data collection. On the one hand, between information-processing costs and the costs of publishing pamphlets, the EPA had spent $26,000 on fuel economy testing and labeling for the 1974 model year.[56] Even the highway test cycle added in 1974 was predicted to cost the agency only $178,000.[57] The National Bureau of Standards, on the other hand, estimated that if it took over fuel economy testing, by the 1976 fiscal year, its costs would be around $3,080,000.[58] Moreover, the Office of Management and Budget was forced to admit that the National Bureau of Standards would probably have to run an emissions test before testing any car for fuel economy, because emissions could not be sacrificed for efficiency. Thus, any other fuel economy program within the federal government was beginning to look extremely redundant.

At the National Press Club on September 20, 1974, the EPA, with the Federal Energy Agency, released its second major fuel economy publication. The news was good: catalytic converters on General Motors and Ford cars had increased fuel efficiency by an estimated 13.5 percent.[59] Emission control was looking a bit better than it had at the beginning of the OPEC embargo. Stork later heard that, by November 1974 the National Bureau of Standards actually put together a "budget request for the staff to plan [a fuel economy] program." The NBS moved at a "geological pace," he wrote. But the "battle [was] long over, and Ash said, forget it."[60] Yet the issue was not completely settled until December 1975 when Congress passed the Energy Policy and Conservation Act, which enshrined the EPA's fuel efficiency testing in the federal automotive fuel economy standards.

Heading toward CAFE

In December 1973, the Senate passed Senator Henry M. Jackson's National Fuel and Energy Conservation Bill, which directed the Department of Transportation to carry out a study of fuel economy regulation. The bill would have to be reconciled with a House version, but six months later both chambers passed the Energy Supply and Environmental Coordination Act. In the end, the law directed the Department of Transportation *and* the Environmental Protection Agency to carry out a 120-day study on how best to improve automotive fuel efficiency—by 20 percent by 1980, if possible.

Responsibility for the study fell to two people, one drawn from each agency to cochair the effort. One cochair was EPA's Eric Stork. The other was Eugene Goodson, a professor at Purdue University who was spending two years as the Department of Transportation's chief scientist.[61] Goodson had earned his PhD in engineering at Purdue in 1963, after which he stayed on as an assistant professor, earning a full professorship in 1970. In 1973 he took a leave of absence to work at the Transportation Department.

As Stork later characterized the matter, "EPA [was] very suspicious of the Department of Transportation and vice versa. . . . At first, [it was] like two porcupines making love."[62] But Goodson and Stork learned to work together. They assembled staff members into seven panels—Technology Assessment, Economic Analysis, Fuel Economy Measurement, Policy Analysis, Truck-Bus Assignment, Safety Implications, and Air Quality Implications.[63] Most panels' "principal members" consisted of one person from the Department of Transportation and one person from EPA—to "combine the two agencies' expertise" no doubt but also so that one could keep an eye on the other.

The Task Force for the 120-Day Study, as it was called, met at the Department of Transportation building in Washington, D.C., on September 9–10, 1974. The group was under considerable pressure. Although its specific charge was to determine whether the auto industry could improve automotive fuel economy by 20 percent by 1980, it had a number of other issues to consider as well.

For example, the task force had to sort out which approach would work best among the several regulatory and other policy tools that could be used to increase fuel economy and, thus, reduce overall automotive fuel use. The group considered a wide variety of options, including "tax on gasoline [crossed out], tax on inefficient cars [gas guzzlers], raise load factors, reduce congestion, reduce auto usage—promote mass transit, and disclosure of autos' fuel economy."[64] Time and again, however, the policy discussions centered on three possibilities. First, the federal government could require a given fuel economy for every new car. This approach was seen as undesirable, however, because it would reduce the variety of vehicles available, including many kinds of cars that consumers—and voters—wanted. The second and third options both involved "averaging"—that is, taking the sum of all the cars a given automaker sold and figuring out the statistical mean fuel economy for all of those vehicles. In the second option, the federal government would require each automaker to increase its average fuel economy by a given percentage. This option had two problems, however. It was harder on manufacturers—especially foreign automakers—who already had a fuel-efficient car fleet. They would be especially pressed to find ways to squeeze the mandated percentage out of their cars' fuel usage. Moreover, from a regulatory point of view, this option was the most difficult to administer. The third option would require the average of each automaker's fleet to meet a certain standard, a specific goal express in miles per gallon (MPG). This option, too, had the problem of favoring companies already producing efficient vehicles. But in the broad sweep of things, this seemed a small price to pay, and it became the dominant model.

The study group's policy papers also mentioned the "free market" as one possible solution to the fuel economy problem.[65] As the final report of the 120-day Study remarked, "The study does not make the judgment of whether a Federal improvement standard is needed inasmuch as a 20 percent improvement goal may be reached solely through the forces of the market.... [Yet] consumer demand for better fuel economy may not be strong enough to induce manufacturers to opt for the substantial improvement possible."[66] Free market solutions would become prominent a little more than half a decade later with the election of Ronald Reagan as president.

The 120-Day Study found that automakers *could* increase fuel economy by 1980—in fact, by much better than 20 percent. "The full range of potential improvements," the task force reported to Congress, "is from 40 to 60 percent."[67] The report had intentionally made no recommendation for an MPG standard, but a number began floating around: a sales-weighted fuel economy average of 28 MPG for each company, or twice the national average of 14 MPG in 1974.[68] This standard found its home in the Energy Policy and Conservation Act, passed on December 22, 1975, a law whose automotive portions largely captured the intentions of the earlier, failed National Fuel Economy Testing Act of 1974.[69] Through a compromise with the auto industry, this number was eventually reduced to 27.5, supposedly by factoring in a 0.5 MPG deduction for energy inefficiencies caused by emission controls.[70] The national fuel economy standards were eventually known as "corporate average fuel economy," or CAFE, standards, though this term appears nowhere in the law. (Staff members who worked for the EPA's Office of Mobile Sources before and after passage of the federal fuel economy standards believe that the acronym *CAFE* was a creation of the automakers, which meant to insinuate that federal regulators were people who spent most of their time hanging out in cafés.)

In the US House of Representatives, the issue of fuel economy had fallen into the hands of the Committee on Interstate and Foreign Commerce. One of the oldest and most powerful committees in Congress, Interstate and Foreign Commerce had its roots in the Committee on Commerce and Manufacturers, created in 1795. One of its subcommittees, the Subcommittee on Energy and Power, took control of the fuel economy issue. And this subcommittee happened to be chaired by a Michigan congressman named John David Dingell Jr.[71] Dingell grew up in Michigan where his father was the US representative for the 15th House district, a position that John Jr. effectively inherited when his father died in 1955. Dingell was and remains a controversial figure. While exceptionally progressive on some issues—he has famously introduced a single-payer health care bill at the beginning of every session of Congress since he was first elected—Dingell has also come under significant criticism for doing the bidding of the auto industry, often shooting down or weakening laws meant to protect consumers because they threatened the Big Three's bottom lines. Dingell is an embodiment of Simon Cameron's epigram "An honest politician is one who, when he is bought, will stay bought."[72]

In truth, the Committee on Interstate and Foreign Commerce had two subcommittees that could have overseen the fuel economy law: Dingell's Energy and Power Subcommittee and the Subcommittee on Public Health and Welfare, which had played a central role in passing the Clean Air Act Amendments of 1970. The division between these subcommittees reflected a division within the executive branch, with Dingell's subcommittee overseeing the Department of Transportation, and the Subcommittee on Public Health and Welfare overseeing the EPA. By ensuring that the Department of Transportation controlled the standards-setting function of the Energy Policy and Conservation Act, Dingell guaranteed that he would have oversight over fuel economy standards.

According to the way the law was written, the EPA would retain the task of measuring fuel economy. Notably, very early in the "energy crisis," Stork had begun jockeying for this position. As the *Detroit Free Press* writer Douglas Williams remarked in an article dated January 18, 1974, "Who told the Environmental Protection Agency to concern itself about how many miles per gallon of gasoline you get in your Brontosaurus V8? Well, nobody. But it's a vacuum the aggressive new government agency chose to fill." He then quoted Stork: "We have no authority in law to enforce fuel economy." Nevertheless, Stork argued that the EPA was the "logical agency for the task." As he explained, "First, we are clearly under attack by people who say emission control has to go for the sake of fuel economy. It is our view that there is no conflict at all between clean air and transportation. . . . [Second,] of all government agencies, we have far and away more data on fuel economy than anyone else in the government."[73] By retaining the fuel economy testing program after passage of the federal fuel economy standards, the Office of Mobile Sources had won the battle it had been fighting for two years. The strange structure of federal fuel economy standards—a structure that remains to this day—stands as a testament to the Office of Mobile Source's skillful gamesmanship.

By the mid-1970s the EPA had established itself as a semi-mature federal agency with a fixed organizational structure. Its experts and administrators had their own interests. Sometimes these interests conflicted with those of regulated industry, but just as often, they led to tensions with other organizations in the executive branch or even with other groups within the EPA. The story of automotive fuel economy measurement and regulation perhaps shows best how at least some of the EPA's work went toward defending its interests and preserving its turf. This defensive posture included both initiating fuel economy measurements to stave off external criticism and protecting the agency's measurement programs from the grasp of other federal agencies. Frequently, the agency succeeded by getting other agencies and nongovernmental organizations to back it—that is, by forming a powerful network of interests. Yet these defensive actions were also productive: they yielded the fuel economy ratings that, with some modification, we still use today. The final structure of fuel economy standards—measured by one agency, administered by another—was abnormal, though not necessarily suboptimal.

At about the same time, in the early 1970s, economists and political scientists increasingly highlighted how public administrators and government experts had their own interests, priorities that did not necessarily mirror those of the public, however that public was to be conceived.[74] Critics of regulation would eventually seize on these ideas to argue that regulators and other government experts overstated their knowledge and capabilities and that they interfered with the proper workings of capitalist markets. Particularly after the election of Ronald Reagan, such ideas gained influence within federal government itself. It is to these critics of government and their concerns about the economic consequences of regulation that we now turn.

CHAPTER TEN

Deregulation and Its Limits

In 1964 the National Aeronautics and Space Administration (NASA) opened a new facility, known as the Electronics Research Center, located in the Kendall Square neighborhood of Cambridge, Massachusetts.[1] The center stirred controversy from the start. President John F. Kennedy and his brother Ted Kennedy had pushed for its creation, fulfilling a vow to "do more for Massachusetts," the state that had just elected Ted as a US senator. Members of Congress attacked the project as obvious pork, but it managed to squeeze through the budget process. By 1970, however, NASA was facing a serious financial crisis, and the research center was scheduled to close. President Richard Nixon's secretary of transportation, John A. Volpe, also from Massachusetts, hatched a plan to make the research center a part of the Department of Transportation. In its new role, the facility and the organization it housed were renamed the Transportation Systems Center (TSC).[2] (It was later renamed again, in Volpe's honor.)

In the mid-1970s, TSC staff, including Richard "Dick" John (who had a doctorate in aerospace engineering from Princeton University), took up an important line of inquiry on fuel economy.[3] The Energy Policy and Conservation Act of 1975 directed the federal automotive safety agency to create fuel economy standards (as recounted in chapter 9). Congress also required an investigation of the expected economic effects of such standards. This concern about economic matters was primarily focused on how fuel economy standards would affect consumers during a period of inflation and stagnant growth. Politicians were wary of affecting auto purchasers' pocketbooks, particularly given that the consumer movement led by the likes of Ralph Nader still held considerable sway. TSC staff members decided to take a broader view, however, and examine how regulations were influencing the automakers themselves. Researchers looked into productive capacity in the United States, the costs of producing each automobile, and the industry's financial state. Along the way, they made a striking discovery: the Chrysler Corporation, a Big Three automaker founded in 1925, was in extremely poor financial health. If economic trends continued, the company would face bankruptcy. And in 1979 it nearly did. The US Congress passed the Chrysler Corporation Loan Guarantee Act, initiating the first "bailout" in the industry's history, and for the next few years, the Transportation Systems Center provided essential information and reports for managing this loan guarantee program.[4]

Federal regulations did not cause Chrysler's financial problems, though some conservative economists would partly blame them. But the Transportation Systems Center's path from studying regulatory standards to uncovering the company's impending failure suggests how interconnected all of these issues were. Thinking about regulatory standards led to thinking about industrial structure at a moment when policymakers, scholars, and ordinary workers worried about the country's economic vitality. By the late 1970s and early 1980s the entire US auto industry—not just Chrysler—was in bad shape. Observers argued that the automakers, once centers of dynamic change, novel products, and innovative activity, had become old, tired, rigid. The automakers had traded innovation for productivity.[5] In the process, they had become locked into old ways. These bureaucratic dinosaurs resisted change during a period when their environment *was* changing. The various "energy crises" of the 1970s led consumers to desire fuel-efficient vehicles, the kinds of cars Detroit was not producing. Younger, upstart Japanese firms were making them, however, and Japanese cars were growing a reputation for being not only kinder at the gas pump but also higher in quality than American models. Many feared these Japanese meteors would drive the Detroit dinosaurs into extinction. A wide variety of individuals and groups in the United States—from industries to legislators, from mainstream economists to labor unions—began to worry about the threat of "international competition," a phrase whose use soared during this period.

This domestic and international context provided rich soil for the emergence of a new governmental philosophy aimed at relieving companies of so-called regulatory burdens. In the 1960s and 1970s, as we have seen, two federal agencies, the Department of Transportation and the Environmental Protection Agency, emerged and began regulating American automobiles. As discussed in chapter 9, these agencies also had their own interests, including maintaining their credibility and gaining and protecting turf. During this same period, conservative economists and others began to question the influence of regulations on the economy and to doubt that regulatory agencies had the public's true interests at heart. These thinkers ultimately aimed to "deregulate" industry. The process of deregulation began under President Jimmy Carter, but Ronald Reagan's election amplified this trend.[6] Reagan brought with him a number of conservative economists who believed that taxes, regulation, and other forms of government intervention hindered economic growth. These individuals, appointed to essential positions in the executive branch, significantly—and enduringly—shifted the course of regulation in the United States, including regulation of the automobile industry.

Experts, including engineers, medical doctors, and lawyers, had influenced the design and use of the automobile throughout its history. In the 1970s and 1980s, economists were the new group to have such an influence. As a participant in this movement later reflected: "The early days of the Reagan administration brought a new breed of civil servant to Washington. Many of them were economists; that is nothing new, but these people were different. They were educated and well-versed in the new economic theory of regulation."[7] This chapter traces the development of the deregulatory ethos and describes the abiding impact this new vision has had on automotive regulation.[8] Yet, if new developments in regulatory practices—such as cost-benefit analysis—had a lasting influence on American government, other deregulation efforts under Reagan, which caused more outrage at the time, were later changed or overturned. Thus, this chapter is a study of deregulation and its limits. First, however, we must account for the historical context that made this deregulatory vision appealing to so many.

Auto Regulation in the 1970s

The Energy Policy and Conservation Act of 1975 created federal fuel economy standards—as of this writing, the last major form of automotive regulation Congress has passed. Yet by the mid- to late 1970s the regulation of automotive safety and pollution control were well-established undertakings managed by mature or semimature bureaucracies—with all of the costs and benefits that came with such bureaucratic development. The Environmental Protection Agency continued to postpone compliance with the automotive air pollution standards that were supposed to have taken effect in 1975, substituting weaker interim standards, but it did so under the working assumption that it was only a matter of time before the tougher, final standards would take effect.

From the perspective of safety advocates at least, auto safety regulation was in direr straits. A 1976 congressional report characterized the National Highway Traffic Safety Administration as in "stagnation." The agency was generating fewer recalls, and recall investigations were taking longer to complete, nearly thirty months in 1976 compared with ten months in 1971 and 1972.[9] Members of Congress found that the agency often dismissed potential defects, siding with industry opinion that they were not dangerous despite considerable evidence to the contrary, leading to worries about the integrity of the process. Moreover, after the agency passed nearly thirty standards between 1966 and 1969, the number of new standards issued fell to sixteen between 1970 and 1974 and only two between 1974 and 1976.[10]

Some of the agency's reticence in creating new standards no doubt stemmed from earlier frustrations and setbacks. In 1969 Richard Nixon had appointed Douglas Toms, Washington State's motor vehicle administrator, to head the federal auto safety agency.[11] One of Toms's preconditions in accepting the appointment was that the agency report directly to the secretary of transportation rather than through the Federal Highway Administration. As a result the agency became more independent and its name was changed from the National Highway Safety Bureau to the National Highway Traffic Safety Administration (NHTSA). In early 1970 Toms announced a new rule, regarding "passive restraints," which was to come into effect in 1973.[12] The idea was that "active" restraints, like seat belts, required active use on the part of drivers and passengers. That is, one had to buckle oneself in. But many, likely most, people were *not* doing that. Active safety technologies thus sat by unused. "Passive restraints" did not require active use—they were effectively replacements for seat belts.

The best-known and most promising passive restraint that emerged during this period was the airbag, which had a history going back at least to the early 1950s. In the 1960s, Eaton, Yale and Town, Inc., a firm based in Cleveland, had further developed the technology and, by 1968, had presented, with staff of the Ford Motor Company, a groundbreaking paper on it to the Society of Automotive Engineers.[13] With an eye on potential profits, the company lobbied government to pass regulations that would require installation of its technology in cars. Like other federal regulations, the National Highway Traffic Safety Administration's efforts regarding passive restraints both drew

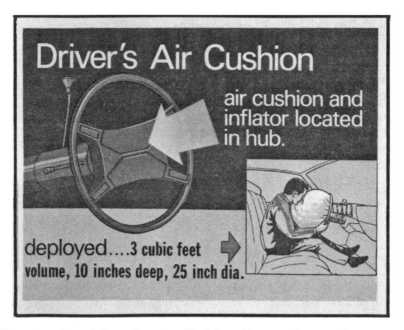

Figure 10.1. General Motors introduced its "air cushion restraint system" as an option in 1973 but removed the systems after its 1976 model year. The introduction and then retraction of these safety systems fit a long automotive tradition going back at least to the 1930s. Airbags remained a priority among safety advocates thereafter but were required in all US automobiles only in 1989. Donald D. Campbell, "Air Cushion Restraint Systems Development and Vehicle Application," SAE Technical Paper 720407, 1972.

on and spurred the development of knowledge, including an international meeting of airbag experts held in 1970.[14]

Toms and other NHTSA staff became fans of the airbag, viewing it as an appealing alternative to seat belts. Yet, federal legislation required the auto safety agency to define standards in terms of performance characteristics, *not* in terms of single technological solutions (which would have been *design* or *technology* standards). "Passive restraints" became a workaround that regulators hoped would effectively push airbags into cars. Toms bought into technology forcing—the idea, most often applied to the Clean Air Act Amendments of 1970, that focused on persuading industry to adopt new technologies and, often in the process, coercing companies to conduct research and development. Very likely Toms was looking to the EPA, which, under the Clean Air Act Amendments, had considerable power to oversee and shape technological change. It was a model that many regulatory administrators sought to emulate. But whether the auto safety agency had the authority to create technology-forcing rules was an open question. Its early efforts to push industry—especially through the creation of Standard 201—had largely been struck down; therefore, its first wave of automotive safety standards had enshrined industry best practices rather than creating substantially new technologies (see chapter 5). Toms confidently moved forward with the technology-forcing passive restraint rule, but the entire undertaking was a gamble.

The automakers, both domestic and foreign, fought the proposed rule vociferously. In 1971 Henry Ford II and Lee Iacocca met secretly with Richard Nixon and his domestic affairs advisor, John Ehrlichman, who became the industry's contact on matters of regulation and pressured Secretary of Transportation John Volpe and safety head Toms to "back off" the automakers.[15] Multiple companies sued the agency, and in 1972 the court returned a mixed ruling: it fundamentally supported the agency's technology-forcing authority, but it claimed that the automotive crash test dummy, which was to form the basis of the performance standard, was not developed enough technologically to create dependable and reproducible results.[16] The ruling mired the agency's efforts, which were already facing resistance along several fronts.

More devastating to Toms's cause was that several automakers at this time opted for ignition interlocks instead of passive restraints. Ignition interlocks kept cars from starting until the driver and front passenger buckled their safety belts. The automakers argued that these devices addressed the same problem passive restraints were meant to solve—people not using seat belts—by forcing users to buckle up if they wanted to go anywhere in their cars. Earlier, Peugeot, Renault, and eventually Ford Motor Company, as well as seat belt makers such as Takata Kojyo and Irvin Industries, had argued that interlock systems were acceptable in lieu of passive restraints.

Toms initially rejected the idea via an "interpretation" issued in the *Federal Register* in May 1971, but five months later—perhaps under pressure from Ehrlichman—he was forced to reconsider, whereupon he issued a rule that automakers could use either interlocks or passive restraints.[17] The ignition interlock became one of the most famous debacles in US regulatory history. The systems were buggy and often malfunctioned, either keeping drivers from starting the car (with the seat belt buckled or not) or, if the car did start, setting off an alarm that could not be silenced until the driver reached a dealership. Users *hated* the systems, and they became a source of considerable complaints to industry and government alike. In the Motor Vehicle and

Schoolbus Safety Amendments of 1974, Congress killed the interlock requirement, substituting a dashboard warning light for seat belts instead.[18] By the mid-1970s, the National Highway Traffic Safety Administration's staff were badly bruised and embarrassed; the agency's credibility, damaged; and its abilities to set tough standards in the future, unclear.

Jimmy Carter entered the White House in 1977 after a campaign that had distanced itself from the corruption of Watergate and the horrors of Vietnam. Carter intended to bring a different sensibility to the presidency. Although both Nixon and Ford had been institution builders in their own ways, each creating several executive branch agencies, they did not relish that role. In the early days of his presidency, Carter symbolized a return to the pro-government, pro-planning ethos of the Johnson administration. He created new agencies, including the Department of Education and the Department of Energy.[19] But Carter signaled his new vision for government perhaps most clearly through the kinds of people he appointed to lead regulatory agencies. Many Americans had long feared that regulated industries would eventually rule, or "capture," the very regulatory agencies that oversaw those industries. The fox would be guarding the hen house. Under Carter's administration, the exact opposite was true. He put members of the consumer movement in charge of consumer protection agencies; he appointed environmentalists to head environmental offices. In the first years of Carter's presidency, the pro-regulation crowd captured the flag of bureaucratic power.

In the realm of auto regulation, a few appointments had notable effects. David Hawkins, a lawyer, became the deputy administrator of the EPA's Office of Air and Waste Management, which contained the Office of Mobile Source Air Pollution Control. After graduating from Columbia University Law School in 1970, Hawkins had gone to the National Resources Defense Council (NRDC), where he had worked until he came to the EPA. Throughout his time at NRDC, Hawkins had acted as a lobbyist for environmental causes. In addition, he testified in favor of strict regulations and enforcement at public hearings, such as the Clean Air Act suspension hearings (described in chapter 8). When Hawkins came to the EPA, he had a vision different from those of his predecessors. His politics and personal management style eventually led to differences with his direct subordinate, the head of the Office of Mobile Sources, Eric Stork. Believing that the EPA needed a change, Hawkins eventually put Stork in a symbolic role, effectively firing him, and replaced him with Michael Walsh, who had previously worked in the EPA's Office of Enforcement. Like Hawkins, Walsh was a true believer in the environmental cause.

When Walsh left the EPA in disgust with the arrival of the Reagan administration, he became the director of the Manufacturers of Emission Controls Association, a trade organization that advocated for air pollution laws around the world. Walsh traveled the globe, delivering sermons on the benefits of clean air. After Reagan came to power, David Hawkins returned to the NRDC.

Carter's choice to lead federal auto safety efforts was even more consequential. He made Joan Claybrook the administrator of the National Highway Traffic Safety Administration. In the agency's early days, Claybrook had acted as Administrator William Haddon's special assistant. After leaving the agency, Claybrook worked for Ralph Nader's Public Citizen. Carter's appointment of an arch–consumer advocate to such a key regulatory position enraged the automakers. Claybrook soon began an ambitious program of rule making and enforcement, just as the industry feared she would. The agency was soon planning, creating, and publishing more standards than it had in years. Claybrook soon became known as the "Dragon Lady," a term that probably said more about Detroit's sexist culture than about Claybrook's character.[20] Auto industry loyalists claimed that Claybrook had accepted the position as agency head only on the condition that Carter's people would promulgate passive-restraint standards, which had been rescinded under the Ford administration; but Claybrook denied the rumor.[21] Still, in 1976, she had written a *Washington Post* article making a strong case for airbags, and as administrator, she did reissue the passive-restraint standard, laying a path to having airbags or other passive devices, such as automatic seat belts, required in all new cars by the early 1980s.[22]

Under Claybrook's direction, the agency also undertook novel efforts in publishing consumer information. It began crashing cars at 35 miles per hour, not for regulatory enforcement (which was primarily based on a 30 mile per hour frontal crash test) but simply to produce data.[23] This information was then published so that consumers could weigh the relative safety of different vehicles when making their purchases. Nader had been advocating for consumer information programs for years. The agency's labeling regime did not differ fundamentally from the National Bureau of Standards' practice in the 1920s and 1930s of publishing which products met federal criteria. By making the results of its tests public, the federal government shaped markets. Claybrook's consumer information program was temporarily discontinued under the Reagan administration, but it became widely adopted around the world.[24]

Claybrook's staff devised a "five-year plan" for the creation of new regulatory standards and the improvement of existing ones. Even these efforts were not enough to meet Nader's high, even zealous expectations, however, and Nader and Claybrook had a falling out that lasted for more than a year.[25] The five-year plan may have reflected naiveté as much as it did wisdom or regulatory acumen. After all, there was no guarantee that the Carter administration would still be in power in five years. Indeed, it was not, and Ronald Reagan's appointees walked in with a completely different agenda. Although Claybrook did not appease Nader, she stood no chance of remaining as agency head in the Reagan administration. Auto industry executives opposed her. And so Claybrook departed, along with Hawkins, Walsh, and many others when Reagan arrived.

By the late 1970s many had begun to focus on the economic costs of regulation. In 1977–78 the Harvard Business School and the Department of Transportation convened a seminar titled "Government, Technology, and the Automotive Future," which brought together representatives from regulatory agencies, legislatures, universities, and the automotive and other industries. The seminar highlighted the basic division in regulatory philosophy that would play out over the coming decades. The goal, the organizers of the seminar presumed, should be to foster innovation and the diffusion of new technologies; and as the organizers later wrote, "depending on one's view," regulation "either stimulated or retarded" innovation.[26] On the one hand, Claybrook, members of her staff, and others defended federal regulation and its ability to generate socially beneficial technological change. The head of the Department of Transportation's Transportation Systems Center, Dick John, and his coauthors argued that fuel economy standards were having a positive impact, though they also cautioned that small manufacturers might face economic hardship if such standards were raised.[27]

On the other hand, more conservative thinkers, primarily legislators and economists, began to level criticisms at federal regulations. Lawrence J. White, a professor of economics at New York University and a member of the Council of Economic Advisors, argued that federal regulations would work better if they were based on "effluent fees," or car model–related taxes related that would come to be known as cap-and-trade.[28] A young US congressman from Michigan, David Stockman, who would go on to influence federal regulation significantly during the Reagan administration, argued that there was no good reason for government to regulate fuel economy. Stockman rejected the idea that the fuel economy standards could adequately address the national security concerns of "energy dependence" or any other "external-ity." (Climate change, or global warming, had not yet emerged as a widely recognized public problem.) The automotive market worked perfectly well, Stockman argued, so if gasoline prices or consumer demand pushed in the direction of fuel-efficient cars, the auto industry would respond in kind.

The seminar "Government, Technology, and the Automotive Future" was deeply interconnected with broader auto-centered developments in Cambridge, Massachusetts. In the late 1970s and early 1980s, researchers at Harvard and MIT began an unprecedentedly thorough examination of the economics and industrial organization of the auto industry. William J. Abernathy, a management professor at Harvard Business School, was one of the leaders of this emerging thought collective. In the mid-1970s Abernathy began a fine-grained analysis of automotive production. The main upshot of his findings, encapsulated in his 1978 book, *The Productivity Dilemma*, was that managers in the auto industry had to choose between productivity and innovation.[29] Managers boosted productivity not by innovating but by locking in specific production regimes, which, in the case of the auto industry, were assembly-line arrangements. The cost of productivity, however, was low innovation, and so productive industries became vulnerable to all kinds of threats in the changing economic environment. And in the late 1970s, of course, Japanese firms were the emerging threat to the US auto industry.[30] Abernathy believed that the US automakers had to increase long-term investment in innovative technologies in order to enhance their competitiveness.

By the early 1980s Abernathy was putting his point in starker terms. In the essay "Managing Our Way to Economic Decline," Abernathy and his co-author pinned American industrial woes on managers who, Abernathy believed, had made a long line of unwise, risk-averse decisions.[31] In another of his cowritten works, *An Industrial Renaissance: Producing a Competitive Future for America*, he used the auto industry as his central example of what had gone wrong and what could make it right again and thereby bring on the industrial revival that he envisioned.[32] Indeed, in much popular economic analysis during this period, the auto industry became a stand-in for the American economy as a whole. Its fate was interwoven with the national self-image. Put baldly, scholars and critics argued that something was wrong with the American spirit. The academic and popular press regularly compared the domestic automakers to the Japanese firms in a negative light. Fear of Japanese dominance would eventually express itself in such works of popular culture as the film *Gung Ho* (1986) and the Michael Crichton novel

Rising Sun (1992), as well as academic works such as Ezra Vogel's *Japan as Number One: Lessons for America* (1979).[33]

Abernathy and other professors at Harvard and MIT began to form a network of researchers in Cambridge, Massachusetts, focused on automotive economics and management. Increasingly, they concentrated on the "just in time" and "lean" production techniques of the Japanese automakers. This work culminated in such popular books as *The Machine That Changed the World*, which examined the Toyota system of manufacturing and held it up as a model for many other kinds of businesses to emulate, not just automakers.[34] The Department of Transportation's Transportation Systems Center was deeply enmeshed in this relatively small, Cambridge-based network of researchers. For example, Dick John and his staff were putting forward a view that fundamentally accorded with the emerging Cambridge orthodoxy when they argued that fuel economy standards forced automakers to re-tool assembly lines and that this changeover put a heavier burden on small firms like Chrysler than on the mammoth General Motors.[35]

While some conservative policy recommendations concerning regulation began mounting in the Cambridge automobile seminars and elsewhere, the Carter administration did not initially act on them. Indeed, in 1978 it renewed the 1969 ban on cooperative research despite protests that it should be lifted to enable more coordination in the suffering auto industry.[36] Moreover, despite loud criticism from conservatives, Carter bailed out the failing Chrysler Corporation.[37] The Chrysler bailout was a move to support the auto industry itself, not the consumer and environmental groups that Carter had favored at the beginning of his term. The company had been struggling for years, but in the recessions of the 1970s, when even the more successful Ford Motor Company and General Motors struggled to maintain profitability and market share, Chrysler fell into bankruptcy. In 1979 the Carter administration, through a complex series of maneuvers, ensured that the company was infused with around $2 billion. The bailout was a success, and the company repaid its loans within a few years. In the process, its CEO, Lee Iacocca, became a celebrity, and his autobiography became a national best seller.

Carter turned against regulations in certain cases, and many of his actions fit with the rising free market ideology. Although he had handed control to consumer and environmental advocates like Claybrook and Hawkins, he turned to a new and different breed in other circumstances. In 1977 the eminent economist Alfred E. Kahn became the chairman of the Civil Aeronautics Board. He immediately set about changing the board's policies by increasing rate competition between the airlines, opening entry to new firms, and allowing companies to create new routes.[38] Embracing Kahn's policies, in 1978 Congress passed the Airline Deregulation Act. Deregulation of other sectors, including the natural gas industry, would soon follow, and on the whole the Carter administration became much more conservative, at least in its economic policies.

For the remainder of his single-term presidency, Carter played an odd balancing act. He rhetorically favored moving ahead with progressive causes, but he rarely used regulation to advance those ends. This balancing act was on full display in Carter's famous "crisis of confidence" speech. On July 15, 1979, Carter sat in front of television cameras in the Oval Office to deliver a message on energy policy, later tagged the "malaise speech." He began by outlining the seemingly intractable problem of energy crises. Carter cast the problem not as one of pure policy but as a moral quandary. He used quotations from people he had talked to in the preceding weeks to suggest that the country's problem was "deeper than gasoline lines or energy shortages, deeper even than inflation or recession."[39] He claimed that one person he had talked to told him, "Mr. President, we are confronted with a moral and spiritual crisis," or a "crisis of confidence," which he likened to a "moral equivalent of war." The path forward, then, was a moral path, "the path of common purpose and the restoration of American values."[40] Carter did outline some new energy policies, including investments in renewable energy sources and rules targeted at utility companies, but he announced no significant auto regulations. The economic climate simply did not allow for such a move. Carter's speech was a political disaster. His opponent, Ronald Reagan, seized on it as a symbol of Carter's pessimism and lack of true leadership. Reagan repeatedly attacked Carter as a president of malaise and gloom and countered that image with his own vision of hope, a vision that won the day in the 1980 presidential election.[41]

President Reagan and Auto Regulation

When Reagan was governor of California, he had—albeit with little enthusiasm—backed a measure tightening the state's automotive emission control regulations, which were more stringent than the federal ones. It was what the people of California wanted. But during the late 1970s Reagan increasingly embraced the free market ideas that were coming into vogue, including deregulation. This focus included an attack on automobile regulations. During the 1970s Reagan had a regular syndicated radio segment. In an episode broadcast in 1977, he went after the regulators who were trying to reshape the car. "Of late we've been told *that* America's love affair with the automobile is over, that *people want* some other method of transportation," he began.[42] "The Generals in the war against the automobile are a 'New Class' of intellectuals, journalists, burocrats [*sic*], and academics who are anti-materialist and opposed to the basic values of American society. Well, which side are you on in this 'War Against The Automobile'?" he asked. Reagan would not put up with such anti-Americanism. *"I'll tell you where I stand—I'm* going out to the parking lot and *give my* gas buggy *an affectionate pat* on the trunk and *tell it how much I care."*

Reagan's radio message relied on B. Bruce-Briggs's recently published book, *The War against the Automobile*, which had begun life as an article in the journal *The Public Interest.*[43] A fellow at the conservative think tank the Hudson Institute, Bruce-Briggs leaned heavily on the notion of the "New Class." Since the 1960s, conservative intellectuals in the United States, such as Irving Kristol (who Bruce-Briggs thanked in his acknowledgments), had been lamenting the rise of a "New Class" of managers and bureaucrats who meddled with core American values.[44] The idea developed thoughts first put forward by the Austrian economists Joseph Schumpeter and Friedrich Hayek, who argued that critical intellectuals who have no knowledge of practical affairs interfere with the proper functioning of capitalism.[45] In Kristol's hands, this anti-expert, antigovernment mindset became a way of critiquing social programs born out of Lyndon Johnson's Great Society, which included the regulation of auto safety and pollution control.

The ideas of Austrian economists, or at least popularized versions of them, provided a foundation for this attack on expertise in other ways as well. In the United States, the ideas of these thinkers—including Hayek, Ludwig von Mises, and other members of the Mont Pelerin Society, along with American disciples such as Milton Friedman, Frank Knight, and George Stigler—came to reside at the University of Chicago and coalesced in the "Chicago School" of economics, as it was known.[46] These economists put forward strong free market visions of society, later described as "neoliberalism," that cast active government as obstructing natural market processes, in part because the huge number of exchanges in any economy simply exceeded the understanding of any social planner.[47]

A corollary of this view came to be called "supply-side economics," also known as "trickle-down economics," a notion that flew in the face of the dominant economic philosophy of the day. On the one hand, followers of John Maynard Keynes believed that politicians could aid the economy by spurring demand (the "demand side") through deficit spending and other mea-sures. On the other hand, supply-side partisans asserted that economic growth would come only if politicians removed barriers—especially taxes, labor unions, *and regulations*—to the entry of new firms and the expansion of existing firms. These ideas found new resonance in the seemingly unstop-pable economic problems of the 1970s—including "stagflation," or simulta-neous economic stagnation and inflation—which put economic consensus based on Keynesian ideas in serious doubt. The active state that would "stim-ulate" the economy and solve "market failures," including those that led to pollution, safety, and other technological risks, was under attack.

Supply-side economics went hand in hand with other theories that emerged from the Chicago School. Nearly all of them were based on a single philosophical anthropology—or picture of human beings—that viewed indi-viduals as selfish, utility maximizers who acted on and reacted to incentives. Public-choice theory applied this thinking to the political process. In this perspective, politicians and executive branch administrators were not "public-spirited individuals benevolently maximizing some well-behaved social welfare function" designed to improve the public good. Rather, they were "rational, self-interested, maximizing agents" seeking to enlarge their own benefits rather than the public's.[48] Viewing regulators and the legisla-tors who supported them as agents in this "behavioral system" led these theorists to believe that regulations, in fact, benefited regulated firms, which curried favor with their overseers. Politicians and administrators act as bro-kers between interest groups, according to this view, and it is interest groups that are served by regulations, not the general public. This "capture theory" of regulation argued, for instance, that regulatory laws and codes created "barriers to entry" that kept new firms out—precisely what established com-panies desired. The adherents of this new theory were *certain* of its truth. As a member of the Chicago School claimed, theirs "was not a philosophical view, it was a hard-edged empirical approach to the world, and it was built on twenty-five years of exacting interdisciplinary academic research."[49]

This theory was based primarily on analysis of economic regulation, such as price controls, which applied to the airline, trucking, natural gas, communications, and several other industries, including public utilities. In the early years, the Chicago School economic view largely overlooked regu-lation of "negative externalities," such as safety, pollution control, and fuel economy; but later it was applied to them, and tendentiously so. For exam-ple, regulatory economist Bruce Yandle, who was executive director of the

Federal Trade Commission under Reagan and later joined the faculty at George Mason University (another free market hotbed), applied the theory to federal automotive fuel efficiency standards.[50] Yandle argued that, during deliberations over where to set fuel economy standards, General Motors pressed for a far higher standard than the one eventually set, *because* a stringent standard would have hurt two ailing companies, Chrysler and American Motors Corporation. Other members of the intellectual movement, as well, blamed Chrysler's bankruptcy on federal regulation.[51] These analyses conveniently set aside gains in safety, reductions in air pollution, and other social benefits that had been won from regulation. Such benefits did not fit the picture of regulation as economic warfare between selfish interests. Yandle's examination of fuel economy standards, for example, could not admit that there was a social benefit to decreasing fuel use. Instead, he argued, the goal of regulation was "forcing fuel-efficient cars on the consumer."[52]

This rising theory of economics thus attacked regulation from several different directions. From the supply side, it argued that regulations burdened the economy, raised costs, and thereby discouraged entrepreneurial action. From the perspective of public-choice and capture theory, it argued that regulations were tools that incumbents used to abuse competitors and raise barriers to entry. A third direction of attack argued that regulations themselves led to perverse, risky behavior. In a famous and controversial 1975 essay, Sam Peltzman, a professor at the University of Chicago School of Business, asserted that automotive safety devices encouraged dangerous driving.[53] Following the Chicago School vision of human beings, Peltzman argued that drivers constantly had to choose—that is, make rational, utility-maximizing choices—between safety and other goods, including thrills and reduced driving time via higher speeds. Safety devices reduced the threat of accidents, thereby rendering speedy, unsafe driving *more desirable*. It was a simple matter of incentives. Peltzman's essay was originally published in the *Journal of Political Economy*, a University of Chicago–based outlet. The American Enterprise Institute, a conservative pro-business think tank, printed a modified version of the essay in a series that included books arguing for the removal of price controls, the deregulation of the pharmaceutical and natural gas industries, a retrenchment in social welfare programs, and the attenuation of antitrust policies.[54] Like capture theory, Peltzman's analysis undermined the credibility of regulation and was, therefore, music to pro-business, antigovernment ears.

Reagan took up these free market ideas when Congressman Jack Kemp joined his team of campaign advisors. Kemp, a representative from a congressional district that included Buffalo, New York, had been considering a presidential run, but he agreed to give up his bid if he had a seat in Reagan's inner circle. The Chicago School had deeply influenced Kemp and his friend and fellow member of Congress David Stockman, who became the head of the Office of Management and Budget under Reagan. In his 1979 book, *An American Renaissance: A Strategy for the 1980s*, Kemp outlined policies based on popularized versions of supply-side economics, including the belief that lowering the tax rate could actually increase tax revenues through increased economic growth and the entry of new firms.[55]

When Reagan became president, he brought these and other free market ideas into office, and he signaled his differences from Carter, as with his symbolic move to take off the solar panels Carter had installed on the White House. The prosperous America Reagan envisioned and preached would not require such icons of limits and thrift. Reagan attacked auto regulations from different angles and on different levels. On April 6, 1981, three months after taking power, the Reagan administration released a plan to aid the automobile industry.[56] The proposal built on a report that Jimmy Carter's secretary of transportation, Neil Goldschmidt, had published just before leaving office.[57] Reagan's staff ignored Goldschmidt's recommendations for aiding the auto industry but used his research and analysis. They took the report's findings and made recommendations "consistent with the philosophy" of the Reagan administration.[58] The Reagan plan had four parts: economic stimulus, regulatory reform, revision of antitrust rules to permit cooperative research, and import quotas for Japanese cars.

Democrats were in general skeptical of the plan but especially about its proposal to lower federal regulatory standards. John L. Burton, a Democratic representative from California, told Reagan's transportation secretary, Drew Lewis, that the White House recommendations contained what Californians used to call a "Gipper gap," the space between Reagan's rhetorical statements and economic reality. "The automobile industry needs some $12 to $15 billion in investment capital," Burton noted, "and the administration program of reduction in safety standards and fuel emissions standards will result in less than $300 million a year over the next five years."[59]

Reagan's turn to so-called "regulatory reform" drew on a long history of US presidents trying to get managerial control over the executive branch. The first move toward regulatory reform may have been the National

Environmental Policy Act of 1969, which required federal agencies to prepare environmental impact statements.[60] The statements featured analyses of the possible environmental consequences of a governmental action. After deeming environmental impact statements a success, staff members of the Office of Management and Budget convinced the Nixon administration to require "quality of life" reviews from the Occupational Safety and Health Administration and the Environmental Protection Agency. The quality-of-life reviews turned the environmental impact statements on their head: instead of requiring government agencies to justify the impact they would have on the natural environment, they required regulatory agencies to justify the impact their rules would have on the economy. Eric Stork, the head of the EPA's Office of Mobile Sources during the 1970s, recalled his staff spending more and more time on such regulatory analyses during the period.[61] Businesses and other anti-regulation forces adored these reviews because they almost always delayed—and often weakened—pending regulations. Moreover, because they required input from other government agencies and the industries they regulated, the reviews provided these interests with "greater private access" to decision makers.[62] Presidents Nixon, Ford, and Carter increased the scope and power of these regulatory assessments, but it was Reagan who brought the movement to its fullest development.[63]

Reagan's work on regulatory reform, including modifications to regulations aimed at the auto industry, began before his full-fledged auto industry plan was released. On February 17, 1981, less than a month into his term, Reagan issued Executive Order 12291, which killed many pending regulations and forced agencies to review existing ones. The auto industry, which Reagan had promised to help, was its primary beneficiary. The executive order affected more than thirty-four potential regulations affecting the automobile, including passive restraints, tighter emission controls, and more stringent fuel economy standards.[64] More important, however, the order required cost-benefit analyses for any future proposed rules.[65] The Office of Information and Regulatory Affairs, which had been created by the Paperwork Reduction Act of 1980 and which sat within the Office of Management and Budget, became responsible for these analyses.[66] So-called desk officers within the Office of Information and Regulatory Affairs who were trained in quantitative methods based in economics reviewed pending regulations to make sure the potential benefits justified the potential costs.[67] Reagan appointed supply-side disciple David Stockman to head the Office of Management and Budget (OMB). In gutting regulations and slashing budgets,

Stockman became known as a fearsome advocate for the Reagan administration's economic philosophy.[68] The OMB rejected or withdrew many regulations, including ninety-five in 1981 and eighty-seven in 1982, a pace that
continued for the rest of Reagan's presidency.[69]

In many ways, these regulatory reform efforts were Reagan's greatest contribution to the history of regulation. Other decisions he made concerning
regulation caused greater outrage at the time, but later presidential administrations undid them. Critics both during Reagan's tenure and afterward
saw his regulatory policies as little more than kowtowing to corporate desires.[70] As we have seen, the actions of the Reagan administration fit with a
longer-running economic philosophy that did not wholly stem from corporate sources, though during this period wealthy donors were pouring more
and more money into pro-business, antigovernment think tanks and publications.[71] Supply-siders argued that regulations burdened the economy in
ways that prevented growth and discouraged entrepreneurship, and public-
choice theorists argued that regulations were little more than weapons that
firms used against one another. Still, the Reagan administration was at times
forced to admit uncomfortably that special interests shaped its policies. Many
of the administration's decisions on auto regulations arose from Vice-
President George H. W. Bush's Task Force on Regulatory Reform. One of the
task force's recommendations was that the executive branch change antitrust rules to allow the automakers to collaborate on research.[72] In the early
1980s no more than 25 percent of any industry was allowed to participate in
research collaboration without facing antitrust suits.[73] Moreover, the automakers had signed a consent decree in 1969 agreeing to cease research collaboration after the Department of Justice brought an antitrust suit arguing
that the automakers were using the joint venture to kill emission controls.
The Reagan administration's suggestion, then, flew in the face of a legal
agreement only twelve years old.

Before Congress, however, Reagan officials were compelled to confess
that Bush's task force included a lawyer who had worked for the auto industry.[74] Before joining the task force, C. Boyden Gray had been a partner at
Wilmer, Cutler & Pickering, home of Lloyd Cutler, who frequently worked
for the Automobile Manufacturers Association, and who had played a role
in weakening the first automotive safety standards in the 1960s and resist-
ing emission control regulations in the 1960s and 1970s. Wilmer, Cutler &
Pickering had not only attempted to fight off the initial 1969 consent decree
but had also lobbied on behalf of the Automobile Manufacturers Association

against an extension of the decree in 1978. Moreover, the Task Force on Regulatory Reform's suggestion to discontinue the consent decree "was made in language nearly identical to that" earlier used by the trade association.[75] Members of Congress suggested that the law firm was using the vice-president's task force to do what it had not been able to achieve through other means. During this period, newspapers also cited auto executives as saying that Secretary of Transportation Lewis had promised a rollback on specific regulations, including passive restraints, *before* regulatory analyses were conducted, making the analyses that eventually emerged seem post hoc justifications for decisions that had been made well before.[76]

In other ways, the Reagan administration's actions never fit the free market ideology it espoused. Import quotas for Japanese cars were perhaps the most obvious example of this discordance. In an attempt to avoid violating free trade agreements, Reagan administration officials met with their Japanese counterparts to work out a "voluntary restraint agreement," which limited imports to 1.68 million cars.[77] In Congress in May 1981, Reagan officials proclaimed that they were working to assist the auto industry from a core "philosophical position, which is basically free trade and the free enterprise system."[78] But Representative Tom Lantos, a Democrat from California, slammed this idea: "It is perfectly obvious that the so-called voluntary restraint by Japan was not anything [like] an action of a free marketplace. . . . We have to sweep away the cloud of mysticism that surrounds much of this dialogue."[79] Over time, free market fundamentalists, including David Stockman, would come to agree that the Reagan administration's vaunted economic philosophy was little more than rhetoric.

Reagan also attacked automotive and other regulations by means of appointments of agency administrators. In many cases, he appointed undertakers—administrators whose chief role was to hamstring or tear apart the very agencies they led. Anne M. Gorsuch, who came to head the Environmental Protection Agency, was a prime example of this trend. (Near the end of her term, Gorsuch married, becoming Anne Gorsuch Burford. To avoid confusion, I refer to her as Gorsuch throughout this discussion.) Having served a term in the Colorado house of representatives, Gorsuch was eminently unqualified for the position of EPA administrator. Upon her arrival at the EPA, she worked hand in hand with Stockman's Office of Management and Budget to systematically deconstruct the agency she led, cutting its budget by 22 percent, downsizing the staff, decreasing regulatory enforcement, and curtailing important regulations, including those affecting pesticides. The last EPA budget under Carter was $1.35 million; by 1984, Gorsuch and Stockman had cut it to $984.6 million.[80] Nearly every program at the agency was affected in one way or another. For EPA staff members, Gorsuch's presence was extremely demoralizing.

Top-tier staff members within the agency went along with Gorsuch's directives, but some lower-level staff members resisted with all their might, beginning a kind of guerilla warfare that included leaking controversial documents to media and environmental groups.[81] One of Gorsuch's methods for bringing the agency to heel was the use of industry-chosen "hit lists" containing the names of career bureaucrats who were to be demoted, fired, or transferred. Gorsuch began carrying out these attacks on individuals, but her methods became controversial when one of the lists was leaked.[82] The leaked lists contained descriptions for each targeted bureaucrat, such as "a Nader on Toxics," "bleeding-heart liberal," and "invidious environmental extremist."[83] Other leaked documents showed that Gorsuch was planning on reducing the 1983 budget to 40 percent of 1981 levels, which would result in a massive—perhaps nearly 50 percent—reduction in force.[84] So much pressure mounted that Gorsuch was compelled to release an open letter promising no further reductions.

The EPA's Office of Mobile Sources in Ann Arbor, Michigan, experienced only mild hardship during Gorsuch's term. While the EPA went through a reduction-in-force, leaders at the Ann Arbor office had to demote a few staff members and move others between divisions, but they did not receive the drubbing that other parts of the EPA did.[85] There were a few exceptions. By the late 1970s, Charles Gray had been put in charge of creating new automotive emission control and fuel regulations in the Office of Mobile Sources.[86] He was known for being a tough, vocal critic of industry. When Reagan came to power, EPA leaders targeted Gray. He moved out of this regulatory position and began his long career of doing R&D at the EPA. In the early 1980s Gray focused primarily on fuels research, particularly the production of methanol, and he cowrote a book on the subject.[87] By the 1990s and the coming of the Clinton administration, Gray had moved on to the so-called Supercar project (described in the following chapter). In general, however, staff members later reported that the Ann Arbor office had the good fortune of being fairly insulated from political upheavals in the EPA during Reagan's first term. The office's geographical isolation from Washington, D.C., might offer one explanation for this experience. Other staff reported that lawmakers also simply liked visiting the Ann Arbor office. The materiality of emissions testing impressed them and gave them a concrete sense that real work was being done there.

Gorsuch's term became increasingly controversial. The storm that ultimately sank her ship involved the fairly young Superfund program, which aimed to clean up hazardous waste sites around the country. Gorsuch and Assistant Administrator Rita Lavelle were charged with colluding with industry and using the funds to influence congressional races and increase the likelihood of Republican victories.[88] A congressional committee requested documents from the program, but Gorsuch moved the documents to the Department of Justice instead and refused to turn them over to Congress. Later investigations, however, found that the Reagan administration was providing these documents to corporate defendants in toxic waste enforcement cases.[89] Congress then subpoenaed Gorsuch, but under advice from the White House, she refused to appear. She became the highest executive branch official to that point in US history to be found in contempt of Congress. The protracted legal squabble that ensued brought bad press both for Gorsuch and for Reagan. Two paper shredders appeared at the EPA—whose leaders claimed that it was destroying "excess copies" of documents.[90]

In 1982 the Meramec River flooded Times Beach, Missouri, causing a hazardous waste spill, apparently as the result of poor oversight by the EPA. When combined with the struggle over Superfund documents, the dioxin disaster in Times Beach made the EPA appear incompetent and possibly corrupt. Reagan fired Lavelle. The Justice Department announced that it would not represent Gorsuch before another contempt of Congress hearing, because its own Federal Bureau of Investigation was beginning to inquire into the Superfund matter. Soon thereafter, Gorsuch resigned when it began to look increasingly likely that Reagan would yield to Congress, release the EPA documents, and enable a firestorm attack on Gorsuch.[91] With Gorsuch gone, Reagan announced that he was bringing back William Ruckelshaus, the first EPA administrator, to his old post to restore the agency's credibility. Ruckelshaus promised, "There will be no hit lists. . . . I intend that EPA will operate forthrightly and honestly."[92] In comparison to Gorsuch's reign, Ruckelshaus's relationship with career civil servants at the agency was "a love fest."[93]

While the Office of Mobile Sources may have experienced relative quiet during the Reagan administration, the ordeal of the National Highway Traffic Safety Administration was diametrically opposite. Reagan appointed Raymond A. Peck Jr. to head the agency. Prior to coming to the NHTSA, he had worked as a lawyer for various companies before becoming the director of regulatory affairs for the National Coal Association, a trade association that lobbied heavily against all forms of regulation that affected the coal industry. Peck brought the auto safety agency's work to a standstill. A staff member later described the agency's rule-making office as a "ghost town." With all rule making halted, staffers spun their wheels, demoralized.

Joan Claybrook and her coauthors, in one of two books about what was happening at the agency after her departure, noted, "Internal reorganizations and re-assignments have resulted in the loss of more than 200 NHTSA staff, a quarter of its workforce."[94] Moreover, in contrast to Claybrook's consumer information campaign, the agency under Peck fell under a "veil of secrecy."[95] Perhaps most important, since federal fuel economy standards were making cars smaller while new regulations died on the vine, cars would be less resistant to crashes without new safety systems to defend them.[96] Probably the most essential standard to fall by the wayside during the Reagan years was that for the side-impact airbag, which had to wait more than a decade to become a reality. The agency's budget was cut from $259 million in 1979 to $211 million when Regan left office in 1989, with the agency's research and development and grant-in-aid programs for states being the most affected.[97]

Yet many agency staff members went along with Peck's programs without protest. By the early 1980s, the makeup of the agency's staff had changed. Haddon, Brenner, and other early members were passionate about the cause of safety. But by the early 1980s many agency officials were "car nuts" more interested in automobiles than in politics.[98] They also did not identify with Joan Claybrook, whom they saw as a "crusader," and they saw the Center for Auto Safety—the main auto safety advocacy organization during this period—as an enemy that attacked them rather than as an ally. Consequently, unlike the staff at EPA and other embattled federal agencies, NHTSA staffers had no desire to leak documents to safety, environmental, or consumer advocates. As a result, external champions of safety and critics of the current status quo were more upset about changes at the federal safety agency than its own staff members were.

The fate of the so-called Research Safety Vehicle (RSV) became emblematic of tensions between safety advocates and the agency during this era. Like the earlier Cornell–Liberty Mutual Safety Car, which had so inspired Ralph Nader, the Research Safety Vehicle was a demonstration project meant to show the state of the art in auto safety equipment.[99] In 1975 the National Highway Traffic Safety Administration contracted for the vehicle's creation with two different groups—Minicars, a safety research firm, and a collaborative research consortium involving Calspan, Chrysler, and Simca.[100] The Research Safety Vehicles were to fit into the changing economic, technological, and environmental context, which included the fuel shortages of the 1970s. The Minicars model aimed to be a so-called S3E car: safe, environment-friendly, efficient, and economical.[101] The research project was completed in 1979. In accord with Claybrook's general enthusiasm for consumer information campaigns, the agency developed a "vigorous program to alert the car-buying public that such injury reducing technologies existed and, through regulatory requirements or public demand in the marketplace, could be incorporated into future new-car designs at reasonable cost."[102] The agency also held press conferences, put the safety cars on public display, and otherwise pressured the automakers into matching the vehicles' safety levels. In 1979 the agency distributed a minute-long public service announcement narrated by actor Lorne Greene to television stations around the country. As Greene intoned, "The RSV—stylish, easy to handle, economical, and absolutely the safest automobile ever created. But there's one slight catch: you can't buy it. Someday soon a car like RSV could be a reality." In other words, the agency followed the script outlined by Nader's 1959 article, "The Safe Car You Can't Buy."

The whole campaign came to a screeching halt, however, when Peck took control of the agency. The RSVs were literally placed out of public view in an "underground government garage."[103] The Lorne Greene public service announcements were withdrawn and replaced by commercials focusing on safe driving. Nader told reporters that Peck was simply doing the bidding of auto firms. "That's his middle name. His name is 'Raymond Give-in-to-Them Peck.'"[104] Peck, for his part, insisted that the RSV was not ready for public consumption and that its benefits had been "overpromoted by its enthusiasts, including Nader." Moreover, Peck claimed, the car drove "like a lawn tractor."[105] He even refused to loan the cars to Nader to film a consumer-oriented commercial. The Reagan administration hoped that, locked away, the RSV would be forgotten.

Yet, while the fury of Claybrook, Nader, and other safety advocates against Reagan's antigovernment obstructionism often hit its mark, it occasionally missed. Claybrook's polemic against the Department of Transportation's pro–seat belt use programs was perhaps the clearest example of safety advocates taking the wrong side. In *Reagan on the Road: The Crash of the U.S. Auto Safety Program*, Claybrook and her coauthors wrote: "The cornerstone of the administration's activities is a $9.6 million safety belt campaign.

SAFETY, ENERGY, ENVIRONMENT AND ECONOMY

Figure 10.2. The 1979 Research Safety Vehicle reflected its time. It included forward-thinking safety devices, such as airbags, but was also fuel efficient, polluted the air less than most cars, and was affordable. Safety and environmental advocates, such as Ralph Nader, saw it as the car of the future and argued that the Reagan administration's decision not to promote the car was a betrayal of public health. Meanwhile, National Highway Traffic Safety Administration administrator Raymond Peck said the car drove "like a tractor." National Highway Traffic Safety Administration, *Research Safety Vehicle, RSV*. Washington, DC: Department of Transportation, 1979.

The agency's 'unofficial' goal is a 20% increase in belt usage. Even if attained it will still leave 70% of all occupants unprotected. . . . Automatic crash protection [such as airbags and other passive restraints] would achieve this goal in about three years and would protect all occupants in 10 years."[106] Claybrook was probably right about the importance of automatic systems like airbags, but she was dead wrong about seat belts. Her bias reflected the belief, going back to the National Highway Safety Bureau's first administrator, William Haddon, that changing technology was more important than changing behavior. Safety advocates saw the focus on seat belts as a reactionary return to the pre-crashworthiness days of driver's education and safety campaigns. Yet one former staff member, who worked for the agency from the 1970s through the 1990s, later observed that, in hindsight, the agency's early, single-minded focus on technological approaches to safety in lieu of any sustained attention to driver behavior was his greatest regret.[107]

During the 1980s the National Highway Traffic Safety Administration increasingly focused on driver behavior. In the early 1980s the agency established a new office, dedicated to behavioral issues including belt use and avoidance of drunk driving. The office had a three-tiered approach: first, in concert with Mothers Against Drunk Driving, the agency pushed for stronger drunk-driving laws, which had an important impact around the nation. Second, the safety agency, in conjunction with private contractors, began public education campaigns for seat belt use and the beloved Vince and Larry crash test dummy public service announcements, which encouraged drivers and passengers to "buckle up." The Vince and Larry commercials precisely inverted the usual meaning of crash manikins: engineers and safety advocates had pushed for crash tests because accidents were ultimately unavoidable and the forces impersonal.[108] The crash test dummy was meant to stand in for everyone. But Vince and Larry—with their tagline, "You can learn a lot from a dummy—

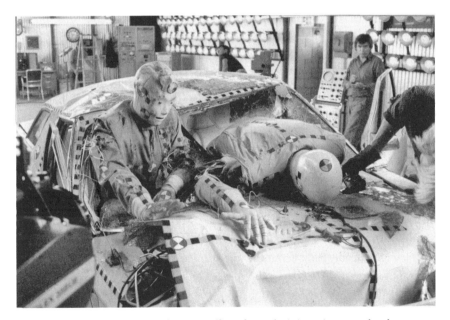

Figure 10.3. The National Highway Traffic Safety Administration's "spokesdummies," Vince and Larry, were the subject of humorous public service announcements in the 1980s and 1990s. With the tagline "You can learn a lot from a dummy—buckle your safety-belt," they fit the era's ethos of once again emphasizing individual responsibility, a message that safety advocates had attacked since the 1950s. Photo from filming of a television commercial, from "Automobile Safety" essay in America on the Move exhibit, Division of Work and Industry, National Museum of American History, Smithsonian Institution

buckle your safety-belt"—involved the remoralization of auto accidents, a return to the pre–World War II message of personal responsibility, which was precisely the idea that safety advocates like William Haddon and Ralph Nader had attacked. Here, "dummy" meant not just manikin but fool.

Third, Elizabeth Dole became the secretary of transportation in 1983. Unlike other Reagan appointees, she was not completely opposed to regulation. She issued a rule requiring a brake light to be placed in all automobiles' rear windows by 1986, an inexpensive solution believed to cut down on rear-end collisions because the higher-positioned light aligned with the driver's line of sight in the car behind.[109] Her greatest innovation, however, was creating an ultimatum for the automobile industry. As she later wrote, "Under the plan we devised, unless by 1986 two thirds of the population was covered by state safety-belt laws that met federal criteria, manufacturers would have to install automatic passenger restraints of their choice in at least 10 percent of all '87-model cars. This figure would rise in successive years until it reached 100 percent by model-year 1990."[110] If the states did pass the laws, the passive restraint rule would be rescinded. This policy effectively made passing mandatory seat belt laws a private-public endeavor. The Department of Transportation promised $40 million toward seat belt education and lobbying for mandatory state seat belt laws; the automakers, which did not desire mandated passive restraints, pledged millions more. Educational kits that explained the benefits of seat belt use soon began showing up at schools and community centers.

In 1984, even before Dole enunciated the policy, New York State created a mandatory seat belt law, becoming the first state to do so.[111] By 1985, nine more states had followed, and by the following year, the total number of states with seat belt laws reached twenty-three. By 1990, thirty-seven states and the District of Columbia had such laws. While only around 11 percent of people had used seat belts in 1980, belt use was hovering around 50 percent by 1990. The rule also encouraged automakers to put passive restraints into vehicles. By 1989 a quarter of new cars had passive restraints.[112] Perhaps Dole was being too optimistic when she later wrote, "Detroit is discovering what we've been preaching for years: safety sells." Still, although Joan Claybrook and other safety advocates had disparaged NHTSA's move toward behavioral interventions, especially its focus on seat belts, these efforts had huge beneficial effects.

In 1989 the highly respected public health expert John D. Graham wrote, "By itself [the federal auto safety agency] did not save the airbag technology.

A new generation of progressive executives at Ford Motor Company followed the European lead of Mercedes-Benz and turned to the airbag in their search for long-run competitive advantage. . . . At Ford, airbags . . . are now elements of long-run corporate strategy to build Ford's reputation as the manufacturer that supplies a quality product that customers learn to appreciate."[113] Graham's point is that, ultimately, private industry and the "market" for safety succeeded in getting airbags into cars, whereas government had failed. But Graham goes too far. This "market" was not virgin land. NHTSA had been trying for several years to build safety-consciousness through seat belt laws, public information campaigns, and the Vince and Larry public service announcements.[114] Historians of business and advertising are always cautious to move from "prescriptive" sources, including advertisements, etiquette books, pamphlets that producers send to consumers in order to inform them, and the like, to descriptions of behavior—and for good reason. We have very little idea how these instruments of suggestion, prescription, and instruction actually affect people's actions, at least at the societal, "in-the-wild" level.[115] Moreover, as we have seen, Elizabeth Dole's initiatives dramatically increased seat belt use and likely played some role in creating a safety mindset around automobiles that eventually led to consumer demand for safety technologies. In sum, we can certainly suggest NHTSA's campaigns played some role in generating consumer demand—enough of a role that the rise of interest in the airbag is more complex than the efforts of a few "progressive executives."

In many ways, the "Reagan Revolution" never was. In 1981 Ronald Reagan took David Stockman "to the woodshed" for being far too honest with a journalist about the administration's lack of progress. He resigned in 1985. A year later, Stockman published his memoirs, titled *The Triumph of Politics*, in which he argued that Washington, D.C.–based political wrangling and Reagan's ambition to succeed in that context had killed the administration's ideals about cutting government spending.[116] Reagan had run a campaign pledging to destroy the Departments of Energy and Education, but when he left office, they were still standing. Stockman wasn't the only one who believed that Reagan had promised more than he delivered. In 1984 an observer noted that the Reagan administration had failed to reform the Clean Air Act or fuel economy standards. Moreover, it had made international trade decisions, such as curbing steel imports, which raised the costs of auto production, and it had increased the gasoline tax, which likely cut into industry profits. The observer concluded, "The Reagan administration's policy toward the auto industry has been far stronger in its stated intentions than in its actual performance."[117] The Reagan administration had halted or killed many regulations, from ozone restrictions to side-impact crash standards. Yet federal agencies instituted these rules in following years. In other words, Reagan's specific interventions in regulatory rules often only postponed them rather than destroying them altogether. As we will see, regulation increasingly became a partisan issue after Reagan's presidency, and the habit of one president undoing the previous president's regulatory decisions became something of a tradition.

The Reagan administration made its biggest and longest-lasting impact by transforming federal rule making to include regulatory review and cost-benefit analysis. The tradition of regulatory review began in the 1960s, but Reagan significantly increased the power of reviewers within the Office of Management and Budget. Later presidents, especially George W. Bush, would strengthen Reagan's regulatory review rules even further. Building the thoughts and tools of economists into federal rule making fundamentally altered the playing field of federal auto regulation. If engineers, doctors, chemists, and lawyers had shaped automotive use and design before 1975, economists increasingly came to play that role thereafter. Economists and their disciples were the new community reconfiguring automotive governance. Their influence was decisive. In time, even left-leaning proponents of environmental and consumer regulations came to frame their arguments first and foremost in the language of cost-benefit analysis. As the following chapter shows, the outcome of these changes has been that it is *much* harder to create new auto regulations, especially when federal agencies face serious resource constraints in terms of both staff and budget. As a result, federal regulation of the automobile since the 1980s has been, at best, uneven.

CHAPTER ELEVEN

Indecision, Regulatory Uncertainty, and the Politics of Partisanship

On January 9, 2002, Spencer Abraham, the US secretary of energy under President George W. Bush, stood at the Detroit auto show with executives from the Big Three and announced a new research program. Called Freedom-CAR, the program would focus on producing hydrogen fuel cell vehicles that would reduce pollution and end the country's dependence on foreign oil. At the announcement Abraham promised, "The gas-guzzler will be a thing of the past."[1] Environmentalists criticized the program, which they saw as a way for automakers to put off near-term improvements in fuel economy, while hydrogen proponents celebrated it, though even they were forced to admit that viable, mass-produced hydrogen technologies were at least decades off. Seven years later, President Barack Obama's secretary of energy, Steven Chu, effectively killed the FreedomCAR program.[2] Chu joked to the *MIT Technology Review* that hydrogen vehicles would require at least four miracles to be viable, whereas human individuals needed to perform only three miracles to become Catholic saints.[3] In other words, it was a technological long shot, and the wiser course was focusing on near-term solutions.

In many ways, these two decisions—one making, the other unmaking—can be taken as a guiding metaphor for automobile regulation in the years following Ronald Reagan's presidency. Reagan focused on beating back regulations and the bureaucrats who created and enforced them. Since 1989, some presidents, mostly Republicans, have shared Reagan's beliefs; others, mostly Democrats, haven't. Some have sought to create new regulations to shape the future of the automobile industry; others have worked to undo those very decisions, arguing that they place unnecessary burdens on business. The open question and point of disagreement between these two visions is how the federal government should use its organizational capabilities and the experts it employs to influence the future development of technology. As shown throughout this book, such regulatory action also guides what problems other experts, particularly those employed in industry, take up and work to solve. The overall effect of this back-and-forth and hemming and hawing between presidential administrations is one of indecision. The work of government and the industries it regulates has lacked definitive direction over the past thirty years.

Growing polarization between the two major US political parties is a primary cause of this wavering and indecision.[4] Since the 1970s, elected Republicans have become more conservative, elected Democrats have become more liberal, and there are fewer elected moderates in either party.[5]

Scholars have laid out many opposing theories to explain this growing divide. One simple and plausible explanation is that the political parties have become better "sorted"—that is, where both parties once had liberal and conservative wings, liberals are now Democrats, and conservatives now Republicans.[6] One result is that the parties' platforms have simplified, and party leaders have standard positions on a wide variety of topics, including abortion, guns, welfare policy, taxation, budget deficits—and regulation.

This growing divide has had significant consequences for regulation. Democrat president Lyndon Johnson created the Department of Transportation, and Republican president Richard Nixon formed the Environmental Protection Agency. Since the 1980s, however, Republicans have increasingly adopted the economic philosophy that undergirded Ronald Reagan's presidency: they see regulation and government action as a problem that impedes business and economic growth. Conversely, Democrats typically have a more positive view of regulation and government more generally. This partisan divide affects not only philosophies of economics and governance but also what pieces of information individuals accept as facts. Most important for automobile regulation, party association is strongly correlated with whether an individual believes human behavior causes global climate change. Concerning global climate change, 79 percent of liberal Democrats believe that humans are causing it, while only 15 percent of conservative Republicans do.[7]

One major consequence of this divide and the back-and-forth between presidencies is what scholars call *regulatory uncertainty*, a term that emerged in the late 1960s and early 1970s to denote the unknowns that those affected by regulation face during periods of regulatory flux. The period was marked by other intense uncertainties as well: scientific; economic, including both gasoline prices and the state of the auto industry itself; and cultural, including worries that machines, such as self-driving cars, would replace humans in many domains. Taken together, uncertainties presented real problems for groups and individuals—including policymakers, the auto manufacturers, other companies and entrepreneurs, safety advocates and environmentalists, and ordinary consumers—who often did not know how to make choices given so many fundamental unknowns. Large industrial firms such as automobile manufacturers work on multiyear planning cycles, and for the past thirty years, they have not been able to predict which regulations will be in force after the next presidential election, never more than four years out. The fact that both political parties have not come to accept human-caused global climate change as a reality, for instance, means that the presence and stability of climate policy are always uncertain.

Government has played some role in forming expert communities during this period, but the work of these communities has not often been translated into policy. When it comes to generating significant technological change, especially that which lies outside the current grasp of industry, the regulations that have emerged during this time, especially safety standards, have been weak. Regulations have been reactive, mandating the use of already-available, widely deployed technologies, rather than proactively setting the agenda.[8]

To examine this period of indecision and uncertainty, this chapter proceeds through three sections. The first examines the uneven development of auto regulation under each of the American presidents since Reagan. Increasingly, particularly since the year 2000, party polarization has shaped automobile and other forms of regulation, with one president after another nullifying his predecessor's decisions. The second section describes the reemergence of industry research collaborations—known as consortia—in the 1990s and explains why US presidents formed these collaborations. The chief finding here is that, in the context of automobiles at least, government-sponsored research consortia have not been nearly so effective as regulation in generating technological change that affects citizens' everyday lives. The final section focuses on the rise of automotive electronics and computing technologies, the most significant changes in automotive technology since the 1980s. By focusing on these technologies, including autonomous, or self-driving, vehicles, we can examine the uneven mixture of corporate strategy and government action that so defined this period.

Regulation after Reagan

From the late 1980s to the present the fate of auto regulation became directly tied, in ways not seen before, to which party held the White House. There were two reasons for this development. First, the US Congress did not pass many new regulatory laws affecting the automobile during this period. This inaction left the creation of new rules and codes up to executive branch agencies and, therefore, ultimately up to the discretion of presidents. Second, Democrats and Republicans became deeply divided during this time. While claims that this is the most partisan era in history are overblown, there are real differences, including the parties' philosophies about government regulation. In part, this divide is a holdover from the presidency of Ronald Reagan, who claimed, "Government is not the solution to our problem. Government *is* the problem." In the years that followed, many Republican politicians took Reagan as a model, believing that he had ushered their party into a new age, and so they—and their wealthy donors—resisted government regulation wherever they could.

In some ways, however, the Republican president who directly followed Reagan did not fit these trends. In June 1989 newly elected president George H. W. Bush announced plans to revise the Clean Air Act, the first such change since the mid-1970s. The Reagan administration had attacked many federal environmental regulations and, for the most part, did not introduce new ones. Bush's air pollution plan was designed to combat urban air pollution and smog, which continued to plague American cities even though significant strides had been made in curbing pollution from automobiles and other sources. Furthermore, environmental concerns in the 1980s had come to focus on "acid rain," a form of pollution that damaged animal life, forests, and buildings.

The most innovative policy in the 1990 Clean Air Act Amendments was the cap-and-trade system, whereby the sulfur dioxide emissions from power plants that were causing acid rain would be curbed by creating a "market" for pollution.[9] The portions of the Clean Air Act Amendments that dealt with automobiles were not based on cap-and-trade but rather in far more traditional pollution control efforts. As Congressman Henry Waxman outlined in a law review article, the new automotive emissions regulations had four elements: "tighter controls on conventional vehicles; requirements to clean up gasoline and diesel fuels; new programs to promote special 'clean-fuel' vehicles; and a first-time program to control emissions from nonroad vehicles such as trains, ships, and construction equipment."[10] The tighter controls included an initial 35 percent further reduction (from 1990 levels) of hydrocarbon emissions, a 60 percent reduction in nitrogen oxide by the mid-1990s, and an additional further 50 percent reduction from those levels by 2003. The law also controlled evaporative emissions—which were especially prevalent on hot days when gasoline would evaporate in cars' fuel containers—and mandated the use of so-called California-type fuel nozzles, which captured evaporative emissions when users refueled at gas stations.

Other parts of the law relied on even more traditional pollution control, namely reforming fuel. Beginning with a successful antismoke campaign in St. Louis, municipal clean air efforts of the 1940s and 1950s had centered on the specification of clean coal types, before such efforts were found not to work on the photochemical smog plaguing Los Angeles (see chapter 3). The most important fuel reformation in the history of automotive regulation had come in the 1970s when the Environmental Protection Agency banned leaded gasoline because it poisoned catalytic converters (see chapter 8). But otherwise regulators had not focused on gasoline when it came to automotive pollution control. The initial Bush plan would have established 100 percent methanol as a standard fuel type (known as M-100) throughout the United States and called for auto producers to produce a million methanol-using vehicles (see chapter 3). Through political wrangling and petroleum industry lobbying, however, the final law set a far less ambitious agenda. The new methanol fuel was reformulated as M-85 (85 percent methanol, 15 percent gasoline), and the law contained no language pushing the auto industry to produce vehicles that used such fuel (see chapter 8). The resulting legislation was toothless, and it did not move automotive technology in any discernible direction.

While parts of the auto industry seemed to recover from its economic low point in the late 1970s and early 1980s, struggles returned in 1990 and 1991, when the companies recorded losses.[11] Journalists chronicled the automakers' constantly changing economic fate, writing books with positive titles like *Reinventing the Wheels: Ford's Spectacular Comeback* and negative ones such as *Rude Awakening: The Rise, Fall, and Struggle for Recovery of General Motors*. The optimistic books lauded auto company executives for their acumen, evident, for example, in their having made a "bold departure from Detroit's stodgy approach to carmaking."[12] The pessimistic ones shamed executives for "hubris, denial, missed opportunities, and self-inflicted wounds."[13] Few of the complimentary volumes or the jeremiads captured the underlying technological and economic structures that led to these boom-and-bust cycles.

Attention to the auto companies' economic health played perhaps the largest role in debates about fuel economy. In 1989, prompted both by ever-increasing importation of foreign oil and concerns about global warming, members of Congress introduced at least six different bills to increase automotive fuel economy.[14] Bush, however, did not favor raising fuel economy standards. John Dingell and his allies were able to halt progress in Congress, and the Bush administration, which could have issued tougher fuel economy rules through the National Highway Traffic Safety Administration, did not do so. Tightening fuel economy threatened the economic well-being of the auto companies, Dingell always insisted. When the United States entered the First Gulf War in August 1990, gas prices spiked, and fuel economy advocates hoped to use the opportunity to pass a law. But these efforts also proved fruitless. The automakers spent millions on lobbying, and the United Auto Workers, which between workers, retirees, and their spouses included millions of people, also opposed the measures.

Furthermore, the automakers mounted a highly successful "astroturf-ing" campaign—in which corporations seek to create the appearance of a grassroots movement.[15] They formed the Coalition for Vehicle Choice, an organization that used newspaper editorials and advertisements in other media to argue that increased fuel economy would limit the kinds of vehicles available on the market, thereby impinging consumer sovereignty. The coalition emphasized its connections to the American Farm Bureau Federation, the Livestock Marketing Association, and the International Professional Rodeo Association, playing up the Jeffersonian ideal of the independent farmer. But it obscured the role of its primary financial backers, the Detroit auto firms. Astroturfing became a common tool in the automakers' quest to resist regulations of all types.

In some ways, the path forward around regulations simply became less clear during this time. As one reason, the goals of regulations often competed—solving one problem created another. Social scientists called this a risk-risk trade-off, and they began to explore the phenomenon in earnest. The first volume dedicated to the topic was published in 1995.[16] One leader in this field was John Graham, a professor and the director of the Center for Risk Analysis at Harvard University's School of Public Health. Graham would go on to become the federal regulation czar during the George W. Bush administration, heading the Office of Information and Regulatory Affairs within the Office of Management and Budget. Graham's early work focused on auto regulation, as did his thinking about risk-risk trade-offs.

Graham's most influential considerations of risk-risk trade-offs zeroed in on the safety risks exacerbated by fuel economy standards.[17] Such standards reduced gasoline consumption less than expected, he argued, both because they encouraged auto users to do more driving and because they discouraged new car sales by leading manufacturers to produce automobiles that American consumers did not want. Moreover, Graham argued, fuel economy standards seemed to contribute less to national security than they were meant to, and they apparently helped the economy and environment barely at all.

On top of all of these perceived problems, fuel economy standards made automobiles less safe, mostly by decreasing their size and weight. Smaller, fuel-efficient cars fared badly in accidents when they collided with larger, heavier vehicles. "The downsizing of the new car fleet" since fuel economy standards were created in the 1970s, Graham argued, "has been associated with increases in occupant injury and fatality risk, compared with what would have occurred if cars had not been downsized."[18] Graham was forced to admit that traffic fatalities had continued to fall during this period, especially because of seat belt and drunk-driving laws. "But highway death rates would most likely have fallen even further" without fuel economy standards, he insisted. Safety advocates never bought into this reasoning, and Joan Claybrook publicly criticized the fuel economy–safety connection.[19] Small cars could be designed that would pass safety tests with flying colors. And if the problem was how they fared when colliding with larger vehicles, the goal should be reducing the number of large vehicles on the road. Yet risk-risk trade-offs remained a real and unavoidable problem. They held true not only for fuel economy standards but for other environmental and safety regulations concerning the automobile, as well. For example, controlling smog-producing chemicals such as nitrogen oxide increased output of carbon dioxide (CO_2)—the primary greenhouse gas emission.[20] Often regulations were inextricably entangled, making any simple, clear decision impossible.

While Washington remained gridlocked over fuel economy standards, the fuel economy of the average vehicle on the road actually became worse during this period. The downward trajectory was due in large part to the emergence of a new vehicle type, the sport utility vehicle, or SUV.[21] Jeep introduced the first SUV in 1984, the Cherokee, which put an automobile-like body and interior on a truck chassis. Jeep marketed the vehicle to urban families as a substitute for station wagons, but the Cherokee's four-wheel drive also promised potential outdoor adventure. Yet the Cherokee was not fundamentally so different from Jeep's other products as the company claimed. The real sign of things to come was the Ford Explorer, introduced in 1990.[22] Within a few years, nearly every auto company was producing an SUV.

SUVs took advantage of a well-known regulatory loophole: because legislators and regulators in the 1970s did not want to put American Motors Corporation, whose only profit-making vehicle was the Jeep, out of business, the Clean Air Act Amendments and fuel economy regulations contained an exemption for light trucks. On the one hand, policymakers wanted to keep American Motors afloat both because it meant jobs and because they had a fundamental antitrust faith that competition was better for consumers. On the other hand, the fact that Jeep produced what many saw as poorly performing vehicles was an argument in favor of the idea that government protection of firms leads to shoddiness. Regardless, the automakers defined SUVs as light trucks, which meant that they were not covered by air pollution or fuel efficiency standards. Increasingly, the SUV became a central part of the automakers' business strategies. They could load up SUVs with all kinds of amenities and sell them as high-priced luxury vehicles.

An iconic moment of the SUV era came when the body builder and film actor Arnold Schwarzenegger bought the first street-legal Humvee, an enormous military vehicle, from a military contractor in Indiana. Schwarzenegger had seen a convoy of the vehicles drive through an Oregon town when he was filming *Kindergarten Cop* in 1989. "Look at those deltoids. Look at those calves," he later said of the vehicles.[23] Soon, he flew to Indiana to meet the contractor's executives, who did not initially want to sell the Humvees to civilians. Schwarzenegger insisted and soon got his way. In 1998 General Motors bought the rights to manufacture a civilian version of the Humvee, which the company called the Hummer.

SUVs were a step backward on nearly every aspect of automobiles that fell under federal regulation in the United States. They consumed more gas and produced more air pollution than smaller cars, in direct contradiction to SUV ad campaigns, many of which cast SUVs as vehicles for nature lovers. They were less safe, both because they were prone to rollovers and because their raised height made them more likely to kill people in collisions with lower-lying automobiles. Again, the danger of SUVs contradicted advertisements that pictured them as bastions of safety and security. Yet the vehicles were hugely appealing to baby boomer parents, especially mothers.[24] While the dangers of SUVs and the regulatory loophole that enabled them were well known and were discussed regularly in major media outlets, federal regulators made few moves to address them.[25]

At the same time, the Bush administration enabled the National Highway Traffic Safety Administration to set rules that had been held up since Reagan took office. The clearest example was the agency's 1990 revision of side-impact standards.[26] The revision added a dynamic, crash test element to a 1973 standard that had applied only static pressure to the side of cars. By the late 1970s, however, regulators knew that side impacts were a significant cause of deaths and injuries, and the agency's administrator, Joan Claybrook, initiated plans for the dynamic test. The Reagan administration killed those plans. Between the late 1970s and the 1990 revision, thousands were killed or injured in side-impact collisions. Throughout the 1990s, then, the National Highway Traffic Safety Administration issued a number of new rules and revised old ones that significantly reduced in-vehicle hazards, though they were rarely tough enough to appease safety advocates such as Ralph Nader and Joan Claybrook. One technological outcome of these new and revised standards—including the side-impact standard and a revision of Standard 201 aimed at making "pillars, side rails, roof headers, and the roof" less injurious during accidents—was that auto designers added an increasing number of airbags to cars.[27] By the mid-2000s, auto owners frequently joked about the many airbags that filled their vehicles.

Initially, environmentalists and safety advocates were elated about the election of Bill Clinton in 1992, which ended twelve years of Republican rule. Clinton promised to increase fuel economy standards to 40 miles per gallon, and his vice-president, Al Gore, well known in environmental circles, envisioned doing away with internal combustion engines altogether.[28] Within months of being sworn in, however, Clinton revised his plans and fuel economy fell by the wayside. The Clinton EPA also announced tough new ozone standards, but through an unfavorable court decision and intense industry lobbying efforts—led again by C. Boyden Gray of the law firm Wilmer, Cutler & Pickering—the ozone standards were considerably weakened.[29] As usual, Democratic representative from Michigan John Dingell stood with the auto industry against his party colleagues. Longtime environmental advocates, including Edmund Muskie, the father of the 1970 Clean Air Act Amendments, went after the federal government for being too weak. "The promise of clean air has been undermined by bureaucratic rigidity at EPA," Muskie said.[30] Ralph Nader and other safety advocates also harshly criticized the National Highway Traffic Safety Administration's lack of progress during the Clinton presidency. Those standards that were passed did little more than enshrine current industry practice, continuing a long national tradition of weak auto safety standards.

As discussed below, rather than further regulating the industry, Clinton and Gore put most of their energy into creating a next-generation vehicle through a public-private research venture. Rhetorically, Clinton and, especially, Gore were the most openly environmentalist leaders in the White House since the 1970s, but like Reagan and other Republicans, they favored nonregulatory means of achieving social goals. For the time being at least, regulation simply did not fit the reigning political philosophy in Washington, D.C.

Aversion to regulation was an even stronger characteristic of the president who followed Clinton. In campaign speeches and presidential debates, George W. Bush idealized Ronald Reagan's presidency more than his own father's. He criticized "big government" and favored deregulation. Both Bush and his vice-president, Richard Cheney, had strong ties to the petroleum industry, and after Bush beat Gore in the contested election of 2000, Bush and Cheney set about creating energy and environmental policies that fit those interests. In January 2001, the month he took office, Bush directed Cheney to form a task force and "recommend a new national energy strategy."[31] In April, Cheney laid out the principles of the report he would produce. Cheney placed greatest emphasis on the need to increase domestic energy production, including mining more coal, drilling in Alaska's Arctic National Wildlife Refuge, building new oil and natural gas pipelines, and constructing more nuclear power plants. "Coal is not the cleanest source of energy, and we must support efforts to improve clean-coal technology to soften its impact on the environment," Cheney said.[32] The phrase *clean-coal technology* was a product of the Reagan years that had fallen into disfavor during the Clinton administration but came back with a vengeance during the George W. Bush administration, which argued repeatedly that there were ways of making coal environmentally friendly.

In his speech, Cheney also mocked Jimmy Carter's "crisis of confidence" speech and derided the political instinct of the 1970s that suggested "we've got to 'do more with less.'"[33] "Conservation may be a sign of personal virtue," he asserted, "but it is not a sufficient basis for a sound, comprehensive energy policy." Ralph Nader and environmental groups such as the National Resources Defense Council attacked Cheney's plan.[34] (During the Bush years, the National Resources Defense Council kept a tally of hundreds of administration policies and decisions that negatively affected the environment.[35]) In an editorial titled "Cheney Tells US to Carry on Guzzling," the left-leaning *Guardian* newspaper argued that the administration's view was that "conservation was for wimps" and that US citizens could go on using "the world's energy resources as if there was no tomorrow."[36] Cheney's only mention of climate change–inducing greenhouse gases came in a passage pushing nuclear power. In May 2002 the Environmental Protection Agency published a report, which had been vetted by the White House and several other executive branch agencies, claiming that humans were responsible for global climate change. White House spokesman Ari Fleischer told journalists that the report admitted that "there is 'considerable uncertainty'" about "the science of climate change" and that "'definitive prediction of potential outcomes" from climate change "is not yet feasible."[37] Bush told reporters, "I read the report put out by the bureaucracy."

The terrorist attacks of September 11, 2001, briefly pushed up oil and gas prices, but greatest influence that the ensuing "War on Terror" had on automotive regulation and many other policy areas was that it pushed domestic affairs to the side, just as other wars had done. Progressive critics argued that the nation's reliance on foreign oil was partly to blame for the attacks and many other foreign policy problems the country faced. For example, Osama bin Laden was enraged when the United States stationed troops in Saudi Arabia during the First Gulf War of 1990–1991, which George H. W. Bush had engaged in partly out of fears that Saddam Hussein's invasion of Kuwait would cut off oil supplies.[38] Furthermore, after the Saudi-led Organization of the Petroleum Exporting Countries (OPEC) announced that it would keep oil prices stable after the 9/11 attacks, the Bush administration chose not to emphasize "evidence that Saudi citizens finance Osama bin Laden's Al Qaeda network and other radical Islamic organizations."[39] After the Bush administration ordered the invasion of Iraq in 2003, arguments only grew louder that the relationship between the United States and Middle Eastern countries with oil resources distorted American foreign policy and increased problems. But such arguments found little resonance with the broader American public, nor did they convince the Bush administration to change its energy policies.

Other unforeseen events also shaped discussions and debates about automobiles. In 2005 Hurricane Katrina devastated New Orleans and damaged many Gulf Coast oil refineries. For this and other reasons, including the rising price of crude and seasonal demand, gasoline prices skyrocketed in the summer of 2006.[40] Prices in general prices fluctuated widely from the 1980s to the 2010s. Consumer desires, policymaker instincts, and auto industry efforts followed these changes. One outcome of high gas prices during the mid-2000s was increasing consumer interest in hybrid and plug-in electric vehicles. Toyota had introduced its hybrid car, the Prius, for sale internationally in 2000, and the car had always been a favorite with environmentalists. But high fuel prices made the vehicles desirable even to individuals who were not passionate about the environment. The Prius led many observers to assert that Asian car manufacturers were continuing to beat American automakers on key measures, including new, innovative products.

High fuel prices and the destruction wrought by Hurricane Katrina went hand in hand with renewed attention to climate change. Al Gore released his book and film, *An Inconvenient Truth*, on the subject in 2006, to both fanfare and scorn.[41] The film sought to demonstrate the reality and threat of climate change to a general audience. Al Gore won the Nobel Peace Prize for *An Inconvenient Truth*, but in many ways he was the worst imaginable spokesperson for global climate change. Conservatives had loathed Gore for years. During the Clinton presidency the conservative pundit Rush Limbaugh called him "Algore," a reference to Victor Frankenstein's lab assistant. Gore was Clinton's lackey, Limbaugh implied. In his book *See, I Told You So*, Limbaugh wrote, "*Algore* is a bona fide tree-hugging, spotted owl–loving, snail-darter-protecting, Gaia-worshipping, radical doomsday prophet who carries water for Greenpeace, the Sierra Club, and every other powerful environmental fringe lobby."[42] Climate change ever-increasingly became a partisan issue, reflected not just in the positions of politicians but also in their constituents' differing scientific beliefs. Corporations, particularly energy companies, led a multimillion-dollar misinformation campaign to persuade citizens that worries about climate change were overblown and were possibly the result of an outright hoax propagated by self-interested scientists looking for more federal research dollars.[43]

Certainly, neither *An Inconvenient Truth* nor any of the other, increasingly despairing jeremiads about climate change led George W. Bush to change course. Indeed, in 2002 the state of California passed a law limiting greenhouse gas emissions as a form of air pollution.[44] Federal law enabled California to create standards that were tougher than federal ones, and traditionally, California decisions had opened a pathway to federal laws and codes. Where California led, the federal government followed. To give but one example, California regulated evaporative auto emissions before federal law did. Yet Bush countered the California greenhouse gas rule by ordering his EPA administrator, Stephen L. Johnson, to deny the state's petition to tighten emission restrictions, arguing that a pending revision of fuel economy standards would address the problem. By not allowing California to enforce its rule, Bush thereby ended a legal tradition that went back to the 1960s, at least temporarily.

Barack Obama ran his first campaign as an anti-Bush candidate. He criticized the war in Iraq and promised action on climate change. Yet, once in office, he inherited the worst economic recession since the Great Depression, a downturn that hit the auto industry perhaps harder than any other. Both General Motors and Chrysler required bailouts, and Detroit became a national, perhaps even global symbol of deindustrialization, as its decaying buildings became the subject of so-called ruin porn. Unlike his predecessors, however, Obama did not see the automakers' financial plight as a reason for avoiding regulation. He frequently argued that regulation induced innovation rather than hindering it. In 2011 he issued an executive order further refining regulatory review. One of the order's subsections made innovation the primary goal of regulation: "Each agency shall also seek to identify, as appropriate, means to achieve regulatory goals that are designed to promote innovation."[45] In 2012 he set fuel economy standards to rise from 35.5 miles per gallon to 54.5 MPG by 2025.[46]

But many of Obama's larger hopes and goals were frustrated. Observers claimed, perhaps with hyperbole, that no other US president had ever faced as much organized opposition as Obama did, both from the conservative political movement known as the Tea Party and from the corporations that supported that movement.[47] Frustrated with congressional inaction, Obama announced in August 2015 that he was issuing a set of rules to be known as the Clean Power Plan in order to limit greenhouse gas emissions.[48] Industry quickly organized against the Clean Power Plan, filing suit against the Obama administration in federal court for overstepping its constitutional authority.

In general, auto regulation during the Obama administration was wavering and lacked any overarching sense of direction. The National Highway Traffic Safety Administration announced that it was exploring some new rules—some of which are described below—but seemed to be without any grander vision. Journalists, policymakers, and safety advocates increasingly focused on "distracted driving," as drivers increasingly paid more attention to their cell phones than to the road. Washington State passed the first law banning texting while driving in 2007. Groups around the country put up billboards with slogans like "Don't Drive InTEXTicated," and the famed director Werner Herzog made a 35-minute film, "From One Second to the Next," about people whose lives had been affected by accidents attributed to texting while driving. The US Department of Transportation held conferences on the subject beginning in 2011.[49] Engineers proposed technological solutions to distracted driving, such as devices that would detect if drivers were using phones, but policies focused on policing and public awareness campaigns.[50]

In the previous chapter, I argue that the most influential outcome of the Reagan administration affecting auto regulation was strengthening the place of cost-benefit analysis in the regulatory process via the Office of Management and Budget. Some developments concerning auto regulation—and in many other regulatory domains—have thrown into doubt whether cost-benefit analysis is always useful and, in particular, whether it invariably reflects the actual lived values of citizens. For example, safety advocates in the mid-2000s began pushing for the installation of rear-view, or backup, cameras in automobiles, mainly to prevent children from being run over. The lead organization pushing for this standard was KidsandCars.org, founded by Janette Fennell.[51] In 1995 Fennell and her family were locked inside the trunk of their car by a kidnapper and were unable to escape the trunk because it had no internal release mechanism. After the family finally escaped, Fennell became an advocate for auto safety, convincing regulatory authorities to require internal trunk releases in all cars.

But Fennell went on to advocate other safety technologies, particularly ones aimed at saving the lives of children. KidsandCars.org pushed the National Highway Traffic Safety Administration to require rear-view cameras in cars, but these efforts encountered a hurdle: in cost-benefit analyses the collective cost of backup cameras outweighed the collective benefits in terms of injuries and deaths prevented. To circumvent this barrier, KidsandCars.org formed a coalition, including parents whose children had died in backup accidents; the group pressured Congress to pass a law requiring the cameras. A law mandating the technology would render executive branch rule making moot. As a result, the Cameron Gulbransen Kids Transportation Safety Act passed in 2008 with more than eighty cosponsors. (The law was named after a two-year-old boy who was accidentally killed in 2002 when his father backed over him in the driveway of their family home.)[52] The act directed the National Highway Traffic Safety Administration to create a backup camera rule by 2011.[53] The NHTSA dragged its feet, however, and released the rule only in 2013, mandating that cameras be put in all new cars by 2018 after KidsandCars.org, Public Citizen, and other organizations sued the agency for failing to comply with the law. Since that time, KidsandCars.org has continued advocating for automotive safety technologies that would save children's lives, including alarms that would sound if children were—accidentally or purposefully—left alone, particularly in hot cars on summer days. Central to the organization's arguments is that cost-benefit analysis often does not reflect our actual moral and political values, a point its leaders believe is evidenced by the fact that the Gulbransen Act flew through both houses of Congress with broad, bipartisan support.

Donald Trump ran a presidential campaign promising to "drain the swamp," a phrase that Ronald Reagan and others had used to describe paring back federal bureaucracy. As president, Trump appointed individuals with strong industry backgrounds and connections to head regulatory agencies, and nonscientists to scientific positions. To head the EPA, he appointed Scott Pruitt, an Oklahoma lawyer (and at the time, the state's attorney general) with a long history of suing the agency and calling for a significant weakening of its powers.[54] Many saw Trump as actively opposed to contemporary science—particularly the notion of human-influenced climate change—and the authority of experts. Activists led the so-called March for Science on April 22, 2017, to protest Trump's science and regulatory policies and stances, with roughly 1.1 million people worldwide taking part in the march or related activities that day.[55] Trump made cutting red tape and diminishing regulation significant themes of his presidency. On December 14, 2017, Trump stood next to two stacks of paper—one giant, about as tall as the six foot, two Trump and much wider; the other considerably smaller. The larger pile represented all of the federal regulations to date; the smaller one, federal regulations in 1960. Trump promised to scale back regulations to pre-1960 levels, saying he could achieve that goal "fairly quickly." In its first year, the Trump administration announced that it would repeal Obama's Clean Power Plan, consider freezing (rather than raising) fuel economy standards, and possibly kill an Obama-era auto safety mandate for vehicle-to-vehicle (V2V) communications systems (discussed below).[56] The regulatory pendulum of presidencies kept on swinging.

To summarize, auto regulation since the 1980s has developed at an uneven pace. These developments have depended largely on the priorities and values of sitting presidents because most new regulations are made through executive branch rule making rather than congressional lawmaking. Growing polarization between the US political parties over such issues as regulation has created a back-and-forth motion: the overall agenda for automobile regulation is less clear than it was in the 1960s and 1970s. Moreover, most of the new rules created since the 1980s have been extremely weak in the sense that they have done little more than codify current industry practices and have not forced new research and development. Finally, federal regulators have built into their processes quantitative methods meant to aid decision making, such as cost-benefit analysis. But these tools are at best imperfect, and even staunch conservatives believe they occasionally lead to perverse results. No simple solution or cure-all method awaits to resolve this complex and contested domain. In this anti-regulation climate, US presidents and policymakers increasingly turn to federal R&D funding meant to spur "innovation."

The Rise of Research Consortia

The period from 1990 to the present featured the reemergence of industry research collaborations, known as research consortia, in the automotive sector. These consortia did not involve coercion: they did not force automakers to adopt the technologies that resulted from the research. Rather, these organizations reflected the belief, increasingly dominant during this period, that a major goal of governance should be to foster "innovation," or the creation and diffusion of new technical knowledge, by funding research and development. As we saw in chapter 7, the automakers had signed a consent decree with the Justice Department in 1969 banning research collaboration on emission controls, after California brought an antitrust suit. Heading into the 1970s, then, the automakers were not betting on any new research collaboration. In general, policymakers viewed industrial research collaborations grimly during the 1960s. Antitrust law forbade more than 25 percent of any industry from participating in research collaborations, which had a long history of disguising cartels.[57]

Yet, in the 1970s, individuals and groups began to question the wisdom of bans on research collaborations, arguing that joint research ventures could help society address difficult technological problems, including the energy crises that the country was then facing.[58] Even more attention was focused on changing the dimensions of international competition for automakers, to address the worry that US car companies were losing out to Japan. In the late 1970s some experts recommended research consortia as a solution to flagging competitiveness, but the Carter administration did not act on such ideas.[59] Indeed, when the automakers moved to resist a ten-year extension of the consent decree banning collaborative research, Carter's people extended it anyway.

Part of the allure of research consortia at this time was that they were thought to be a key to Japan's success. An article on research consortia asked, "Can United States industry beat the Japanese at their own game?"[60] In the quest to discover what factors contributed to Japanese industries' stellar success, a great deal of attention fell on Japan's Ministry of International Trade and Industry (MITI), the focus of an academic obsession during this period. Among many other tasks, MITI organized research consortia in targeted industries, including "steel, automobiles, microelectronics, TVs, machine tools, satellites, biotechnology, and artificial intelligence."[61] Chalmers Johnson published *MITI and the Japanese Miracle* in 1982, in which he argued that research institutes were one of MITI's primary methods for fostering growth and technological change. Research consortia began to seem less a threat to antitrust barriers than a means of keeping up on the global stage. To compete with Japan, US policymakers wanted to use a Japanese technology of governance—the joint research venture.

Ronald Reagan sought to end the ban on research consortia as one way to help the auto industry.[62] Reagan took this step because he bought into an economic philosophy that did not see monopoly as a problem, thereby removing the intellectual basis of the ban on collaboration.[63] In addition, the membership of Vice-President George H. W. Bush's Task Force on Regulatory Reform, which suggested rescinding the consent decree, included lawyer C. Boyden Gray, a partner at Wilmer, Cutler & Pickering, the firm that frequently worked for the industry trade association the Automobile Manufacturers Association. Not only had the firm attempted to fight off the initial 1969 consent decree, but it had also lobbied for the Automobile Manufacturers Association against an extension of the decree in 1978. Thus, Reagan's decision arose from a murky mixture of economic philosophy and corporate interest.

Reagan's decision opened the door for research consortia in the 1980s. Moreover, Congress passed the National Cooperative Research Act of 1984, which assured corporations that the Justice Department would not bring antitrust suits against them for entering into such collaborations.[64] Research consortia emerged in a number of industrial sectors in the 1980s, especially in the semiconductor industry, with the most famous being SEMATECH. Yet, even though the Reagan administration explicitly enabled the automakers to form research consortia, no major automotive consortium emerged. In part, the automakers did not act on this opportunity because their primary concerns were not the kinds of fundamental scientific research that consortia typically focused on. Rather, in the face of Japanese competition, they were preoccupied with learning from their Japanese rivals in order to reform their production systems. General Motors formed a partnership with Toyota known as New United Motor Manufacturing, Inc. (NUMMI), which focused in part on the transfer of Japanese production techniques to the United States.[65]

The first major automotive research consortia after 1969 was the US Advanced Battery Consortium, formed in 1992 at the behest of the George H. W. Bush administration.[66] Bush hoped to aid the auto industry with pending state-level regulations. If the federal government was not making strides toward fuel economy under Bush, the state of California was. The 1990 Clean Air Act Amendments preserved California's right to set emission control standards tougher than those in the rest of the country, and it also allowed other states to adopt California's standards if they so chose. California lawmakers used this opportunity to create an initiative eventually known as the Zero Emissions Vehicle (ZEV) program.[67] The requirement mandated that 2 percent of vehicles sold in the state in 1998 would have no tailpipe emissions whatsoever; the number was to rise to 10 percent by 2003.

The Bush administration formed the Advanced Battery Consortium, then, to soften the blow of these regulations. The battery program was the brainchild of Michael Davis, an official at the Department of Energy who proposed pooling federal monies with a significantly smaller amount from the automakers. Although the Big Three automakers put forward only 16.5 percent of the battery consortium's overall funds, they had enough control over the program that it could "essentially define goals for itself that enabled an indefinite postponement of progress."[68]

In California, the automakers began protesting the Zero Emissions Vehicle program, claiming that current battery technologies simply could not meet the state's requirements. They also argued that consumer demand was not great enough to justify making electric cars. Again, the automakers and petroleum companies mounted astroturfing campaigns to simulate grassroots resistance to the program. A memo from the American Automobile Manufacturers Association that was leaked to the media in 1995 outlined a plan: "The AAMA is conducting a search for a qualified contractor to manage a statewide grassroots and educational campaign in California to create a climate in which the state's mandate requiring automakers to produce a fixed percentage of electric vehicles beginning in 1998 can be repealed."[69] Oil companies created the astroturf organization Californians Against Utility Abuse to foment citizen outrage over the prospect of electric vehicle charging stations driving up electricity rates.[70] Environmentalists, electric vehicle fans, and other advocates insisted that the technology was much closer than Detroit claimed and that consumer demand was higher than anyone expected. They drew parallels between the automakers' claims about electric vehicles and the companies' earlier (false) denials about the near-term viability of airbags.[71]

The effectiveness of the Advanced Battery Consortium can be difficult to judge, though by any measure it was not as successful at transforming the average American automobile as earlier *regulatory* measures. The consortium still exists, and it continues to award grants for research. In 1998 participants wrote a report congratulating themselves for some success, and perhaps some of the research the consortium supported has ended up shaping automotive technologies.[72] Even this report admitted to real limits and problems, however. The industry-controlled consortium took nearly two years *just to set its goals.* In the end, however, California significantly weakened the ZEV rule through a memorandum of understanding in 1996, and the automakers withdrew and destroyed the electric cars they had produced.[73] The regulatory pressure that led to the Advanced Battery Consortium's original formation disappeared, as did the incentive to get battery technologies into cars. In 2015 there were about 291,000 plug-in electric vehicles on American roads, but that was out of a total of 256 million registered vehicles, or about 0.1 percent. A plausible counterfactual looks like this: if California had not rescinded its Zero Emission Vehicle requirement in 1996, by 2003, 10 percent of the new cars sold in the state might have been electric cars. Since the 1960s the federal government has historically followed California's lead and adopted its standards. If the federal government had eventually adopted the ZEV standard, 10 percent of the cars automakers sold would be electric (or some other ZEV—the standard does not mandate which option the market should choose), rather than a mere 0.1%.

The subsequent two federally funded research consortia are easier to judge in that they came to naught. While the US Advanced Battery Consortium continued working throughout the 1990s, President Bill Clinton and Vice-President Al Gore searched for other solutions after they entered the White House in 1993. Both men, but especially Gore, believed that technological change could solve, or at least improve, social problems. A member of the White House Office of Science and Technology Policy called the Environmental Protection Agency's Charles Gray, a tough regulator who had written about the potentials of methanol in the 1980s.[74] Gray told the White House advisor that addressing fuel economy was fairly straightforward, and over the coming days Gray and his staff began estimating to what extent fuel economy could be taken into account. They put forward a goal of reaching 80 miles per gallon. The automakers initially resisted the plan, but the project, soon called the Partnership for a New Generation of Vehicles, or the "Supercar" program, was pitched as a public-private research consortium, rather than as a binding technology-forcing regulation. Clinton and Gore announced the program in a Rose Garden press conference in 1993. The Supercar project initially made slower-than-expected progress but caught its stride in the late 1990s.[75]

When George W. Bush took office in 2001, he ended some government initiatives aimed at improving the environment. In the auto world, the chief victim was the Supercar program—which was only months away from delivering planned prototypes.[76] On January 9, 2002, Bush's secretary of energy, Spencer Abraham, announced that the Supercar Program would be ended and replaced with a new program called FreedomCAR.[77] Under Supercar, the Big Three automakers had all made diesel-electric hybrid vehicles that got over 70 miles per gallon, though only General Motors' car reached the program's goal of 80 miles per gallon. But none of the companies wanted to move ahead into actual production with these automobiles, which they argued would cost over $7,500 more than conventional cars and for which, they believed, there was little consumer desire.[78] The Bush administration was not going to pressure the automakers. The program died.

Bush's FreedomCAR program focused on the creation of hydrogen vehicles. The "hydrogen economy" was a pet project for Bush, who announced a $1.2 billion hydrogen fuel initiative in early 2003.[79] Environmentalists admitted that moving to hydrogen fuel was a worthy ideal, but they criticized the plan for failing to include any short- to medium-term solutions. Daniel Becker, the director of the Sierra Club's Global Warming and Energy program, claimed that FreedomCAR and related federal hydrogen programs "serves as a shield" for the auto industry against increasing fuel economy standards and other policies that would lead to short-term improvements.[80] Ashok Gupta, the head energy economist at the National Resources Defense Council, said, "The FreedomCAR is really about Bush's freedom to do nothing about cars today."[81]

When Barack Obama became president, his secretary of energy, Steven Chu, promptly killed the FreedomCAR program in favor of technologies that would be available in the near term.[82] "We asked ourselves," Chu said, "'Is it likely in the next 10 or 15 or 20 years that we will convert to a hydrogen economy?' The answer, we felt, was 'No.'"[83] For a decade, then, automotive research consortia became White House pets—as Supercar was for Clinton and Gore and FreedomCAR was for Bush. Then, shortly after his inauguration, the next president would strangle his predecessor's pet. The result was intense political uncertainty, a lack of direction, and little technological progress. These government-sponsored research consortia certainly formed, re-formed, and consolidated expert communities, as the trail of official reports makes clear. Because the consortia lacked the coercive power of regulation, however, there was no pressure for the automakers to deploy widely the technologies that resulted from the research. This fact is most obvious with the US Advanced Battery Consortium. When California greatly weakened its Zero Emission Vehicle goals, the automakers lost incentives to deploy electric vehicles. This is not to say that the original California law was a good one—or that it is wise to "pick winners," as conservative critics like to put it—but only that force is often essential for getting the companies to actually use costly technologies that address negative externalities. Automotive research consortia emerged as a way for US presidents to induce technological change while avoiding regulation. Instead of issuing more regulations, they would support "innovation" through research and development. But when the outcomes of automotive research consortia are compared with continuing revisions of the Clean Air Act—which have reduced some regulated auto emissions by 99 percent from 1970 levels—automotive research consortia look extremely weak indeed.

Vehicle Electronics and Computerization

Throughout history, several expert communities—including automotive and illuminating engineers, actuaries, psychologists, medical doctors, chemists, lawyers, and economists—have remade aspects of the automobile, its users, and its system. Since the mid-1980s, experts in computing—or, more accurately, information and communications technologies—have altered many components of cars. But contrary to the rhetoric of the age that is focused on "revolution" and "disruption," this transformation, too, has been incremental—an evolution rather than a revolution. Computers and communication technologies, such as Global Positioning Systems (GPSs), have contributed to the incremental technological change that has marked the automobile industry since at least the 1920s. Moreover, automotive electronics and computing have arisen from a mixed medley of business strategy and government action, especially regulation. In this way, tracking the history of these technologies allows us to examine the factors—including government—that have induced technological change during this period.

Engineers employed by automakers and supplier firms began considering the use of electronics in cars long before it was feasible. The First National Conference on Automotive Electrical and Electronics Engineering was held at Wayne State University in September 1964. Four universities and the main electrical and electronics engineering society, the IEEE (Institute of Electrical and Electronics Engineers), cosponsored the event.[84] Early papers from engineers at such companies as the Radio Corporation of America and Westinghouse Electric Corporation envisioned using electronic components—transistors, resistors, solenoids—for basic functions like ignition and braking. It took another decade for a real automotive electronics community to emerge, however. The first International Colloquium on Automotive Electronic Technology was held in 1974, convening every two years thereafter. Around automotive electronics a research and design community formed that envisioned "nothing less than a total redesign of the internal combustion engine."[85] The colloquium was conceptualized in terms of the "convergence" between electronics or computing and automobiles.[86] Convergence was a common framing of the diffusion of computing technologies during this period. The term was also applied to the marrying of computing and communications that eventually led to networking systems such as the Internet.[87] Slowly, an expert community emerged around automotive electronics, led by engineers at Ford Motor Company, General Motors, Bendix Motor Corporation, Allen-Bradley Company, and the Society of Automotive Engineers.

During this early period, these visionaries struggled with two fundamental limiting factors in electronics: cost and reliability. Given these cost concerns, it is not surprising that government regulation was perhaps *the* force moving these technologies into cars: regulations coerced firms into taking on costs they would otherwise avoid. And early proponents of automotive electronics were perfectly aware of the government's role, which they realized was creating opportunities for their skills and ideas to be used, opportunities that may not have otherwise existed. Trevor O. Jones, an engineer at the General Motors Corporation and a founder of the automotive electronics community, outlined the centrality of government in his 1974 essay, "The Challenge of Automotive Electronics in the USA." Jones believed that three primary factors were driving the "period of accelerated growth of automotive electronics": the "consumer protection movements, computer and space age technology, and inflationary economics."[88] Jones included automotive pollution laws under the umbrella of consumer protection, and he listed four laws enacted in the 1960s and 1970s that regulated automotive safety, pollution, noise control, and cost savings. Jones realized not only that regulations were providing opportunities for automotive electronics but also that many of these technologies were spin-offs from the military and space races. That is, automotive electronics were deeply entangled in a web of industry and government that could not be separated. Early electronics were expensive, but Jones had the vision to see that through constant improvement and cost reductions—a phenomenon enshrined in Moore's Law—electronics would eventually make auto production cheaper rather than more expensive. In the context of stagflation and energy crises, Jones saw great potential in automotive electronics. "The need for efficient use of materials and energy resources," he wrote, "will demand increased application of advanced technology."[89] Government loomed large.

Indeed, "no across-the-board electronics devices emerged [in automobiles] until the seat belt interlock was mandated on 1974 model year passenger cars."[90] As discussed in the previous chapter, this mandate was short-lived, but regulation also induced the next wave of application. The original use of on-board computers, known as electronic control units, in cars stemmed from the requirements of the Clean Air Act Amendments of the 1970s. The automakers could not meet those pollution standards without finely controlling engine functions. By 1979 the companies had teamed up with Motorola, Intel, Delco Electronics, Siemens, American Microsystems, and other electronics firms to install microprocessors in cars.[91] Originally, the Big Three used computers only to control carburetion (later fuel injection) and, thereby, combustion, but applications soon broadened. In 1988 the California Air Resources Board required on-board diagnostics— better known as the "check engine" light—on all cars sold in the state, and the systems became standard nationwide even before the Clean Air Act Amendments of 1990 mandated it.[92] Over the following decade, auto designers began tapping the unrealized potential of computers, using them to control more and more of the automobile's workings, including consumer conveniences, the power train, body control, and entertainment, communications, navigation, and safety systems (particularly the ever-increasing number of airbags). The automobile was remade as an information and communications technology.

With prices dropping and technologies improving, automotive electronics and computing increasingly became a part of corporate strategy. Automakers could not only replace mechanical functions with cheaper electronic ones but also introduce new features, such as entertainment and navigation devices, as sales points in annual model changes. Electronics became enshrined in the stagecraft of automobile marketing. As a result, the vast majority of electronics in cars have come about through company choice rather than government coercion. The number of sensors—originally important components of airbag and emission control systems—exploded in cars. Auto users would visit mechanics after check-engine and service lights began shining on their dashboards, only to find that it was the sensors rather than some mechanical system that had failed. Auto maintenance increasingly depended on the use of computers, making traditional at-home maintenance and repair impossible and mechanics' lives difficult.[93]

Regulation after 1990 was incremental and uneven. Government tended to follow industry's lead. Some regulations, such as requiring on-board diagnostics, certainly forced industry to adopt technologies that it had little interest in. But most federal regulations mandated the deployment of systems that were already well developed and widely used. In the realm of safety, for example, the National Highway Traffic Safety Administration eventually required the adoption of anti-lock brakes, electronic stability control, and backup camera systems. But these technologies were already ubiquitous in new cars. The safety agency did little to spur their development. Auto companies had been working on anti-slip and anti-lock brake systems since the 1960s, and the safety regulations of that decade, along with Ralph Nader's attack on the industry for not making progress on this front, influenced this work.[94] The federal auto safety agency played only a minimal role in the technology's evolution, however, and mandated it only after it was mature and widely available. Again, if anything, federal regulations first coerced firms into putting these technologies into all of their models, not just their luxury lines, and then ensured that the companies would continue using the systems, precluding any backsliding that might occur because of cost considerations or design fads. Only rarely did the federal government proactively encourage work on technologies not already deployed, such as when the National Highway Traffic Safety Administration announced in August 2014 that it was considering mandating vehicle-to-vehicle (V2V) communications systems.[95]

Automotive electronics and computers also occasionally undermined existing regulations. As chapter 7 shows, so-called defeat devices, which circumvented federal standards, often relied on the use of electronic sensors. Computerization increased opportunities for cheating. In 1995 General Motors paid a $45 million fine for using computers to shut off emission controls in Cadillacs when drivers turned on the air conditioner.[96] In September 2015 the Environmental Protection Agency announced that Volkswagen had been using computer-based defeat devices on so-called clean-Diesel cars for more than a decade.[97] Sensors in the defeat devices detected when cars were being tested and, under those conditions, changed the vehicle's operation to meet test criteria. Subsequent investigations discovered that other companies were using cheating systems much like Volkswagen's, and less than a year later the German government's study found that a number of companies, including Jeep, General Motors, and Mercedes-Benz, were using systems supposed to protect cars from engine damage to flout emissions rules, thereby producing between five and twelve times the amount of nitrogen oxide allowed.[98] Moreover, hackers repeatedly demonstrated that they could use computers to take control of a car from the driver and remotely run its functions, including brakes and transmissions, making it likely that cybersecurity experts would be yet another community to shape automotive design.[99]

In the 2010s, futurists and technology boosters increasingly focused on the promises of autonomous, or self-driving, vehicles. While media coverage tended to highlight fully autonomous vehicles such as the Google Car, automakers had been introducing computer-, laser-, and radar-based systems for controlling vehicles functions, including the throttle and brakes, since the 1990s. Over the following years, they sold or experimented with a wide variety of such systems, including adaptive cruise control, forward collision and lane departure warning systems, and vehicle-to-vehicle and vehicle-to-infrastructure (e.g., "smart roads") communication systems. Popular accounts of experimental autonomous vehicles proclaimed how incredible they were—near miracles—but given the current state of technology, the number of experts working on the problem, and the amount of money being spent by rich corporations, the development of autonomous vehicles was a perfectly ordinary story of knowledge communities developing, testing, and refining ideas. It was an unremarkable example in the history of technology—evolution rather than revolution.

The federal government did not form a large research consortium around autonomous technologies, but it did play a role in initiating and shaping the research field. In 2003 the US Department of Defense's Defense Advanced Research Projects Agency (DARPA) announced the DARPA Grand Challenge, a prize competition to spur the design of autonomous vehicles.[100] The prize would go to the individual or group whose self-driving car would be the first to cross a rough 200-mile track in the Mojave Desert in less than ten hours. By the mid-2000s, such prize competitions, whose historical roots went back at least to the eighteenth century, had become a fad. Policymakers and government administrators searched for noncoercive, market-based ways to incentivize research and development. Or as DARPA put it on its buzzword- and euphemism-filled website, the goal of the competition was "to leverage American ingenuity to accelerate the development of autonomous technologies that can be applied to military technologies."[101] The competition came at a time when the US military and intelligence services were relying increasingly on unmanned aerial vehicles, or drones. Moreover, in the war in Iraq, insurgents frequently used improvised explosive devices (IEDs) to injure and kill American soldiers out on patrol. That military leaders would want armed autonomous vehicles on the ground was unsurprising.

The first Grand Challenge attracted the attention of universities and corporations as well as hobbyists involved in the robot combat craze of the late 1990s and early 2000s, which had been both captured and spurred on by television shows such as *Battlebots, Robot Wars,* and *Robotica.*[102] But the first competition ended in failure: no team finished the course. The vehicles created at Carnegie Mellon got the farthest, traveling a little over seven miles before getting stuck on a rock. DARPA continued holding competitions, however, and autonomous technology went further each year. Stanford University's robot car, Stanley, completed the course and won the 2005 competition. Carnegie Mellon won DARPA's 2007 competition, which was focused on urban driving.

The influence of the Grand Challenges on current military technology is unclear (in part because it is still classified), though military officials in 2014 spelled out a twenty-five-year vision that featured increasing reliance on autonomous ground vehicles.[103] But the competitions incited action in the private sector. Among the more publicized efforts was Google's self-driving car program, which drew researchers from the Grand Challenge teams of Stanford, Carnegie Mellon, and other universities.[104] In this way, the autonomous vehicle was a technology developed first for military uses that later spilled over into domestic uses—just as much of impact biomechanics had moved from military aviation safety to domestic auto safety.

Advocates for self-driving cars frequently touted the technology's potential contributions to areas of concern and regulation regarding the automobile, including safety, emission control, and fuel economy. Sebastian Thrun, the computer scientist and roboticist who led the development of Stanford's Stanley before becoming the head of Google's self-driving car program, wrote on a company blog, "We believe our technology has the potential to cut [the number automobile deaths] "perhaps by as much as half. We're also confident that self-driving cars will transform car sharing, significantly reducing car usage, as well as help create the new 'highway trains of tomorrow.' These highway trains should cut energy consumption while also increasing the number of people that can be transported on our major roads."[105]

In these and similar proclamations, the Google team and other autonomous vehicle advocates voiced fantasies that went back at least to General Motor's Futurama exhibit at the 1939 New York World's Fair and the company's 1956 film, *Design for Dreaming*. Some self-driving car enthusiasts saw the technology as part of a much larger dynamic—the eventual obsolescence of human beings as operators. Given the long history of dangerous motorists—extending back to the Auto Fiend of the 1900s and the accident-prone auto user of the 1920s and 1930s—it is not surprising that some of us see ourselves as the problem, that some of us dream of casting human drivers into the dustbin of history.

Not everyone was enthusiastic about self-driving vehicles, however. Some doubted their safety. Others believed that Americans loved the *act* of driving too much to ever turn control over to computers. But autonomous vehicles also fit within broader anxieties in contemporary culture about the role of computers, robots, and artificially intelligent machines—anxieties especially focused on how machines may replace workers. The phrase *technological unemployment* was a product of the Great Depression, though the underlying fear extended back at least to the Luddites of the early nineteenth century.[106] In the 2010s, however, with economic inequality on the rise since the 1980s and wages that remained stubbornly static, fretting over technological unemployment experienced a new vogue. Authors penned a wave of new books about how robots and artificial intelligence would permanently displace labor, including Erik Brynjolfsson's and Andrew McAfee's *The Second Machine Age: Work, Progress, and Prosperity in a Time of Brilliant Technologies* (2014), Nicholas Carr's *The Glass Cage: Automation and Us* (2014), and Martin Ford's *Rise of the Robots: Technology and the Threat of a Jobless Future* (2015). Journalistic bastions of liberal anxiety, such as *The Atlantic*, published articles with titles like "The End of Work" and taglines like "Technology will soon erase millions of jobs."[107]

While these authors acknowledged that worries about technological unemployment were not new, they constantly reiterated that *this* was the moment when such nightmares would come true, especially because of the new capacities of information and communications technologies.[108] They repeated the futurist creed "This time is different." Chunka Mui, a writer for *Forbes.com* and coauthor of the e-book *Driverless Cars: Trillions Are Up for Grabs*, has argued that autonomous vehicles are the ultimate "disruptive innovation" because they will eliminate not only professional drivers (including taxi and delivery drivers and big-rig truckers) but also eventually "all the jobs related to a car accident," including "the ambulance driver, ER clerk, nurse, doctor, technician, claims adjuster, insurance call center operator, body shop repair crew, replacement parts factory worker."[109] Meanwhile, David Mindell, an expert in the history and development of robotics for undersea and space exploration, has argued that, given our experience in these other domains, the deployment of fully autonomous vehicles is quite *un*likely.[110] How the technology will develop is unclear. As an eminent statistician once put it, "There are no facts about the future."

The National Highway Traffic Safety Administration has not played any fundamental role in generating autonomous vehicle technologies or pushing automakers to do so. Many firms around the globe—both traditional automakers and younger, digital technology companies, such as Google, Tesla, and Uber—are racing against one another to produce the first viable self-driving cars.[111] If anything, the federal safety agency attempts to play a coordinating role vis-à-vis autonomous vehicles, helping firms get on the same conceptual page, in a way that Herbert Hoover would have endorsed nearly one hundred years earlier. As has so often been the case in the history of automobile regulation, standards have been central to such coordinating efforts.

Often, the first step in standardization is creating systematic nomenclature. In 2013 the National Highway Traffic Safety Administration defined five levels of autonomy, ranging from traditional nonautonomous vehicles at level zero to fully self-driving vehicles at level four.[112] In 2016 the agency adopted a similar classification system, created by the Society of Automotive Engineers, which added a fifth level. This classification split full autonomy (level four of the previous system) into two levels: a kind of autonomy in which drivers would have to take control in certain limit conditions, such as driving on dirt roads, and a "steering wheel optional" level in which human intervention is never required.[113] These classification systems have had a desired outcome in that they have come to inform many conversations and debates. For instance, experts disagree considerably about when, or if, it will ever be possible to move from level-four autonomy to level-five.

Fitting the Trump administration's focus on removing "regulatory burdens," in late 2017 the National Highway Traffic Safety Administration announced that it would ask concerned parties to identify "any unnecessary regulatory barriers to automated safety technologies."[114] The agency was clearly endorsing the potential of self-driving cars, but would clear the way for such technologies instead of pushing them forward. The NHTSA also released "Automated Driving Systems 2.0: A Vision for Safety," its report outlining "voluntary guidance" for companies working on autonomous vehicle technologies. The document encouraged companies to conduct and publish "Voluntary Safety Self-Assessments" to "demonstrate to the public (particularly States and consumers)" that the companies were thinking about safety, communicating with government agencies, "encouraging the self-establishment of industry safety norms" for autonomous vehicles, and "building public trust, acceptance, and confidence through transparent testing and deployment."[115] It was a vision of industry self-regulation, made even clearer by a box of text in the report that read, "Entities are not required to submit a Voluntary Safety Self-Assessment nor is there any mechanism to compel entities to do so. While these assessments are encouraged prior to testing and deployment, NHTSA does not require that entities provide disclosures nor are they required to delay testing and deployment. Assessments are not subject to Federal approval." The document also noted that the agency "encourages collaboration and communication between Federal, State, and local governments and the private sector as the technology evolves."[116] In many ways, the federal auto safety agency had retreated to a 1920s vision of associational government, embracing coordination and eschewing direct influence.

~

Regulation and other efforts to reduce automotive problems took no decisive direction from the late 1980s to the late 2010s. Rather, they operated by fits and starts. With few exceptions, most of the regulations created during this period simply codified current industry practice—sometimes the state of the art, but more often technologies and practices that automakers could easily have achieved years earlier. While from this perspective the regulations were weak, they had the advantage of forcing the technologies into all production lines, including budget models, and ensuring that companies did not backslide by later removing the features. When political leaders leaned away from regulation, they often leaned toward federal research-and-development funding and the formation of research consortia. But these consortia have accomplished little. In the absence of regulation, corporations lack incentives to implement costly solutions to negative externalities. When negative externalities are present, regulation is much more efficient at generating innovation than research funding is.

Electronics and computerization were the most significant developments in automotive technology during this period, and they also arose from the admixture of corporate strategies and government actions typical of the times. Regulation drove electronics into cars, particularly to deal with emission control, but since the 1990s, auto companies have put increasing numbers of computers, sensors, and other electronic systems into their cars on their own accord. When government has regulated, it has followed industry rather than leading it. The federal government did encourage the development of autonomous vehicles—not through a research consortium but through a DARPA-sponsored prize competition. Increasingly, self-driving car advocates assert that the technology will greatly reduce the automobile's harmful risks, including safety, pollution, and fuel consumption. Autonomous vehicles themselves present a kind of risk, however. Focusing too much on how this magic bullet technology will eventually save us might distract us from mandating more-ordinary technologies that could save lives in the near term. Unsurprisingly, Silicon Valley is the center of self-driving car enthusiasm. The valley's denizens famously embrace a free market philosophy—sometimes called techno-libertarianism—that eschews government regulation. One writer who spent time in Silicon Valley interviewing techno-libertarians concluded that many corporate executives in the region "take it on faith that corporate capitalism, unchecked just a bit longer, will bring about an era of widespread abundance."[117] Applied to the self-driving car, this belief suggests that the best thing government can do is get out of the way. Yet, if this book has shown anything, it is that government has been involved in mitigating the automobile's problems throughout that technology's entire history. These thoughts invite us to consider what conclusions we might draw from the foregoing examination of the history of auto regulation in the United States from 1893 to the present.

Conclusion

In 2012 the National Highway Traffic Safety Administration estimated that since 1960, "vehicle safety technologies associated with Federal Motor Vehicle Safety Standards" had saved more than 600,000 lives in the United States. By 2009, automobiles sold in the United States produced 99 percent less of some regulated emissions than did cars sold in the 1960s. And in 2015, vehicles sold in the United States, both cars and light trucks, reached an average fuel economy of 24.8 miles per gallon, over twice what their counterparts got in the early 1970s. Government regulation led to most of these technological improvements and, just as important, kept the automakers from sliding backward after introducing them. Many problems still remain, some of them seemingly intractable, even hopeless. More than thirty-seven thousand people died in auto accidents in 2016, up 6 percent from the year before. Motor vehicles are still the primary contributors to air pollution, including urban smog. And American vehicles still guzzle huge amounts of gasoline—at an average of 19.53 million barrels of petroleum per day in 2015. Still, in terms of safety, pollution, and fuel economy, the American automobile has objectively improved over the past fifty years and stands as a leading example of how government can influence socially beneficial technological change.

This book examines the history of automobile regulation in the United States from 1893 to the present. In doing so, it traces a path between two dominant ways of thinking about the relationship between government and technology. On the one hand, conservatives and libertarians often argue that the best thing government can do to encourage technological change and other forms of innovation is get out of the way. The University of Chicago economist Sam Peltzman, for example, argues that regulation is not needed and that benefits such as improved safety come about through the "natural progress of opulence," or economic growth and prosperity.[1] On the other hand, liberals and progressives often argue that government itself is, in fact, an essential source of technological change, and they therefore advocate for more government research. Today, this argument is perhaps most associated with Mariana Mazzucato's 2013 book, *The Entrepreneurial State: Debunking Public vs. Private Sector Myths*, which examines how government-funded research lies behind many successful technological advances.[2] Mazzucato finds that devices such as the iPhone—an icon of the digital economy—relied on government R&D, particularly that of the military.

This book lays out an alternative path to either leaving markets alone or having government do the work: regulation draws together experts and gets them to focus on solving specific problems. This path is superior to the libertarian and progressive visions for at least two reasons. First, the libertarian perspective cannot account for the federal government's influence on the automobile and the systems associated with it, an influence extending back not only to the 1960s but all the way to the 1920s. Second, as seen in chapter 11, research funded or conducted by the federal government has not been nearly as successful as regulation at generating meaningful technological change in American automobiles. Regulation forces open possibilities and potentials that neither free markets nor (nonregulatory) government activities generate on their own. Without such pressures, firms often fall into routines and shy away from significant changes, even if such changes would greatly benefit the public.[3]

For a variety of reasons, including the purpose of structuring future law and policy, understanding this history is crucial. Surveying the long history of government-business relations concerning the automobile, we see that regulatory processes have changed markedly. Have these changes improved government performance? Have they increased the scope of government action? Have they "disciplined" the market so as to ensure innovation in automotive safety, emissions, and efficiency? This conclusion to the book has three aims: first, to put forward a general picture of regulation and technological change; second, to show how this overarching pattern has played out differently across four historical periods going back to the late nineteenth century; and third, to offer some final thoughts about the future of both automobile regulation and the use of government performance standards to shape beneficial technological change.

Regulations as Focusing Devices

Historians and social scientists have described numerous parallel cases in which regulation shaped technological outcomes, but so far we lack a general picture of regulation and technological change. The account developed here focuses on the social organization of human attention and problem-solving abilities. Since at least the 1920s, trained experts have played a dominant role in national discussions about problems associated with the automobile. Indeed, the history of the automobile from the late nineteenth century through the present maps onto what might be called the age of the expert, and this history of automotive regulation is, in some ways, a history of the trials and tribulations of expertise.[4] Put simply, regulations are most effective when they motivate experts to focus on specific problems with technologies—that is, effective regulations compel attention and demand problem solving. A problem, according to one definition, is a "matter or situation regarded as unwelcome or harmful and needing to be dealt with and overcome."[5] Regulation involves a dialectic between two dynamic forms of attention: first, how experts come to focus on and take up problems and, second, how problems come to widespread public attention and thereby become what social scientists call public problems.[6] Regulation is one central way whereby public concern shapes expert problem selection.

This process takes place against a wider backdrop of the ways knowledgeable, skilled individuals choose problems to "solve," sometimes creating profound forms of technological change in the process. In many ways, regulation depends on these wider and older ways of organizing problem solving. In capitalist societies, markets—or more accurately, perceptions of markets' potential—have "suggested" problems to inventive individuals and groups. Eighteenth-century inventions, such as James Watt's steam engine and Edmund Cartwright's power loom, addressed fundamental "bottlenecks" in industry and production and enabled the emergence of significant industries.[7] Similarly, mass production arose as knowledgeable individuals moved to new industries—from armaments to sewing machines, from bicycles to automobiles—to capitalize on opportunities they perceived in new applications of emerging technologies.[8]

Over time, organizations, voluntary associations, and businesses further organized attention and problem solving. The emergence of scientific, engineering, and related disciplines, as well as the scientific and professional societies that went along with them, were part and parcel of modernity.

These groups communicated about, and organized efforts to address, well-known scientific and technical problems and often diffused real and proposed solutions to them. In the United States, for instance, beginning in 1826, the *Journal of the Franklin Institute*, a Philadelphia-based organization dedicated to the promotion of the mechanic arts, published articles on research problems and findings regarding (what became) mechanical engineering.[9] The number of such engineering societies expanded greatly in the late nineteenth and early twentieth centuries. Of course, practicalities and profits alone never wholly determined which problems individuals and groups found interesting. Niftiness, fascination, and play always affected selection, and social dynamics within disciplines and professional communities, including simple faddishness, also made problems become fashionable or "sexy."[10] In the late nineteenth century, railroad companies and other firms began encouraging engineers and scientists to address specific problems, a trend that culminated in the creation of formal corporate research and development laboratories.[11] General Motors was the first automotive firm to form an R&D lab, as we saw in chapter 2.[12]

The key group of experts in the history of the automobile and of auto regulation consists of the engineers who work for the automakers and their professional organization, the Society of Automotive Engineers. After its founding in the first decades of the twentieth century, the SAE quickly became an important social space for the creation of technical standards and the sharing of technical information.[13] As we have seen, however, crucial moments of technological change—which often went hand in hand with and were induced by regulation—frequently featured experts who were not automotive engineers, from chemists to medical doctors and from psychologists to computer scientists. Such interdisciplinary groups of experts—what the late historian Ann Johnson called "knowledge communities"—seem to be the norm for particularly difficult and hairy technological problems.[14] Still, given that automotive engineers control the design and production of automobiles, even technologies originally invented or shaped by others must eventually pass through their heads, hands, and technical publications.

Most of the technical and scientific problems taken up by the Society of Automotive Engineers have arisen from corporate prerogatives and strategies as the automakers have competed to win consumers. A few examples suffice. Soon after closed-body cars became the dominant automotive design, the automakers came to focus on what was called "riding comfort," which had some connections to auto safety in that vibrations and other discomforts were thought to cause driver fatigue.[15] Automotive engineers broke down "riding comfort" into a series of interrelated technical problems focused especially on the design and material science involved in tires, springs, shocks, suspensions, and other subsystems of the chassis.[16] Similarly, when General Motors opened its Art and Color Section in 1927, it initiated a boom in automotive styling that involved the technical problems associated with making stylish design (once the domain of one-off, bespoke handicraft) fit the requirements of mass production.[17] The same was true of the so-called horsepower race that emerged after World War II. The engineers at American automakers focused on designing and developing high-compression engines and other technologies that lived up to advertisements featuring speed and performance, much to the chagrin of safety advocates (see chapter 4). In summary, the dictates of industrial and consumer capitalism have selected the overwhelming majority of problems in automotive engineering.

Occasionally, auto companies and the engineers they employed turned to problems of concern to consumers and the public. As chapter 2 reveals, automakers in the 1930s removed protruding interior knobs and made other changes in the name of safety. But just as often, the automakers dropped such features and effectively backslide. Moreover, before World War II the companies never voluntarily addressed automotive air pollution; nor were the firms much involved with the creation and diffusion of traffic rules, signs, signals, and other measures meant to deal with auto safety. Therefore, these changes—and the attention and problem solving that undergirded them—came from outside the auto industry.

Problems do occasionally rise to the level of public awareness, and such awareness sometimes leads to legislative and administrative change. Some problems are private, and others are public.[18] An individual may have a credit problem after having too much fun shopping online or buying a house beyond his or her means, but individuals' access to easy credit becomes a public problem only when it is widely discussed, plays out in the mass media, or becomes the subject of government investigation—all of which came into effect after the financial crisis of 2008, for example. Similarly, a family may experience a tragic car accident that forever changes its members' lives, but automotive safety, pollution, and fuel economy only sometimes rise to the level of collective attention.

Problems typically become public via three well-known paths and the interplay between them: media organizations choose them; interest groups put them forward; or policymakers select them. All three of these pathways have shaped the history of automobile regulation. There are real limits to public awareness about problems, however. Since the late 1960s, scholars have tracked what they call the issue-attention cycle. As the originator of the term explained, "American public attention rarely remains sharply focused upon any one domestic issue for very long—even if it involves a continuing problem of crucial importance to society. Each of these problems suddenly leaps into prominence, remains there for a short time, and then—though still largely unresolved—gradually fades from the center of public attention"[19] What we see in the history of auto regulation are waves of intense public awareness followed by long troughs of relative inattention.

Only rarely has public interest lead to meaningful political change that actually shaped technology. This book defines regulation as any activity aimed at addressing technological problems, from informal efforts, such as holding meetings or conferences, to formal laws and governmental codes. What it has found again and again is that regulation acts as a kind of *focusing device* that—through enticement, coercion, or some combination thereof—persuades expert communities to focus on specific problems.[20] As described above, the ways by which scientific and engineering disciplines, professional societies, and corporations organize the efforts of experts form a central component of modernity. In this context, regulations work by inducing organization and reorganization around public problems that, for whatever reason, are not being addressed.[21] Regulation has encouraged experts to focus on problems in several different ways, and some forms of regulation dominated during particular periods, which is the primary reason we examine different eras of regulation below. Since the 1910s, however, a dominant policy tool for regulating automobiles has been the *performance standard*. That is, regulations for technologies have consisted of tests, such as those for headlights, crashes, and emissions, that manufacturers' products must meet, along with criteria for passing those tests. For the most part, regulations have avoided specifying how companies should meet those criteria, and instead mandated a specific solution, known as a design or technology standard. Technologies thus meet the standards as long as they pass the tests. Regulations are often most effective when they create new markets or economic opportunities around problems, as has occurred again and again in automotive history.[22]

The Early Period of the Automobile, 1900–1920

Automobiles became a public problem soon after they appeared, when newspapers began covering accidents, injuries, and deaths. As we saw in chapter 1, the first cars were *expensive* objects for the wealthy, and early news coverage of automotive hazards thus tended to focus on the behavior of wealthy auto owners and their chauffeurs. Some of the earliest regulatory efforts concerning the automobile were private, or at least nongovernmental, as when rich residents of Oyster Bay, Long Island, New York, hired a private force to police recently constructed roads that so-called Auto Fiends were tearing up by racing their cars over them (see chapter 1). In the first years of the twentieth century, only a decade after the Duryeas constructed the first automobile in the United States, state and municipal governments began regulating roads, primarily through speed and traffic laws. The first individuals who worked on auto safety, including the self-appointed traffic guru William Phelps Eno, were not experts in any formal sense. Eno did not have a college degree, and the "science" he brought to bear on traffic management was little more than organized common sense.

Even these early laws—though not aimed specifically at changing technology—influenced the technically minded. For instance, inventors designed and marketed speedometers largely in response to speed laws (see chapter 1). In this context, speedometers were a means for drivers to ensure that they kept their speed under the posted limit, or even to challenge traffic cops who pulled them over. Automotive magazines and catalogs during the period teemed with products—from bumpers to horns, from goggles to headlights—that promised to make driving safer. In this way, when independent inventors and automotive supply companies focused on creating safety technologies, however basic, they did so largely in response to perceived opportunities in the automotive aftermarket. As discussed in chapter 1, Ford Motor Company, and likely other firms, also encouraged auto dealers to sell their vehicles on the basis of safety.

Experts with formal training and advanced degrees first began to work seriously on automotive problems during the 1910s. At that time, illuminating engineers, some of whom had connections with the auto industry and the SAE, took an interest in the problem of glare, which was a growing public concern with the ever-increasing use of electric lights (see chapter 2). States and municipalities began to regulate headlights during this period, but illuminating engineers appear to have taken up the headlight glare problem mostly of their own accord. Solving the general glare problem was important to the continued public acceptance of lights, and it was also a matter of professional pride—particularly for illuminating engineers who made much of the connection between the metaphor of light and the reason-centered idea of Enlightenment. Indeed, one function of professions is to resolve social problems in their domains in the absence of government, and such actions are sprinkled throughout the history of technology, though with some obvious limits.

As states and municipalities began to regulate technologies like headlights and brakes, however, experts in such organizations as the Illuminating Engineering Society and the National Bureau of Standards assisted in codifying and improving technical standards and tests for these technologies (see chapter 2). The decision in *MacPherson v. Buick*, which established that the automakers were liable for defects, encouraged automobile firms to take product inspection and materials testing more seriously. In the late 1910s, experts within the SAE began paying more attention to these issues and promulgating technical standards addressing them. Moreover, all these developments in the 1910s took place within a more general safety movement that emerged with the birth of the National Safety Council in 1913. The council sought to apply the lessons of factory safety to public problems through a variety of promotional efforts and publications such as *Sure Pop and the Safety Scouts*. By World War I, many of the pieces that would constitute a national automotive safety movement existed, but only after the war were these disparate pieces brought together.

The First National Auto Safety Movement, 1920–1950

In the early 1920s the growing carnage of automotive accidents regularly made front-page news across American cities. In 1916 an estimated 8,200 people died in auto accidents in the United States, but by 1923 that number had more than doubled to 18,400.[23] To meet this challenge, individuals formed the first national auto safety movement. As historians have long noted, the First World War vastly expanded the size and capabilities of organizations in the United States, and many of the organizations that came to shape auto safety efforts in the 1920s and 1930s either expanded or had their origin in that struggle. The National Bureau of Standards, for instance, significantly expanded its efforts concerning automobiles during and just after the war. The Council of National Defense, a civic organization that coordinated the domestic war effort, first became aware of the problem of the lack of uniformity in traffic laws while using trucks to move agricultural goods to market—and to transport the aid that US Food Administration head Herbert Hoover organized to feed starving Europeans. By the mid-1920s, now known as the Great Engineer, Hoover was fighting to make traffic laws uniform across the United States. Similarly, the American Engineering Standards Committee, the most important standards body when it came to safety efforts in the 1920s, began life as a "war baby."

The emerging national movement focused on connecting trained experts to well-known public problems. The most important coordinating effort in automobile regulation before World War II was Herbert Hoover's National Conference on Street and Highway Safety. As described in chapter 2, the Hoover conferences fit a more general mode of governance known as associationalism, which Hoover popularized during this period.[24] The Hoover conferences formed networks and created opportunities. They created a market for ideas in both the literal and the figurative sense. Some groups sought financial gain by marketing solutions. Others sought nontangible benefits, such as increasing the status of their professions. Significantly, William Phelps Eno was invited to the first national auto safety conference, but he did not attend, nor did he take part in later ones. Increasingly, trained professionals dominated the field of auto safety; the day of wealthy amateur reformers was done.

Municipal laws and safety efforts also opened a significant market for traffic control devices, including traffic signals, and other technologies. Moreover, as regulation of brakes and headlights was established, firms began producing standardized testing instruments sold to local garages that conducted annual inspections. Attempts to reduce automotive hazards opened up markets of all varieties. Agencies of the federal government also sought to shape markets by producing credible, reliable knowledge and to inform policymaking at the state and local level as well as individual consumer decisions. From the 1910s through the 1930s, the National Bureau of Standards led these efforts affecting automobiles and published findings on a range of automotive topics.[25] For these endeavors, the federal government hired its own experts and assigned them to focus on particular problem domains. Experts in the National Bureau of Standards also took up problems through their own entrepreneurial initiative.

The insurance executive Albert Whitney was another crucial player in fostering the national auto safety movement. Whitney acted as the head of the American Engineering Standards Committee (AESC; later named the

American National Standards Institute); he was an active member of Herbert Hoover's various safety efforts, including the National Conference on Street and Highway Safety; and he coauthored the first American driver's education textbook, *Man and the Motor Car* (see chapter 2). Through his leadership, both as head of an insurance trade organization and the AESC, Whitney guided a wide variety of experts in focusing on safety. To give one example, Whitney created fellowships at Columbia University's Teacher's College that led to a number of novel studies of safety, including the first dissertation on the topic, Ruth Streitz's *Safety Education in the Elementary School: A Technique for Developing Subject Matter* (1927).

The Hoover conferences and Whitney's efforts yielded several important automotive safety outcomes. These efforts included improving roadways, laws, signs, and signals to increase coordination among drivers; improving technologies, such as brakes and headlights, that enhanced driver mastery; and reforming drivers themselves, eventually through licensing tests and driver's education. By 1930, engineers taking part in the Hoover conferences had standardized signs and signals, including the three-color traffic light. The conference also issued uniform traffic codes, which proved influential but were never universally adopted across the country. In many ways, what we consider the standard road system today was developed during this early period, and the Hoover conferences played no small part in coordinating these efforts.

There were also real limits to these efforts, however. Legal uniformity, for instance, depended on the willingness of state and local officials who often had no incentive to join in. More important, it was more or less universally accepted during this period that drivers caused accidents and, therefore, that regulatory efforts should focus on improving driving. Regulations affecting brakes and headlights put some pressure on automakers to ensure the quality of these technologies, but even these devices were aimed at enhancing *driver mastery*. The automakers had few incentives to address problems that economists call "negative externalities," or social costs, such as accidents and pollution, that affect third parties but do not factor in firms' balance sheets.

The Rise of Federal Automobile Regulations, 1950–1975

After the rise of the first national safety movement in the 1920s and 1930s, the next significant development in the history of automobile regulation was the creation of federal regulations between 1965 and 1975. Yet both federal automotive safety and air pollution standards relied on science and engineering developed earlier, as early as World War II but blossoming in the 1950s. Most significant, the sciences of automotive safety and air pollution *reframed* the perceived problems of and around automobiles. The emergent field of impact biomechanics, or crashworthiness, argued that the goal should be to make cars safer *in* accidents, rather than to avoid accidents in the first place. Auto users had always known that cars produced carbon monoxide that could be harmful in enclosed spaces, but the emergent science of automotive air pollution and smog established that cars were hazardous in previously unrecognized ways.

The sciences of crashworthiness and automotive air pollution contrasted greatly both in how they came into being and in how they came to public consciousness. These differences had lasting consequences for their incorporation in federal regulation. Public uproar over regular bouts of smog in Los Angeles, California, led to the science of automotive air pollution. When Caltech's Arie Haagen-Smit discovered that automotive emissions were one of the primary causes of photochemical smog, he was doing research for the Los Angeles Chamber of Commerce (see chapter 3). Repeatedly, it was commercial concerns in California, including agriculture and real estate, that drove experts to focus on air pollution. Air pollution regulation developed first at the local, regional, and state levels, none of which fully addressed the overall problem. The federal government entered the field slowly, initially by using its experts in the Public Health Service to generate credible science and by holding associational conferences to bring experts together. But well into the 1960s, research networks focused on automotive air pollution remained weak, dispersed, and uncoordinated. Most important, automakers and others had made little progress in developing meaningful emission controls by the late 1960s. Legislators in the US Congress avoided active intervention in air pollution, which was framed as a state and local problem. It took *years* of automaker inactivity to change politicians' minds.

Unlike air pollution science, research on crashworthiness took place largely outside public awareness. Researchers were attracted to crashworthiness for several different reasons. For one, the biomechanical research of Elisha Gurdjian and Herbert Lissner at Wayne State University arose largely from Gurdjian's heartbreaking experiences as an emergency room doctor and his frustration with the lack of knowledge about concussions and other biological responses to physical trauma, including automobile accidents. Yet a dominant line of crashworthiness research, including the early work of Hugh DeHaven, originated in the study of military aviation hazards. Most important, Colonel John Stapp played a fundamental role in consolidating the research community around *crash safety* (see chapter 3). In many ways, the science of crash safety was a spillover from the military to the domestic sphere. By the mid- to late 1950s, the basic outline and practical upshots of crashworthiness research were well defined.[26] Ford Motor Company used some of these ideas in its Lifeguard Safety Package in 1955, only to retract it months later. Crashworthiness research came into public view only slowly; for instance, when journalists covered development of the Cornell–Liberty Mutual Safety Car. By 1959 the frustrated young lawyer and reformer Ralph Nader complained, in his article "The Safe Car You Can't Buy," that these safety technologies were unavailable to consumers.

Between 1965 and 1975, Congress passed several major laws regulating air pollution, safety, and fuel economy, including the National Traffic and Motor Vehicle Safety Act of 1966 and the Clean Air Act Amendments of 1970. Congress debated and enacted both of these laws in the context of intense public attention, even outcry. In the case of safety, the discovery that General Motors had hired private investigators to spy on Ralph Nader—and the acute fallout from that discovery—guaranteed virtually unopposed passage of the law. Similarly, in 1970 Edmund Muskie used public concern over environmental problems, including the first Earth Day, and mounting frustration with automakers' inaction on emission controls to shepherd through one of the toughest regulatory laws in US history. Both of these laws required automakers to address the automobile's public problems and forced them to internalize the costs of these negative externalities.

These two landmark laws had widely divergent outcomes, however. Many of these differences and divergences were rooted in how the laws were structured, but some of them stemmed from differences in administrative leadership at the relevant regulatory agencies. Put another way, scientific and engineering expertise played a crucial role in the history of automobile regulation, but *administrative expertise and leadership* were also vital. On the one hand, the first leaders of the federal auto safety agency, William Haddon and Robert Brenner, labored under a weak law that required them to go through cumbersome federal rule-making procedures. Yet they also seemed to have little or no understanding of the adversarial relationship they were entering with the industry; nor did they have any experience managing a regulatory agency. As discussed in chapter 5, the automakers embarrassed federal staff members during rule-making procedures, thereby leading to significantly weaker standards that had long-term consequences for federal safety regulation. On the other hand, Eric Stork, who took control of the Environmental Protection Agency's automotive emissions program around 1970, worked under a clear, powerful law, and he also had significant regulatory experience. He understood and embraced adversarial struggles with regulated firms. As chapters 7 and 8 make clear, Stork reoriented the agency to focus on technology assessment, which helped ensure that experts within the automakers worked diligently on pollution control with measurable outcomes.

Federal auto safety standards did not initially push automotive engineers or other experts to generate substantial new knowledge (see chapter 5). By the mid-1960s, the research networks associated with Stapp's auto safety conferences had built a considerable body of biomechanical knowledge that the automakers *were not using*. While the Society of Automotive Engineers had begun creating safety standards after legislation was threatened around 1960, they were not put into effective practice. Thus, federal safety standards involved the codification of the existing state of the art. As focusing devices, the standards pushed automotive engineers to translate existing scientific and medical knowledge into concrete design, which involved some learning, but such translation did not necessitate any considerable R&D. As later chapters show, particularly chapter 11, this tradition of relatively weak safety standards continues in the present: in general, federal automotive safety standards in the United States codify current practice rather than setting ambitious technology-forcing performance standards. If the standards have any significant outcomes, they tend to be twofold. First, the standards ensure that automakers build the affected safety technologies into all of their vehicle lines, instead of just luxury lines, where new technologies are typically introduced. Second, they guarantee that automakers will not backslide by introducing and then removing safety features, as they frequently did before the advent of federal safety regulations.

Federal emissions laws had a much more profound influence on knowledge production following passage of the 1970 Clean Air Act Amendments. As discussed in chapters 6 and 7, earlier federal automotive pollution control laws were weak and largely relied on the federal government's own, underfunded R&D programs. The 1970s law, however, was *technology forcing*: it mandated that the automakers attain emission reductions that were not possible with the current state of the art. As a focusing device, the regulations forced the automakers to build new R&D programs and to form new knowledge communities, which eventually resulted in the introduction of the catalytic converter (see chapter 8). Moreover, the 1970 law opened up a significant new market, which drew attention from other industries, such as chemical firms and manufacturers of advanced materials, including Corning. In 1976 these corporations formed their own trade association, the Manufacturers of Emission Controls Association, to share credible information about emission controls and advocate for pollution laws around the globe. In the end, the Clean Air Act Amendments of 1970 created some of the most successful pollution control regulations in the US history.

Safety and pollution control laws also created incentives for perverse action, however, especially in the case of firms that used their attention-focusing and problem-solving capabilities to attack regulations and find ways to work around them. As highlighted in chapter 5, the automakers may have put more thought and creativity into combatting and weakening the first proposed federal safety standards than they did into inventing safety technologies themselves. In this way, regulation engenders resistance. Automakers poured serious resources, both money and the time of engineers, scientists, and lawyers, into criticizing the government's proposals. Performance standards also engendered a long and proud history of automakers cheating the tests—a time-honored form of creativity, problem solving, and innovation. Cheating extends back to the standards' earliest days and continues in the twenty-first century, as evidenced in the Volkswagen diesel emissions scandal of 2015. The advent of computers in cars has exacerbated these problems by making it easier for automotive engineers to design around tests.

To date, the federal automotive standards created between 1965 and 1975 and amended thereafter have been the most successful method of convincing car companies to address negative externalities and preventing them from backsliding on technological progress. The laws have never been perfect and always have had serious limitations. By the late 1970s, however, critics argued that regulation itself had become a significant public problem that needed to be addressed.

From Deregulation to Inaction, 1975 to the Present

From the 1950s through the 1970s, economists and social scientists developed theories arguing that regulation damaged the economy and that civil servants acted in their own self-interest rather for any public good. They also asserted that regulatory agencies were typically "captured" by the very interests they were meant to oversee. As chapter 9 demonstrates, regulators at the EPA fiercely defended the credibility and legitimacy of their work. The agency's fuel economy measurement program arose as a way of dealing with external criticism of the agency and its enabling laws. Moreover, as fuel economy became a fashionable topic after the OPEC oil embargo, the EPA defended its turf around fuel economy measurement from other organizations in the federal government that tried to claim it, often at the behest of the auto industry. But civil servants would argue—rightly and justly, I think—that such forms of self-defense were necessary to defend the soundness of the Clean Air Act, which did greatly benefit the public good.

The push for deregulation reached its apotheosis in the Reagan administration. As chapter 10 notes, struggles over the economic impact of regulation have drawn other kinds of experts, particularly economists, into federal administration. Efforts toward "regulatory reform" ensconced cost-benefit analysis and other economic methods in federal law. Experts within the Office of Management and Budget have used these techniques to regulate the regulators ever since, and the rules have had significant consequences for the creation of new regulatory standards. Almost universally the rules have weakened proposed regulations. Clearly, regulations create opportunities for all kinds of experts, including nontechnical ones like public administrators, economists, and lawyers. Experts both within and outside the federal government have devoted increasing attention to justifying new regulations in the face of cost-benefit hurdles. Critics argued that cost-benefit analysis often placed too heavy a burden on the regulatory process and that—as in the case (described in chapter 11) of backup cameras to avoid running over children and other persons—the outcomes of cost-benefit analysis often do not align with our actual moral values. Moreover, cost-benefit analysis is subject to manipulation by experts and often ignores intangibles that cannot be easily quantified.

Since Reagan left office, the path of federal automobile regulation has been both uneven and uncertain. Typically, the federal government has followed industry rather than leading. Moreover, the US government has been far less effective at generating meaningful and socially beneficial technological change in the automobile through its own research and development efforts than it has through regulation. Indeed, when federal pollution control standards depended on federal R&D during the 1960s, the standards made very little progress. Only with the coming of the tough Clean Air Act Amendments of 1970 did auto emissions decrease significantly. Similarly, US presidential administrations since the 1980s have been fond of the collaborative R&D

ventures known as research consortia, often featuring public-private part-nerships. Consortia have been attractive in part because they seem to by-pass the need for unfashionable coercive regulation. But most of the research consortia focused on problems posed by automobiles have ended in failure and generated little meaningful change (see chapter 11). The experts involved in these undertakings produced some new knowledge, but the automakers were under no pressure to implement it—and so they did not. Only regula-tion has induced such action. Since the 1990s, there has rarely been a clear vision for how the federal government should use its own experts in auto-motive regulation, let alone shape the attention-getting and problem-solving skills of experts within automotive firms.

Performance Standards and Their Future

In contrast to the hopeful, passive prodding of the federal government through public-private partnerships, regulations, particularly those struc-tured using performance standards, can be powerful tools of technological transformation. All regulations act as focusing devices, though they can have widely divergent outcomes. The weakest require firms simply to act on existing knowledge that they are currently ignoring—translating current know-how into design, diffusing knowledge into products. The strongest force firms to apply problem-solving skills in new ways and often to form new or reform existing knowledge communities.

Frequently, solutions have come from outside the auto industry itself. In other words, regulation often induces the movement of knowledge between sectors.[27] In itself, this fact is not surprising. Invention and innovation stud-ies almost always find that significantly new ideas come from outsiders, not established, or incumbent, firms or communities deeply invested in the cur-rent system.[28] Auto safety efforts from the 1910s through the 1930s involved the large-scale movement of know-how and technological devices, such as signal lights, from railroads to streets. Crashworthiness was first developed for military aviation before being applied to cars. Air pollution control drew on scientific methods and technical solutions, such as catalysis, originally ap-plied in other domains. And, since the 1970s, regulations have increasingly influenced the relationship between the automakers and information and computing technologies. Regulation often induces the production and move-ment of knowledge by opening up new markets. Performance standards require technologies to meet socially determined norms without either designing or choosing specific technologies. As the late economist Steven

Klepper wrote, "One of the great virtues of capitalism is the decentraliza-tion of decision making."[29] Performance standards allow for—and even harness—this virtue. Regulations work best when they tap the potential of capitalism rather than shutting it down.

Successful regulation, however, requires a knowledgeable government—one that has reliable knowledge (often produced "in house") and substantial expertise and organizational capabilities, which is to say that regulation is quite fragile. US presidential administrations have frequently appointed industry-friendly administrators who allow regulatory agencies to be captured, and seek to rescind agencies' previous and pending regulatory work.[30] Safety, environmental, and consumer advocates perennially attack the "revolving door" between industry and regulatory agencies. Reduced budgets undermine organizational capabilities and regulatory enforcement, and hostile environments and low morale within agencies cause deeply knowledgeable experts to flee. What is done can be undone. Progress isn't guaranteed.

Performance standards have played two key roles in the history of auto-mobile regulation. First, since the late nineteenth century, technical standards have become the primary means of coordinating scientific and engineering practice across space, time, and organizational boundaries. The number of technical standards created by governments, standards bodies, and engineer-ing and science societies since the late nineteenth century is staggering. In this way, federal performance standards comport with broader processes for rationalizing production, and they have drawn on already-existing prac-tices and standards within the auto industry and the Society of Automo-tive Engineers.

Second, performance standards have fit the liberal political culture of the United States. They do not control the daily operations of corporations or put much constraint on consumers. To understand and appreciate this point, let us briefly explore alternative paths not taken, or counterfactuals. The primary alternative to performance standards would be mandating that corporations use specific technologies—sometimes called design, or tech-nology, standards. Use of such standards in US automotive regulation has been extremely rare. By using performance standards, lawmakers and public administrators have given corporations the leeway to solve problems in whatever way they wish, as long as they pass the test. As Georgia congress-man James MacKay said during hearings on the 1966 auto safety law, "I have proposed that the standards be for performance of the automobile, not for its total design, because this kind of test places the least restriction on the industry and yet fulfills our need."[31] Such a view fits comfortably with the values and political ideologies widely held in the United States. In this way, performance standards are *a liberal technology of governance*.

Performance standards, then, offer a specific way of defining business-government relations around technology-rooted public problems, but their relationship with consumers is just as important. In the end, performance standards demand little from consumers beyond raising the price of new cars and, in some cases, requiring annual inspections (which can be costly and have a greater impact on low-income auto owners). This lack of burden placed on consumers is an important reason that such standards succeed politically. Moreover, all public problems associated with the car, including safety, emissions, and fuel consumption, could be reduced *if people drove less*. That is, another alternative to performance and design standards would take the form of rules that tightly restrict driving and lower driving mileage. Indeed, some readers may be disappointed that this book ignores ideas for moving away from automobiles altogether. Yet the simple reality is that historically, discussions about ending automotive culture or seriously curtailing auto use and restricting drivers rarely entered mainstream discussions of the problems examined in this book, especially at the federal level. Legislators, public administrators, and other interested parties who pushed for automotive regulations gave central place to consumer sovereignty. Mirroring their culture, they put faith in what scholars call the technological fix—that is, they assumed that public problems could be solved by changing technologies rather than by constraining the habits and desires of free citizens. Such a vision implicitly assumed that the goal of regulation should be technological change rather than cultural or social change.

It seems unlikely that US legislators and policymakers will use performance standards to further significantly reshape automobiles anytime soon. The largest barrier to such change is obviously political gridlock and polarization. But there are also significant technological uncertainties, with no dominant vision for how automotive technologies should be developed. Advancing automotive air pollution and fuel economy regulations would most likely involve moving toward a wider array of alternative energy vehicles, such as hybrids and plug-in electrics. Moreover, many Americans still like big cars, SUVs, and trucks, and many others do not want or simply cannot afford to pay more for hybrid or electric vehicles.

When it comes to safety, many individuals currently place their greatest hopes in autonomous, or self-driving, vehicles. But this technology is surrounded by serious hype and unrealistic promises. When hype-mongers like Elon Musk, who consistently fails to meet deadlines and basic production quotas while spinning visions of science fictional tech utopias, tell us that the arrival of self-driving cars is imminent, we should be more than skeptical. We should not blindly count on the verity of the prophets of fully autonomous vehicles who promise a zero-accident, zero-fatality future and greatly decreased auto pollution through the hyperrational application of this technology. I have argued elsewhere that focusing on full autonomy risks blinding us to available and near-term solutions, such as lane departure warning systems and radar-actuated cruise control and braking, that could save many lives and reduce injuries.[32] I believe that regulators should be more aggressive about getting these available systems into new cars. Autonomous vehicle systems are currently too underdeveloped at present to be the basis of extensive, technology-shaping regulations. Some experts have argued that we should go *much* further in pushing automotive safety, which never went through the technology-forcing process that emission controls did that led to regulated pollutants being reduced by more than 90 percent. Some scholars have proposed using a similar percentage-based criteria as a means of regulating auto safety.[33] How would automakers transform their products if we mandated that they must reduce the number of automotive fatalities in new cars by, say, 40 percent within ten years?

Performance standards also have much wider applicability and potential, though, again, the current political climate makes such application unlikely in the short term. Our historical experiences suggest that we have powerful, time-tested tools for addressing technological problems, including one of our greatest contemporary concerns: anthropogenic climate change.

While nationalists pride themselves on the "American spirit of invention and entrepreneurship" or whatever, there is, in fact, a widespread hesitancy in the United States to prod firms to be more innovative or to force consumers to change their behaviors, wants, and desires. Contrast this outlook with China and several European countries, which have announced plans to phase out gasoline- and diesel-powered vehicles completely within the next twenty-five years.[34]

This book presents a case for limited optimism. Since the 1960s, both conservatives and progressives have grown skeptical, even cynical, about expertise. Yet, as the foregoing account shows, there are domains in which experts have made great strides. Regulation has spurred the accomplishment of social gains. This view is not Pollyannaish. Experts often *are* self-interested. Regulations *can* lead to perverse results or exact costs that markets cannot bear, and addressing one risk *can* exacerbate others. But we must resist the contemporary fashion that attends to such ironies and trade-offs while ignoring the more general trajectory, which overall is *positive*. The shortcomings just listed simply mean that policymakers and regulators must be creative and aware. At the same time, this book also sounds notes of caution about the limits of regulation, about the ability of corporations to resist meaningful change, and about our unwillingness to regulate our own behavior. Often, the greatest hope lies not in changing technology but in shifting our cultural patterns and individual habits; yet we show little inclination toward such changes.

In the end, however, the central argument of this book is that we must change the way we think about regulation. We must examine it in finer detail, with more historical fidelity. For many years, regulations have been cast as uniformly stultifying, rigid, even Kafkaesque. But when regulations are well crafted, the exact opposite is true—they unleash creativity and channel the human capacity for problem solving toward our greatest objects of concern.

Any book that takes as long to write as this one mounts far too many debts to acknowledge fully. Parts of this book began as a dissertation in history and policy at Carnegie Mellon University, where I had an excellent committee in David Hounshell, Jay Aronson, John Soluri, Joel Tarr, and Steve Usselman. My graduate work was supported by two National Science Foundation research grants, one that created the Climate Decision Making Center and the other titled "The Socio-political Construction of Technologies under 'Technology-Forcing' Regulations: A Tale of Two Automotive Technologies." Thanks to colleagues in the Department of Engineering and Public Policy for always inspiring through their cogent policy analyses.

A fellowship at the Lemelson Center for Invention and Innovation fundamentally transformed the nature of this project. Thanks to Eric Hintz for making my time there worthwhile and to the great Roger White for always helping my research. For five years my colleagues at the Stevens Institute of Technology created a supportive environment for continuing to rework this manuscript. Special thanks to Jim McClellan, whose constant advice to his young charges was that we should be ourselves. The Department of Science, Technology, and Society at Virginia Tech gave me two semesters free to work on this and other projects. Many thanks to my dear colleagues here in beautiful Blacksburg. Thanks also to the many librarians and archivists who made this work possible and to the former and present civil servants who shared their stories and papers with me, especially Eric Stork, Joe Somers, and Karl Hellman. I am grateful for the good work of the editors and staff members at Johns Hopkins University Press.

Much of what is good about this book is attributable to three mentors: Steve Usselman, whose work inspired me to enter the study of history and technology, has stood with me from before graduate school to put finishing touches on this manuscript. I have greatly benefited from his exacting comments. Richard John, too, has offered his wise guidance for years now, from the time I acted as his research assistant before grad school to the time he acted as the series editor of this book. I will always be grateful for the way he has shared his encyclopedic knowledge and pushed me to go further and further. Finally, no single human has shaped my thinking, including the thinking in this book, as much as my doctoral advisor, David Hounshell. I thank him for taking me on, even though I'm sure he regretted it *many* times. If there is a single readable sentence in this book, it's because David taught me how to write it.

I wrote this book in the bosom of two professional societies, the Society for the History of Technology and the Business History Conference, and I have benefited immeasurably from their support. I also gained a great deal of insight from seminars at the Fashion Institute of Technology, the Hagley Museum and Library, New York University, and the University of Pennsylvania. Thanks to Roger Horowitz, Matt Stanley, and Dan Wilk. My research for this book benefited from innumerable conversations with colleagues and friends, including Ed Balleisen, Jeremy Blatter, David "C is for Cutie" Brock, Kevin Brown, Larry Burke, Lindsey Cormack, Mags Curnutte, Garry Dobbins, Yulia Frumer, Hugh Gorman, Meg Graham, Barbie Hahn, Dan "Piles and Ditches" Holbrook, John Horgan, Kristyn Karl, Samantha Kleinberg, Patty Cakes McCray, Costa Samaras, Phil Scranton, Susie Lee Spellman, Joel Tarr, Dom Tobbell, Heidi "Headbanger" Voskuhl, Benji Waterhouse, Dan Wilk, and JoAnne Yates. Arwen Mohun staged an intervention in the life of chapter two that greatly improved its ways, though she may still be unhappy with them. I wish Ann Johnson was alive, so we could argue about what I've written. I asked Andy Russell to read and comment on more drafts of more chapters of this book than is moral really, and he always did it without complaint. The guy's a saint.

My parents and siblings supported me even as I pursued a career they often did not understand. My in-laws, Bob and Linda Middleton, provided emotionally and financially and made my family's life possible. My daughter, Henrietta, and son, Alban, were born during the second half of this project. They have never known a world in which I wasn't working on this book, and before they came along, I never realized how vastly better my world would be with them in it. The Baron and the Gypsy were a calming presence because they never gave one single lick about my work—not one. Abigail, the publication of this book is another milepost on our long, strange trip together. Thank you for always being there, Shugs.

Introduction

1. "1959 Chevrolet Bel Air," Old Ride, http://www.oldride.com/library/1959 _chevrolet_bel_air.html; "2009 Chevrolet Malibu," *U.S. News & World Report,* http://usnews.rankingsandreviews.com/cars-trucks/Chevrolet_Malibu/2009/specs /Malibu-LT-1LT-300503/.

2. "1959 Chevrolet (USA) Bel Air 4-Door Sedan," Automobile Catalog, http://www .automobile-catalog.com/make/chevrolet_usa/full-size_chevrolet_5gen/full-size_bel _air_3gen_4-door_sedan/1959.html; "2009 Chevrolet Malibu," fueleconomy.gov, U.S. Department of Energy, https://www.fueleconomy.gov/feg/bymodel/2009_Chevrolet _Malibu.shtml.

Chapter 1 • The Auto World Gets Organized

1. "Speeding of Automobiles," *New York Times,* August 30, 1902; Flink, *Car Culture,* 25.

2. Bromley, *William Howard Taft,* 16.

3. Bromley, *William Howard Taft,* 16.

4. Flink, *America Adopts the Automobile,* 19.

5. As quoted in Flink, *America Adopts the Automobile,* 31.

6. Flink, *America Adopts the Automobile,* 29–30.

7. On the Newport scene, see Cohn, *Combustion on Wheels,* 1–4. See also Edwin Emerson Jr., "Newport in the Lead," *Automobile Magazine* 1, no. 1 (October 1899), 5–16, which Cohn draws upon without attribution.

8. Emerson, "Newport in the Lead," 12–13.

9. Emerson, "Newport in the Lead," 10–11.

10. Emerson, "Newport in the Lead," 16.

11. "New Street Menace," *Chicago Daily Tribune,* January 20, 1900.

12. On early automobile drivers, see Buerglener, "Creating the American Automobile Driver, 1898–1918."

13. Bowden, *Our Destiny and Other Poems,* 101.

14. Borg, *Auto Mechanics,* 17–18.

15. "The Licensing and Education of Chauffeurs," *Town & Country,* January 6, 1906, 16, clipping in Warshaw Collection, Automobile Industry, Box 20.

16. "What Some Missourians Think of Autos," *The Automobile* 15, no. 25 (December 20, 1906), 890.

17. This film is now widely available on the internet.

18. "Automobile Legislation in the United States," *Horseless Age* 10, no. 19 (November 5, 1902), 505–511.

19. Ward and Warren, *Silent Victories*.

20. Eno, *Story of Highway Traffic Control*, 1–2.

21. "Rules of the Road," *New York Times*, January 20, 1920.

22. "Eno Says Welsh Bill Is Bad Traffic Measure," *New York Times*, April 1, 1917.

23. Traffic Law Ordinance of the City of Nashville Regulating Traffic on the Streets (Nashville: Higginbotham, 1914), 11, Frances Loeb Library, Harvard University.

24. Flink, *America Adopts the Automobile*, 144–145.

25. Quoted in Flink, *America Adopts the Automobile*, 145.

26. For a listing and detailed description of a wide variety of suits and penalties, see "Some Leading Automobile Suits," *Horseless Age* 10, no. 19 (November 5, 1902), 512.

27. "What Is Class Legislation?" *Horseless Age* 19, no. 9 (February 27, 1907), 300.

28. Franz, *Tinkering*.

29. Ballou-Wright Automobile Supplies Catalog, 1906, 34.

30. Catalog for 35% Automobile Supply Co., 89, Division of Work and Industry, National Museum of American History, Smithsonian Institution.

31. Corn, *User Unfriendly*.

32. Ballou-Wright Automobile Supplies Catalog, 42.

33. Corn, *User Unfriendly*.

34. Orson Welles's film *The Magnificent Ambersons* (1942), which among other things examines the emergence of the auto industry during the early twentieth century, contains a humorous scene that demonstrates the difficulties of driving early automobiles over snow- and ice-covered roads.

35. Brown, "Status of the Automobile," 225.

36. Brown, "Status of the Automobile," 226.

37. Hounshell, *From the American System to Mass Production*, 220.

38. Flink, *Car Culture*, 29.

39. Donovan, *Wheels for a Nation*, 63.

40. Hounshell, *From the American System to Mass Production*, 224.

41. Albert, "The Nut behind the Wheel," 364.

42. Cain, "Fatalities and Injuries in Motor Vehicle Accidents, by Functional Category of Roadway: 1980–1995," table Df434–447.

43. The formation of the Society of Automotive Engineers and other groups dedicated to sharing engineering and technical knowledge about the automobile are tied up with the complicated history of the Selden Patent, which I have chosen to elide in this narrative for the sake of economy. For the story of the Selden Patent, see Flink, *The Automobile Age*, esp. 51–55; and Thompson, "Intercompany Technical Standardization."

44. Sinclair and Hull, *Centennial History of the American Society of Mechanical Engineers*; Usselman, *Regulating Railroad Innovation*; Noble, *America by Design*.

45. "History of Automotive Standardization," 699.

46. Usselman, *Regulating Railroad Innovation*.

47. Thompson, "Intercompany Technical Standardization," 8.

48. Mohun, *Risk*, 137.

49. Aldrich, *Safety First*.

50. Bailey, *Sure Pop and the Safety Scouts*. For another reading of *Sure Pop*, see Mohun, *Risk*, 155–157.

51. *MacPherson v. Buick*, 111 N.E. 1050 (N.Y. 1916), 1055, quoted in Clarke, *Trust and Power*, 64.

52. See E. F. Roberts, "General Inspection," *Automotive Industries* 28 (June 26, 1913), 1306–1308; A. B. Browne and E. H. Lockwood, "The Practical Testing of Motor Vehicles," *S.A.E. Transactions* 10, no. 1 (1915), 68–91; Ralph H. Sherry, "Metallurgy in the Automotive Industry," *S.A.E. Transactions* 12, no. 2 (1918), 323–350.

53. Russell Huff, "Factors of Safety," *S.A.E. Transactions* 11, Part 2 (1917), 70–87. On Huff, see also "To Push Standards Work of Auto Engineers," *New York Times*, January 9, 1916.

54. Belden, "Future Passenger Car," 147.

55. Horwitz, *Transformation of American Law*.

56. Cohn, *Combustion on Wheels*, 9–10.

Chapter 2 • Standardization Is the Answer

1. This and the following quotations come from Stevens, *Adventure*, 9–10.

2. On Whitney, see Russell, *Open Standards and the Digital Age*; and Mohun, *Risk*.

3. Stack, *Safety for Greater Adventures*, 31–32.

4. Stack, *Safety for Greater Adventures*, 33.

5. Stevens, *Adventure*, 7

6. This is according to Google's Ngram tool, which visualizes usage patterns of words and phrases over time.

7. Stack, *Safety for Greater Adventures*, 42.

8. Stack, *Safety for Greater Adventures*, 45–46, excerpting a talk, titled "Why Safety and Recreation Belong Together," that Whitney read before the Eleventh Annual Congress of the Playground and Recreation Association of America in Atlantic City on October 17, 1924.

9. Using published and archival sources, I created a social network diagram of individuals and organizations that were influential in automobile safety in the 1920s and 1930s. The organizations included various associations, engineering societies, state and federal agencies, the National Research Council, and the National Safety Council. Of the nine individuals plotted, Whitney had six known connections to safety-focused organizations, while all of the others had no more than three known connections. In this and other ways, it is easy to argue that Albert Whitney was the best connected and most influential individual in the automotive safety movement during this period.

10. See, for example, Norton, *Fighting Traffic*; Miller, "'Death Car' Reckoning," (PhD. Diss., Northwestern University, 2012), 11–12; McShane, "Origins and Globalization of Traffic Control Signals"; Mohun, *Risk*.

11. "Motor Vehicle Accidents, Death Rates, and Deaths, by Type of Accident: 1913–1996," *Historical Statistics of the United States*, table Df448–456.

12. "Motor Vehicle Registrations, by Vehicle Type: 1900–1995," *Historical Statistics of the United States*, table Df339–342.

13. Miller, "'Death Car' Reckoning," 11–12.

14. Sinclair, *Philadelphia's Philosopher Mechanics*; Usselman, *Regulating Railroad Innovation*.

15. Corn, *User Unfriendly*.

16. Norton, *Fighting Traffic*, 95–99.

17. A one-page brochure for the Glaroff Amber Vacuum is in the automotive section of the Warshaw Collection, National Museum of American History, Smithsonian Institution Archives.

18. Marks, "Tenth Anniversary of the Illuminating Engineering Society," 597.

19. The full committee consisted of Perley G. Nutting (chairman), Nelson M. Black, F. H. Gilpin, M. Luckiesh, F. K. Richtmeyer, and F. A. Vaughn. Miscellaneous Notes, *Transactions of the Illuminating Engineering Society* 9, no. 7 (1914), 2.

20. "General Report on Glare," *Transactions of the Illuminating Engineering Society* 10 (1915): 987–999.

21. "Automobile Headlights," Report No. 11 of the Committee on Glare, *Transactions of the Illuminating Engineering Society* 11 (1916): 29–35.

22. The members of the 1916–1917 committee were P. G. Nutting (chairman), C. O. Bond, William Churchill, J. R. Cravath, E. J. Edwards, W. B. Lancaster, W. F. Little, and S. C. Rogers.

23. Illuminating Engineering Society, *Specifications of Laboratory Tests for Approval of Electric Headlighting Devices*.

24. "Eastern Motor Vehicle Department Heads Organize for Co-operation," *Motor Truck News* 11 (December 1921): 16.

25. "Single Effort by Hartford Dealers Makes Better Show," *Motor Age* 39 (March 3, 1921): 31.

26. Bureau of Public Roads, *Motor-Vehicle Traffic Conditions in the United States*, part 4, p. 4.

27. "Texas Headlight Laws Ridding Roads of Glare," *Automotive News*, October 27, 1925.

28. California Department of Motor Vehicles, *Table of Approved Headlight Devices*.

29. "Abstract of Discussion," in J. R. Cravath, "The Headlight Glare Problem," *Transactions of the Illuminating Engineering Society* 12 (1917): 190.

30. On the number of headlights approved by the motor vehicle administrators, see Clarke, *Trust and Power*, 155–156.

31. Clarke, *Trust and Power*, 156.

32. Advertisement for School of Automotive Electricity, *Popular Mechanics* (January, 1922), 93.

33. H. M. Johnson, review of *Psychology and the Motorist*.

34. "State Crusades Inspire Sales of Headlights," *Automotive News*, October 19, 1925.

35. The Weaver Manufacturing Company has told me that it has no company archives, but it has put a helpful historical page on the Safety Lane online: "Weaver History Museum," Castle Equipment Company, http://www.castleequipment.com /Museum/safety_lane_weaver_history.htm.

36. The best history of the American Engineering Standards Committee to date is Russell, *Open Standards and the Digital Age*, ch. 3. See also Knowles, *Disaster Experts*, 139.

37. Stack, *Safety for Greater Adventures*, 35.

38. The details of this paragraph and the following one are taken mostly from Stack, *Safety for Greater Adventures*, 11–17. This work includes both a biography written by Whitney's colleague Stack and a compilation of excerpts from some of Whitney's speeches and writings.

39. A few historians of technology have recently (and independently, I think) rediscovered Whitney as an important historical actor. See Packer, *Mobility without Mayhem*; Mohun, *Risk*; and Russell, *Open Standards and the Digital Age*.

40. On the Merritt Committee, see Knowles, *Disaster Experts*.

41. Aldrich, *Safety First*.

42. Stack, *Safety for Greater Adventures*, 92.

43. Hord, *History and Organization of Automobile Insurance*, 1.

44. Friedman, *American Law in the Twentieth Century*, 364.

45. Hord, *History and Organization of Automobile Insurance*, 3.

46. Friedman, *American Law in the Twentieth Century*, 264–265.

47. Usselman, *Regulating Railroad Innovation*; Burke, "Bursting Boilers and the Federal Power."

48. Whitney, "On Assuming the Chairmanship of the Workmen's Compensation Service Bureau" (1914), quoted in Stack, *Safety for Greater Adventures*, 29.

49. Whitney, "How the National Bureau of Casualty and Surety Underwriters Developed from Early Rating Bodies" (1939), quoted in Stack, *Safety for Greater Adventures*, 19.

50. Whitney, "Application of the National Bureau of Casualty and Surety Underwriters for Consideration in Respect to the Award to Be Made by the American Trade Association Executives" (1931), quoted in Stack, *Safety for Greater Adventures*, 33.

51. Whitney, quoted in Stack, *Safety for Greater Adventures*, 19.

52. Stack, *Safety for Greater Adventures*, 81–82.

53. Whitney, *Place of Standardization in Modern Life*.

54. McShane, "Origins and Globalization of Traffic Control Signals."

55. "Proposed Standardization of Colored Signal Lights," 206. Rudd is a largely unexplored character in the history of engineering and technology, though Steven W. Usselman cites a few letters from Rudd in the signals and safety chapter of *Regulating Railroad Innovation*.

56. John O'Brien, "Flashlight Signal Control of Traffic," *Transactions of the Illuminating Engineering Society* 17, no. 5 (May 1922): 249–250.

57. Harriss, "Signal Control of Traffic," 246–247. Interestingly, there seems to be some confusion about how the New York signals actually worked, though Harriss's account can be taken as authoritative since he built the systems.

58. These materials are found in a folder marked "Traffic Signals" in the Transportation Library of the Smithsonian Institution's National Museum of American History.

59. *Year Book of the American Engineering Standards Committee 1923*, 33.

60. *Year Book of the American Engineering Standards Committee 1923*, 15; Stack, *Safety for Greater Adventures*, 82.

61. *Year Book of the American Engineering Standards Committee 1923*, 15.

62. Whitney, *Man and the Motor Car.*

63. Hoover, *Memoirs of Herbert Hoover*, 42.

64. On the criticism that standardization leads to conformism and other problems, see Russell and Vinsel, "Dynamic Interplay between Standards and Routines."

65. Hoover, quoted in Whyte, *Hoover*, 274–275.

66. For a more complete story of the role of the National Bureau of Standards in automotive regulation during the 1920s and 1930s, see Vinsel, "Virtue via Association."

67. Another division, dedicated to electric light, contributed to formulating headlight standards and standardizing traffic signals, but it did so in conjunction with the Heat Division.

68. National Bureau of Standards, *National Directory of Commodity Specifications.*

69. Hoover, *Memoirs*, 67–68.

70. Burgess quoting a letter from Whitney in one of the essays that Burgess delivered before the American Engineering Standards Committee; "Promoting the Use of Specifications Certification Plan," 1, National Bureau of Standards (hereafter NBS), General Correspondence 1925, IDS, Box 116, National Archives and Records Administration, College Park, Md. (hereafter NARA).

71. Mohun, *Risk*, 180.

72. Monthly Report, March or April 1920, p. 4, NBS, Heat Division Monthly Reports, Box 1, NARA.

73. Monthly Report, July 1921, p. 2, NBS, Heat Division Monthly Reports, Box 1, NARA.

74. Monthly Report, December 1921, p. 3, NBS, Heat Division Monthly Reports, Box 1, NARA.

75. S. Von Ammon, "Developing a Method for Testing Brake-Linings," SAE Technical Paper 220014 (1922).

76. National Bureau of Standards, *Standards Yearbook*, 196–197.

77. A. B. Barber to Vernon Kellogg, June 23, 1925, "AG&Depts: Commerce: National Conference on Street and Highway Safety, 1924–1926," Committee on the Psychology of the Highway, National Research Council, Archives of the National Academy of Sciences, Washington, D.C.

78. On Barber, see Norton, *Fighting Traffic*, esp. 187–193.

79. Hawley, "Herbert Hoover,." For a recent updating and restatement, see Balogh, *The Associational State.*

80. The full list of supporters for the 1924 conference and the amount each gave are as follows: the American Automobile Association, the American Electric Railway Association, the American Mutual Alliance, and the American Railway Association each gave $2,500. The US Chamber of Commerce, the National Automobile Chamber of Commerce, and the National Bureau of Casualty and Surety Underwriters each gave $4,000. The National Association of Taxicab Owners gave $500, and the National Safety Council, $1,500. Later, the National Research Council chipped in $1,000. All of these figures come from a document in the archives of the National Research Council, in which someone had recorded the various contributions of the groups in a pamphlet titled "Questions and Answers Based on the Report of the

National Conference on Street and Highway Safety" in "AG&Depts: Commerce: National Conference on Street and Highway Safety, 1924–1926."

81. Carson, "The Science of Merit," 182. See also Kevles, "Testing the Army's Intelligence."

82. National Conference on Street and Highway Safety, *Report of the Committee on the Causes of Accidents*, 8.

83. Sidney Williams discussed the form's creation with the psychologist Alvhh Lauer in the early 1930s. Alvhh Lauer, "Report of the Interview with the National Safety Council," September 23, 1928, "Committee on the Psychology of the Highway— Work at Ohio State University," Records of the Committee on the Psychology of the Highway, National Research Council, Archives of the National Academy of Sciences, Washington, D.C. (hereafter NRC). A government report published later contained an example of an advertisement from Oklahoma that described that state's use of the NSC's standardized accident reporting form and suggested its use to others. Bureau of Public Roads, *Motor Vehicle Traffic Conditions in the United States*, Part 3: *Inadequacy of State Motor-Vehicle Accident Reporting*, 9.

84. Moss, "Standardized Tests for Automobile Drivers"; Snow, "Tests for Chauffeurs"; Viteles, "Transportation Safety by Selection and Training"; Weschler, "Tests for Taxicab Drivers."

85. Burnham, *Accident Prone*.

86. Weiss and Lauer, *Psychological Principles in Automotive Driving*.

87. For a more complete account of the NRC psychologists and their relationship to automobile safety, see Vinsel, "'Safe Driving Depends on the Man at the Wheel.'"

88. R. S. Woodworth requested funds in a letter to Whitney dated August 18, 1925. Whitney replied to Woodworth, August 24, 1925. Both in "Committee on Psychology of Highway, 1925–1926," Records of the Committee on the Psychology of the Highway, NRC.

89. Whitney, quoted in Follows, "Safety and Welfare Work in the Engineer's Education," 547.

90. Dewey, *Human Nature and Conduct*.

91. Whitney, *How Knowledge Driveth Away Fear*, 5.

92. Whitney noted in an address on driver's education, "A habit is something that gets rooted for the most part in the reflex nerve centers." Elsewhere he wrote: "We must begin with safety as it appears on the lower, simpler plane of physical well-being, where the response from the child comes not from the upper reaches of the mind but from the lower depths of instinctive feeling." Whitney, "The Place of Classroom Instruction in Driver Education," excerpted in Stack, *Safety for Greater Adventures*, 55; Whitney, *The Inner Meaning of the Safety Movement Particularly in Relation to the Problem of Education* (New York: National Safety Council, 1928), excerpted in Stack, *Safety for Greater Adventures*, 43.

93. Whitney, *How Knowledge Driveth Away Fear*, 13.

94. Whitney, *How Knowledge Driveth Away Fear*, 15.

95. Whitney, *Inner Meaning of the Safety Movement*, 43.

96. Space doesn't allow me to spell out how Whitney's Darwinian model connects to Dewey and the pragmatist tradition. How pragmatism was in part an extension of Darwin's ideas to human thought and psychology is a major theme of Louis Menand,

The Metaphysical Club: A Story of Ideas in America (New York: Farrar, Straus and Giroux, 2002).

97. Whitney, *Place of Standardization in Modern Life*, 5. I owe my awareness of this quotation to Andy Russell, who used it in another context in an essay we cowrote, "Dynamic Interplay between Standards and Routines."

98. Whitney, *Inner Meaning of the Safety Movement*, 43.

99. Aldrich, *Safety First*.

100. Dewey, *The Quest for Certainty*, 7.

101. The book also mentions how proper street lighting greatly reduced accidents. Whitney, *Man and the Motor Car*, 93.

102. Stack, *Safety for Greater Adventures*, 29, 11.

103. Based on querying the exact phrase "safety movement" in Google's Ngram tool, which uses the database of all books scanned into Google Books to track historical trends in word and phrase usage.

104. Bureau of Public Roads, *Motor Vehicle Traffic Conditions in the United States*, part 4, p. 4.

105. Bureau of Public Roads, *Motor Vehicle Traffic Conditions in the United States*, Part 1: *Uniformity of State Motor-Vehicle Traffic Laws*, 10.

106. Mohun, *Risk*.

107. Joel W. Eastman, *Styling vs. Safety. The American Automobile Industry and the Development of Automotive Safety, 1900–1966* (Lanham, Md.: University Press of America, 1984), 181.

108. Harry Armand, "Make the Automobile Safer!" *Safety Engineering* (March 1938), 35.

109. Harry Armand, "Safety and the New Cars," *Safety Engineering* (December 1938): 32–34.

110. On the Automotive Safety Foundation, see Norton, *Fighting Traffic*, 247–248, 251–254; and Eastman, *Styling vs. Safety*, esp. 139–145.

111. Hoffman, *Seven Roads to Safety*.

Chapter 3 • The Creation of Crashworthiness

1. [John Updike], "Packaged," *New Yorker*, September 29, 1956. The piece was in the "Talk of the Town" section, which did not feature bylines at that time, but the *New Yorker*'s digital archive claims that Updike wrote it, and the sarcastic, ironic prose certainly *sounds* like Updike.

2. The best work on Truman's use of safety conferences is Richard E. Weingroff's long, book-length essay, "President Dwight D. Eisenhower and the Federal Role in Highway Safety," which is available on the Federal Highway Administration webpage, http://www.fhwa.dot.gov/infrastructure/safety.cfm.

3. This and the following two paragraphs draw heavily on Gangloff, "Medicalizing the Automobile."

4. DeHaven, "Mechanical Analysis of Survival in Falls."

5. DeHaven, "Mechanical Analysis of Survival in Falls," 546, 539.

6. "Obituary: Neurosurgeon Dr. Elisha Gurdjian was a Pioneer in His Medical Field," *Detroit Free Press*, December 8, 1985; Wayne State University Press Release, November 30, 1985; E. S. Gurdjian CV; L. M. Thomas, "Elisha S. Gurdjian, M.D.: A

Biographical Sketch," *Clinical Neurosurgery: Proceedings of the Congress of Neurological Surgeons* (1971); xvii-xix, all in Wayne State University Archives, clipping files.

7. "Prof. Lissner Dies, Taught for 26 Years," *Inside Wayne*, June 23, 1965; Herbert Lissner curriculum vitae, both in Wayne State University Archives, clipping files.

8. Colonel John P. Stapp, "Past, Present, and Future of Biomechanics at Wayne State University," 391, Wayne State University Archives, clipping files.

9. On the history of materials testing, see, for instance, Usselman, *Regulating Railroad Innovation*; and Slaton, *Reinforced Concrete and the Modernization of American Building*.

10. "Simmons and the Strain Gage," 20; Nielsen, *From Locomotives to Strain Gages*.

11. All quotations in this paragraph are from Gurdjian and Lissner, "Mechanism of Head Injury," esp. 393.

12. Perry and Lissner, *Strain Gage Primer*, 245.

13. Gurdjian and Lissner, "Mechanism of Head Injury," 394; Patrick, "A Look Back on My Stapp Conferences,".

14. Lissner, "Human and Animal Impact Studies in U.S. Universities."

15. Patrick, "A Look Back on My Stapp Conferences," xi.

16. Roach, *Stiff*, 96–97.

17. Patrick, Lissner, and Gurdjian, "Survival by Design," 494–495.

18. This and the following quotations in the paragraph are from "The Fastest Man Alive," *Time* (September 12, 1955), 80–88. See also Ryan, *Sonic Wind*.

19. "The Fastest Man Alive."

20. "Conference Founder," in *Proceedings of Tenth Stapp Car Crash Conference*, v. See also Bushnell, "Administrative History of the Aeromedical Field Laboratory."

21. "The Fastest Man Alive."

22. Siegel, *Forensic Media*, 162.

23. Siegel, *Forensic Media*, 162.

24. "The Fastest Man Alive."

25. As quoted in Bushnell, "Administrative History of the Aeromedical Field Laboratory," 24.

26. "The Fastest Man Alive."

27. "The Fastest Man Alive."

28. Ryan, *Sonic Wind*, 198–200, 207.

29. Siegel, *Forensic Media*, 164.

30. Patrick, "A Look Back on My Stapp Conferences," ix.

31. On the role of witnessing in scientific culture, see Shapin and Schaffer, *Leviathan and the Air Pump*.

32. Col. Robert Muldrow, "Welcome," in *Proceedings of Tenth Stapp Car Crash Conference*, ix.

33. Eastman, *Styling vs. Safety*, 191. Eastman's suggestion that industry pressure ultimately halted Stapp's research appears to be based on an interview with Stapp.

34. I base this claim on keyword searches in Google Books, JSTOR, and other academic databases.

35. Yoganandan, Nahum, and Melvin, *Accidental Injury*, 84.

36. *The Elmer A. Sperry Award for 1967 to Edward R. Dye, Hugh DeHaven, and Robert A. Wolf* (Detroit: Society of Automotive Engineers, 1968), 12, (booklet available

online at http://www.sperryaward.org/awardbooklets/1967%20Sperry%20Award%20Booklet.pdf).

37. Gangloff, "Medicalizing the Automobile," 180.

38. Nader, *Unsafe at Any Speed*, 66.

39. Eastman, *Styling vs. Safety*, 221.

40. Gangloff, "Medicalizing the Automobile"; Eastman, *Styling vs. Safety*, 220–221.

41. "Notes on Planning Conference," December 16–17, 1952, Papers of the Crash Injury Research Project, Box 1, Folder 2, Cornell Medical College Archives, New York City. The groups represented at the conference included Ford, Chrysler, General Motors, Packard, Studebaker, the National Safety Council, the Automobile Manufacturers Association, U.S. Army Ordnance, the Office of Naval Research, the Army Research and Development Command, the Metropolitan Life Insurance Company, the Association of Casualty and Surety Companies, the Farm Bureau Mutual Automobile Insurance Company, the Cornell Aeronautical Laboratory, Motor Vehicle Research, Dunlap and Associates, and state police from Indiana, Virginia, Connecticut, and New Hampshire.

42. *Elmer A. Sperry Award for 1967*, 14–15, 21–22.

43. Friedman, *American Law in the Twentieth Century*, 364–365.

44. [Updike], "Packaged."

45. *Elmer A. Sperry Award for 1967*, 21–22.

46. Nader, *Unsafe at Any Speed*, 191–192.

47. Severy, Siegel, and Mathewson, *Statement on Crashworthiness of Automotive Seatbelts*, 1. On the rise of crash safety research at UCLA, see Siegel, *Forensic Media*, 175–187.

48. Paul Niquette, *101 Doozies I've Met* (1996), http://niquette.com/books/doozies/dooz101.html.

49. Bierman and Larsen, "Reactions of the Human to Impact Forces Revealed by High Speed Motion Picture Technique."

50. Quoted in Siegel, *Forensic Media*, 177.

51. Severy, Siegel, and Mathewson, *Statement on Crashworthiness of Automotive Seatbelts*, 2.

52. Siegel, *Forensic Media*, 187.

53. Stonex and Skeels, "Summary of Crash Research Techniques Developed by the General Motors Proving Grounds."

54. Stonex and Skeels, "Summary of Crash Research Techniques Developed by the General Motors Proving Grounds," 8–9.

55. This and the following two quotations are from Stonex and Skeels, "Summary of Crash Research Techniques Developed by the General Motors Proving Grounds," 9.

56. Stonex, "Roadside Design for Safety"; Stonex, "How Roadside Ditches and Slopes Can Be Designed for Safety"; Cschowski, Skeels, and Hawkins, "Appraisal of Guardrail Installations by Car Impact and Laboratory Tests"; Lundstrom and Skeels, "Full-Scale Appraisals of Guardrail Installations by Car Impact Tests."

57. All quotations in this paragraph are from Gandelot, "Engineering Safety into Automobile Bodies."

58. Nader, *Unsafe at Any Speed*, 87.

59. This quotation and the following one are from Mehling, "Big Three Fight over How Safe to Make Cars." For further discussion of Gandelot's response, see Luger, *Corporate Power, American Democracy, and the Automobile Industry*, 62; and in Byrne, *The Whiz Kids: Ten Founding Fathers of American Business—and the Legacy They Left Us* (New York: Doubleday, 1993), 261.

60. See Byrne, *The Whiz Kids*, ch. 17, esp. 256–264. Byrne's account of Ford, McNamara, and auto safety is problematic in that it relies almost wholly on interviews with McNamara even though Byrne concedes at several points that McNamara's memory was often faulty. It is, however, still the best account we have of internal Ford workings regarding safety during this period.

61. "Alex Haynes, 89, Automotive Safety Pioneer," *New York Times*, September 3, 1999.

62. Eastman, *Styling vs. Safety*, 222–223.

63. Eastman, *Styling vs. Safety*, 222–225.

64. Revealing images of one of these steering wheels, pre- and post-impact, can be seen in Ford Motor Company, *Freedom of the American Road*, 103.

65. Byrne, *The Whiz Kids*, 258. This fact is also mentioned in the Errol Morris film *The Fog of War* (2003).

66. Byrne, *The Whiz Kids*, 259.

67. Byrne, *The Whiz Kids*, 260. Video of the Mary Martin and Ethel Merman Ford Fiftieth Anniversary Show can be found online.

68. Mehling, "Big Three Fight over How Safe to Make Cars," 40.

69. Byrne, *The Whiz Kids*, 261.

70. Nader, *Unsafe at Any Speed*, 88.

71. As with many sayings passed around the auto industry, the ultimate source of this phrase is hard to pin down. Byrne (*The Whiz Kids*, 262) points out that McNamara incorrectly attributed it to a headline in the industry newspaper *Automotive News*. The primary public expression of this phrase may have been an article in *Fortune* Magazine, though the writer credited it as "a favorite crack making the rounds at GM." See Cutlip, *The Unseen Power* (unpaginated digital copy).

72. Byrne, *The Whiz Kids*, 262.

73. Eastman, *Styling vs. Safety*, 241. Eastman's source is an oral history interview he conducted with Roberts.

74. Siegel, *Forensic Media*, 164.

75. See, for example, U.S. Senate, Committee on the Judiciary, "Compensation to Mrs. Cecil N. Broy for Husband's Death in Automobile Accident While in Foreign Service"; U.S. Senate, Committee on the Judiciary, "Compensation for Civilian Injuries and Death from Collision of Army Truck and Automobile."

76. U.S. Senate, Committee on Labor and Public Welfare, "Motor Vehicle Safety Responsibility Act of District of Columbia." July 10, 1953. P.L. 83-365.

77. The other original members of the subcommittee were Walter Rogers (D-Texas), John V. Beamer (R-Indiana), Samuel N. Friedel (D-Maryland), and Paul F. Schenck (R-Ohio). US House of Representatives, "Highway Traffic Safety," Report of the Special Subcommittee on Traffic Safety, 84th Congress, 2nd sess., January 3, 1957. Within a year, the subcommittee was joined by Alvin R. Bush (R-Pennsylvania) and J. Carlton Loser (D-Tennessee).

78. U.S. House of Representatives, *Traffic Safety*, 7.

79. U.S. House of Representatives, *Traffic Safety*, 429.

80. For Douglas's comments on horsepower, see U.S. House of Representatives, *Traffic Safety*, 4. Douglas also submitted a number of editorials and magazine articles criticizing the horsepower wars.

81. Many of the witnesses at the traffic safety hearings mentioned horsepower. It was a *known* problem.

82. See Chayne's charts comparing 1941 and 1956 vehicles as to horsepower and top speeds, in U.S. House of Representatives, *Traffic Safety*, 340–341.

83. U.S. House of Representatives, *Traffic Safety*, 355.

84. Eastman, *Styling vs. Safety*, 99–100.

85. Eastman, *Styling vs. Safety*, 146–148; Luger, *Corporate Power, American Democracy, and the Automobile Industry*, 59–60; Nader, *Unsafe at Any Speed*, 203–215.

86. U.S. House of Representatives, *Traffic Safety*, 60.

87. U.S. House of Representatives, *Automobile Seat Belts*, 6.

88. U.S. House of Representatives, *Research Needs in Traffic Safety: Hearing before a Subcommittee of the Committee on Interstate and Foreign Commerce, House of Representatives*, 85th Congress (1958).

89. Luger, *Corporate Power, American Democracy, and the Automobile Industry*, 66.

90. John M. Hanson, deputy assistant commissioner, Motor Equipment Division, General Services Administration, estimated that the US government owned 224,000 vehicles, with 116,213 in civil agencies and 108,743 in the military. U.S. House of Representatives, *Motor Vehicle Safety*, 47–48.

91. U.S. House of Representatives, *Motor Vehicle Safety*, 53–57.

92. I take these numbers from the historical profits of Fortune 500 companies compiled at archive.fortune.com.

93. This and following quotations were transcribed by the author from "John O. Moore," NYPR Archive Collections, *WNYC*, August 25, 1959, https://www.wnyc.org/story/john-o-moore/.

94. One other aspect of Moore's appearance on this program is worth noting. One of the students pointed out that many people did not use seat belts when they were available. Moore replied that this was a criticism not of the seat belt but of human behavior and likened it to the polio vaccine, which prevented 95 percent of polio cases, yet half the population of New York City had not received it. He was happy to say that a number of organizations, including the Public Health Service and the National Safety Council, had agreed to a long-term public education effort encouraging people to buy and use seat belts.

Chapter 4 • From Movement to Government Agency

1. McCarry, *Citizen Nader*, 21.

2. Buckhorn, *Nader*, 13–15.

3. Buckhorn, *Nader*, 16.

4. Buckhorn, *Nader*, 19.

5. Hodgson, *The Gentleman from New York*, 54.

6. US Senate, Committee on Government Operations, *Federal Role in Traffic Safety*, vol. 1, p. 251.

7. Kearney, "A Safety Expert Looks at the '59 Cars," 90.

8. Kearney, "A Safety Expert Looks at the '59 Cars," 90.

9. "Medical News," 1597.

10. Gangloff, "Medicalizing the Automobile," 291.

11. "Medical News," 1597. On Haddon's alcohol research, see Barron H. Lerner, *One for the Road: Drunk Driving since 1900* (Baltimore: Johns Hopkins University Press, 2011).

12. Woodward, "Medical Criticism of Modern Automotive Engineering." See also Siegel, *Forensic Media*, 157–158, 161–162.

13. Woodward, quoted in Siegel, *Forensic Media*, 158.

14. Francis, "Cars *Can* Be Safer." The artist Stewart Rouse's concept car drawing is on page 83.

15. Woodward, "Medical Criticism of Modern Automotive Engineering," 631.

16. Eastman, *Styling vs. Safety*, 194–196.

17. In her dissertation on Hugh DeHaven, Amy Gangloff argues that his and the other Cornell researchers' work constitutes a form of "medicalization," a term that first emerged in sociology to describe how nonmedical phenomena come under medical authority. Childbirth and social deviance are two classic examples of things that became medicalized over time. Gangloff's use of the medicalization concept is fruitful, insightful, and valuable. My own understanding, however, is that the Cornell researchers and others in their community drew on medical metaphors in their rhetoric *when it seemed helpful to do so*. At other times, they ditched medical metaphors and relied, for instance, on ones derived from design or engineering.

18. Walsh McDermott to August Heckecher, September 2, 1955, Records of the Crash Injury Research Project, Box 1, Folder 1, Cornell University Medical College.

19. Moynihan, "Epidemic on the Highways." Later, as the director of the Joint Center for Urban Studies of the Massachusetts Institute of Technology and Harvard University, Moynihan led and wrote a lesser-known study on highway safety: US Department of Health, Education, and Welfare, *Report of the Secretary's Advisory Committee on Traffic Safety*.

20. Moynihan, "Epidemic on the Highways," 17.

21. McCarroll and Haddon, "A Controlled Study of Fatal Automobile Accidents in New York City"; Haddon et al., "A Controlled Investigation of the Characteristics of Adult Pedestrians Fatally Injured by Motor Vehicles in Manhattan"; Haddon, "Research with Respect to Fatal Accident Causes." As the head of the epidemiology research program, Haddon also published on a number of non-auto-related public health topics, including skiing injuries and the relationship between fluoridated water and dental caries and osteoporosis. See, for example, Ellison et al., "Skiing Injuries"; Carlos, Gittelsohn, and Haddon, "Caries in Deciduous Teeth in Relation to Maternal Ingestion of Fluoride"; Goggin et al., "Incidence of Femoral Fractures in Postmenopausal Women."

22. Haddon, "A Note concerning Accident Theory and Research with Special Reference to Motor Vehicle Accidents." A good example of Haddon's explicitly methodological work is Levison and Haddon, "The Area Adjusted Map."

23. See Nader, Unsafe at Any Speed.

24. As quoted in Lerner, *One for the Road*, 57. Lerner has written the best account we yet have of Haddon, though it is necessarily focused more on Haddon's work on drunk driving than it is on other aspects of the epidemiologist's career. Lerner is likely also the only historian so far who has examined Haddon's private papers, which remain in the possession of his family.

25. Gangloff, "Medicalizing the Automobile," 292–293.

26. Ullmann, "Some Economic Aspects of Automobile Safety Devices," 263.

27. Moynihan, "Legal Regulation of Automobile Design," 265.

28. Moynihan, "Legal Regulation of Automobile Design," 271.

29. Moynihan, "Legal Regulation of Automobile Design," 278–279.

30. Moynihan, "Legal Regulation of Automobile Design," 279.

31. At a deep level, this belief that traditional views of accidents were caught up in folklore and superstition was tied to science's suspicion of what would later be called indigenous knowledge. This suspicion went hand in hand with top-down Cold War–era "Third World" development initiatives, and it came under heavy fire from academics during the ascendancy of postmodernism.

32. Haddon, "Changing Approach to the Epidemiology, Prevention, and Amelioration of Trauma," 1433.

33. Improvements in ambulances and other factors in emergency care and medicine have been critical in reducing automotive death and injury since the 1960s, and many of these changes have resulted from policymaking. See Andrew T. Simpson, "Transporting Lazarus."

34. Haddon, "Changing Approach to the Epidemiology, Prevention, and Amelioration of Trauma," 1436.

35. On the linear model, see Stokes, *Pasteur's Quadrant*; and Godin, "Linear Model of Innovation."

36. Quoted in O'Connell and Myers, *Safety Last*, 217–218.

37. O'Connell and Myers, *Safety Last*, 218.

38. US Senate, Committee on Government Operations, *Federal Role in Traffic Safety*, Vol. 1, p. 201.

39. Franklin Floete to Sec of Interior, October 18, 1960, Record Group 269: 36/1/4, Central File, 1959–1961, Box "Vehicles," Folder 1, Records of the General Services Administration, NARA.

40. The participating groups are listed in US Senate, Committee on Government Operations, *Federal Role in Traffic Safety*, Vol. 1, p. 204.

41. Nader, Unsafe at Any Speed, 226.

42. On the seventeen devices, see David R. Jones, "U.S. Agency Spurs Car Safety Drive," *New York Times*, January 27, 1965; and Jones, "Safer Autos," *Washington Post*, August 18, 1965.

43. R. Bazzell, "The SAE Manikin Interior Space Evaluator," SAE Technical Paper 630176 (1963).

44. One example of criticism of chrome and other literally flashy and glaring design features is John Keats's 1958 book, *The Insolent Chariots*, discussed below.

45. As quoted in Nader, *Unsafe at Any Speed*, 137.

46. Nader, *Unsafe at Any Speed*, 226–231.

47. Nader, *Unsafe at Any Speed*, 228.

48. Miller, quoted in Luger, *Corporate Power, American Democracy, and the Automobile Industry*, 67.

49. "Expert at Hearing Condemns GSA Auto Safety Standards," *Washington Post*, March 30, 1966.

50. Nader, *Unsafe at Any Speed*, 229.

51. "Auto Safety Panel to Aid GSA Buying," *Washington Post*, May 21, 1966. The other members of the committee were two medical doctors and the chief of special projects in the Republic Aviation Division of Fairchild Hiller Corporation.

52. Robert Irvin, "Auto Makers Troubled by Federal Safety Rules," *Washington Post*, October 23, 1966.

53. Buckhorn, *Nader*, 3–5.

54. Buckhorn, *Nader*, 5; Robert Darnton, "Accidents Called a Major Disease," *New York Times*, December 20, 1964.

55. Buckhorn, *Nader*, 7.

56. Bowen, *Ralph Nader*, 32.

57. On Katz, see Lemov, *Car Safety Wars*, 24–26.

58. Katz, "Liability of Automobile Manufacturers for Unsafe Design of Passenger Cars," 864.

59. As quoted in Lemov, *Car Safety Wars*, 26, drawing on an interview Lemov conducted with Nader.

60. Lemov, *Car Safety Wars*, 30.

61. Lemov, *Car Safety Wars*, 30.

62. The article is reprinted in *The Ralph Nader Reader*.

63. Hodgson, *Gentleman from New York*, 62.

64. Lerner, *One for the Road*, 44.

65. Moynihan, "The Legal Regulation of Automobile Design," 277.

66. Hofstadter, *Anti-intellectualism in American Life*, esp. chapter 13.

67. Keats, *Schools without Scholars*. On Nabokov and Keats on progressive education, see Miller and Nowak, *The Fifties*, 256–257.

68. As quoted in Miller and Nowak, *The Fifties*, 257.

69. Tenney, *Highway Jungle*.

70. Tenney, *Highway Jungle*, 156.

71. Nader, *Unsafe at Any Speed*, 219.

72. The Nader quotation on the Corvair comes from archival footage in the documentary film *An Unreasonable Man* (2006).

73. The address was later published as "Taming the Corporate Tiger," reprinted in *The Ralph Nader Reader*.

74. Nader, *Unsafe at Any Speed*, vii.

75. Holsworth argues that Nader's focus on the individual makes his environmental arguments incoherent. Nader cannot argue for environmental conservation beyond pollution's effect on human health, and the role that consumerism plays in environmental degradation is a blind spot for him. This conflict is best captured in the title and substance of a subsection of Holsworth's book: "Self-Interest and Scarcity: Nader's Contradiction." Holsworth, *Public Interest Liberalism and the Crisis of Affluence*, 53–56.

76. Nader, "Taming the Corporate Tiger," 136.

77. Nader, "Taming the Corporate Tiger," 137.
78. McCarry, *Citizen Nader*, 81; Lemov, *Car Safety Wars*, 94.
79. McCarry, *Citizen Nader*, 81–82.
80. McCarry, *Citizen Nader*, 81.
81. McCarry, *Citizen Nader*, 81.
82. US Senate, Committee on Public Works, *Highway Safety Act: Hearings on S. 3052*, 89th Cong., 2nd sess. (1966), 248. By the "idealized norms of science," I am referring to the so-called Mertonian norms of science first put forward in Robert K. Merton, "The Normative Structure of Science."
83. US Senate, Committee on Public Works, *Highway Safety Act*, 248.
84. "The White House Message on Transportation," p. 1, Finding Aid UD–UP–3, Box 1, Folder 2, Record Group 416, Records of the National Highway Traffic Safety Administration, NARA (hereafter cited as Records of the NHTSA).
85. Brenner, "Analysis of Traffic Flow Using 'Time' as the Dependent Value."
86. The records of this group, the Federal Highway Agency Task Force, are contained in "FHA Task Force Proposed Organization," Finding Aid UD–UP–3, Box 2, Folder 11, Record Group 416, Records of the NHTSA.
87. On RAND, see Hounshell, "The Cold War, RAND, and the Generation of Knowledge"; Jardini, "Out of the Blue Yonder"; and Light, *From Warfare to Welfare*.
88. Brenner, "Analysis of Traffic Flow Using 'Time' as the Dependent Value," 1.
89. "Proposed Statement of Mission, Federal Highway Administration, Department of Transportation," in "FHA Task Force Proposed Organization," 3.
90. Brenner, "The Future of Highway Research."
91. Brenner, "The Future of Highway Research," 59.
92. Thus, where Haddon advocated the linear model, Brenner championed what Donald Stokes calls Pasteur's Quadrant. Stokes, *Pasteur's Quadrant*.

Chapter 5 • The Limits of Federal Automotive Safety Regulation

1. McCubbins, Noll, and Weingast, "Political Origins of the Administrative Procedure Act."
2. I have heard rumors, perhaps apocryphal, of a flow chart used in the Environmental Protection Agency to explain the agency's standards-making process. After illustrating the steps mandated by the Administrative Procedure Act, the chart moves to the final step, wherein at least one regulated firm sues the agency, claiming that the rule is unworkable and unfair and that it will most certainly destroy the industry. Thus, the courtroom becomes a de facto step in the regulatory process. As Robert A. Kagan has shown, regulation in the United States, both during the rule-making process and through post-rule-making legal challenges, always takes place in an adversarial environment. Kagan, *Adversarial Legalism*.
3. "Report on the Development of the Initial Federal Motor Vehicle Safety Standards," iv, NHTSA Microfilm, Reel 2, Records of the NHTSA.
4. "Car Safety Rule Termed Illegal," *New York Times*, March 7, 1967.
5. National Highway Safety Bureau (NHSB), "Brief in Support of Proposed Findings of Fact: Before Presiding Officer Russell A. Potter," 7, NHTSA Microfilm, Reel 2, Records of the NHTSA.

6. On the history of consensus standards, see Russell, *Open Standards and the Digital Age.*

7. "Report on the Development of the Initial Federal Motor Vehicle Safety Standards," iii.

8. "Report on the Development of the Initial Federal Motor Vehicle Safety Standards," iii.

9. US Department of Commerce, National Highway Traffic Safety Agency, "Report on the Development of the Initial Federal Motor Vehicle Safety Standards Issued January 31, 1967," March 17, 1967, p. 6, NHTSA Microfilm, Reel 1, Records of the NHTSA (original quote in *Congressional Record*, August 31, 1966, p. 20600).

10. NHTSA, "Report on the Development of the Initial Federal Motor Vehicle Safety Standards Issued January 31, 1967," 58, 110. Standard 201 was based on GSA standards 515/2a—Padded Instrument Panel and Visors for Automotive Vehicles; 515/3a—Recessed Instrument Panel Instruments and Control Devices for Automotive Vehicles; 515/18—Window and Door Controls; 515/19—Ashtrays and Lighters; 515/20—Armrests; and 515/21—Padding for Automotive Seat Backs. The GSA published its finalized standards in the *Federal Register* on June 30, 1965, pp. 8319–8327.

11. The tests remain largely unchanged today. See, for instance, the SAE website for the standards, https://www.sae.org/standards/.

12. Bazzell, "The SAE Manikin Interior Space Evaluator" (paper presented at the National Automobile Meeting, Detroit, March 19–21, 1963), 1. Bazzell was an engineer in Chrysler's Engineering Division. The difference between Chrysler's role in auto safety and its role in emissions control was partly related to the difference in their relationship to the firm's core: Chrysler designed its own bodies and, thus, had vested interests in playing some part in shaping the industry standard, but the firm had much fewer capabilities affecting chemistry and chemical processes and often relied on suppliers, including GM in the case of emissions control, for developments in these fields.

13. Kaptur and Myal, "The General Motors Comfort Dimensioning System," (paper presented at SAE International Congress of Automotive Engineering, Detroit, January 1961).

14. Hansen, "SAE Test Procedure for Instrument Panels," 1. Hansen worked as the supervisor of material development in Chrysler's Engineering Office.

15. The NHSB never spelled out the intended purpose for the larger and smaller manikins in its publications. But, if the expectations were not plain enough, at a hearing, one NHSB attorneys, Robert O'Mahoney (discussed below), asked one of the auto industry's engineers, "Let me ask you, sir: Would it be your understanding of the definition which uses as one outer limit the 95th percentile male and as another limit the fifth percentile female, that it was intended to protect the segment of the population that has sizes between the 95th percentile male and 5th percentile female?" To this question, Edward C. Adkins, a representative for General Motors replied, "I object to this line of questioning. This came up several times. The intent of the standard is not in issue here. I think that such questions are irrelevant and particularly so since this witness is asked to give an opinion as to what he might

conceive to be the intent of the standard." ("Transcript of Proceedings in the Matter of Reconsideration of Initial Motor Vehicle Safety Standard 201," May 22, 1967, Warren, Mich., p. 91, NHTSA Microfilm, Reel 2, Records of the NHTSA).

16. National Center for Health Statistics, *Weight, Height, and Selected Body Dimensions of Adults.*

17. NHTSA, "Report on the Development of the Initial Federal Motor Vehicle Safety Standards Issued January 31, 1967," ii.

18. Lowell K. Bridwell, "Order in the Matter of the Petitions for Reconsideration of Initial Motor Vehicle Safety Standard No. 201," April 21, 1967, p. 2, NHTSA Microfilm, Reel 1, Records of the NHTSA.

19. Lowell K. Bridwell to B. R. Allen, April 20, 1967, NHTSA Microfilm, Reel 1, Records of the NHTSA.

20. "'Breakdown' Cited in Air-Sea Rescue," *New York Times*, October 24, 1947.

21. "3 on 33-Foot Sloop Lost on Trip to Virgin Islands," *New York Times*, December 28, 1968.

22. Lowell K. Bridwell to B. R. Allen, April 20, 1967, NHTSA Microfilm, Reel 1, Records of the NHTSA.

23. Russell Potter, Federal Highway Administration, "Reconsideration of Initial Motor Vehicle Safety Standards: Docket FHA-1: Recommended Findings (Report of Presiding Officer)," June 23, 1967, p. 22, NHTSA Microfilm, Reel 2, Records of the NHTSA (hereafter cited as "Report of Presiding Officer"). In all likelihood, Potter cowrote this document with a number of other people, but since no record exists of the identity of these other contributors, I will refer to him as the author.

24. Robert M. O'Mahoney and David Schmeltzer, "Statement of Position in the Matter of Petitions for Reconsideration of Initial Motor Vehicle Safety Standard No. 201," April 29, 1967, p. 1, NHTSA Microfilm, Reel 1, Records of the NHTSA.

25. In chapter 8, I describe how similar disagreement arose over the terms *good faith* and *feasible* in the law that regulated automotive emissions.

26. In his "Recommended Findings" report, discussed in greater detail below, Potter pointed to two General Motors employees, Edwin Klove and George Ryder, as being very persuasive about the standard's lack of clarity and objectivity as well as about other points. George Ryder earned his bachelor's in engineering at Michigan State University in 1949, and in 1966 he became the assistant chief engineer of GM's Buick Motor Division, the position he held during the 1967 Standard 201 hearings. As assistant chief engineer, Ryder had "complete responsibility for translation of advanced body designs into a usable production design." Much of his career had been spent moving between models and renderings in the production of objects in the automotive and aircraft industries, including the "translation of an aircraft plywood design into sheet metal" and multiple projects on car bodies and "automotive glass development." Later, in 1981, Ryder formed an automotive consulting firm. Less is known about Klove, but he received a number of patents as co-inventor of interior features. "Direct Testimony of George R. Ryder" in "Direct Written Testimony of General Motors Corporation," May 12, 1967, p. 1, NHTSA Microfilm, Reel 1, Records of the NHTSA; Edwin H. Klove Jr., Joseph J. Magyar, Marvin A. Packett, Detroit, and Robert E. Meshew, Birmingham, MI, assignors to General Motors Corporation (GMC), "Safety Belt Buckle," US Patent 3,256,576, filed April 29, 1964, and issued

June 21, 1986; Edwin H. Klove Jr., Warren, and James L. Noll, Livonia, Mich., assignors to GMC, "Closure Latch," US Patent 3,423,117, filed August 17, 1967 and issued January 21, 1969; Edwin H. Klove Jr., Bloomfield Hills, and James L. Noll, Livonia, Mich., assignors to GMC, "Inflatable Occupant Restraint," US Patent 3,618,977, filed May 11, 1970, and issued November 9, 1971.

27. "Statement of Position of General Motors Corporation," n.d., p. 3, NHTSA Microfilm, Reel 1, Records of the NHTSA.

28. For the most part, however, industry representatives would not just come out and say that they wanted a two-dimensional standard. Most of their rhetorical work was devoted to attacking the H-Point and the other terms. For instance, they questioned whether the H-Point was knowable. If a manikin was placed on the driver's seat, its weight would depress the seat cushion. The automakers argued that slight variations in the manikin's and the seat's tolerances would mean that the H-Point would always be at a slightly different place, rendering the tests unrepeatable.

29. "Verified Statement of Derwyn M. Severy," p. 27, NHTSA Microfilm, Reel Two, Records of the NHTSA.

30. "Affidavit of Colonel John P. Stapp USAP MC," pp. 5–7, NHTSA Microfilm, Reel Two, Records of the NHTSA.

31. Patrick's curriculum vitae was submitted along with his testimony and is contained in NHTSA Microfilm, Reel 1, Records of the NHTSA. Additional biographical information can be culled from Mary Roach, "I Was a Human Crash-Test Dummy," *Salon.com*, November 19, 1999, http://www.salon.com/health/col/roac/1999/11/19/crash_test/print.html.

32. As Potter wrote, "John P. Danforth . . . testified that the Wayne State research was a cooperative venture with General Motors." "Report of Presiding Officer," 32–33.

33. See, for example, Roberts's statement in "Report of Presiding Officer," 33.

34. "Verified Statement of Derwyn M. Severy," 22.

35. "Affidavit of Colonel John P. Stapp USAP MC," 5–7.

36. See, for example, Hilgartner, *Science on Stage.*

37. For a partial biography of Cutler, see his obituary, "Lloyd N. Cutler, Counselor to Presidents, Is Dead at 87," *New York Times*, May 9, 2005.

38. Nader quoted in "Lloyd N. Cutler, Counselor to Presidents, Is Dead at 87."

39. "Environmental Currents."

40. "Transcript of Proceedings in the Matter of Reconsideration of Initial Motor Vehicle Safety Standards 201," hearing held May 22, 1967, Warren, Mich., NHTSA Microfilm, Reel 2, p. 9, Records of the NHTSA (hereafter cited as Hearing Transcript).

41. Hearing Transcript, 12–13.

42. Hearing Transcript, 13.

43. Adkins received his law degree from the University of Michigan and worked for the law firm of Arthur & Hadden in Cleveland before becoming corporate counsel for General Motors. At GM, Adkins took part in the successful defense against Ralph Nader's civil suit. Obituary, *Star Beacon* (Ashtabula, Ohio), August 18, 2000.

44. Hearing Transcript, 42.

45. Hearing Transcript, 47.

46. Hearing Transcript, 50.

47. Hearing Transcript, 135.
48. Winner, "Do Artifacts Have Politics?"
49. "Report of Presiding Officer," 22.
50. Kisiel, "Soichi Kawazoe." Kawazoe received his bachelor of science from Dayton University in 1930 and his master of science from MIT in 1931. He worked for General Motors' and Ford's Japanese branches before moving to Nissan.
51. "Report of Presiding Officer," 24.
52. The preceding three quotations are all from Hearing Transcripts, 136–137.
53. Hearing Transcripts, 140–141.
54. Hearing Transcripts, 149.
55. "Report of Presiding Officer."
56. "Report of Presiding Officer," 6–7.
57. "Report of Presiding Officer," 7.
58. "Report of Presiding Officer," 7.
59. "Bureau Counsel Exceptions to Recommended Findings," NHTSA Microfilm, Reel 2, p. 3, Records of the NHTSA.
60. "Bureau Counsel Exceptions to Recommended Findings," 5. At the heart of the matter lay the question of whether the proceedings should have been quasi-legislative or quasi-judicial. The NHSB's disagreement with Potter on this point began early in the process. O'Mahoney had called Potter on May 15, before the hearings, to clarify how the hearings would proceed. Potter suggested to him that the support materials that the NHSB had submitted for the hearings (written testimony, engineering studies, etc.) were inadequate. As O'Mahoney reported, Potter suggested that he would not be asking the NHSB any questions during cross-examination, because it was not his role to "help it out at all." O'Mahoney took this to mean that Potter saw his role as being close to a judge, while he believed that the purpose of the hearings should be to "establish fact," which the agency would then take into consideration in revising the standard. Rather, Potter had claimed in the May 15 phone call that one of the questions was whether "the requirements in Standard 201 [were] reasonable, considering costs, appearance of the automobile, and saleability of the automobile." That is, he would be judging the standard itself. As O'Mahoney wrote, "It seems clear from this conversation . . . that we will have difficulty with Mr. Potter." Robert M. O'Mahoney, Memorandum to "The Files," "Telephone Conversation with Hearing Officer," May 15, 1967, NHTSA Microfilm, Reel 2, Records of the NHTSA.
61. "Bureau Counsel Exceptions to Recommended Findings," 5.
62. "Auto Makers Sue on Safety Rules," *New York Times*, April 1, 1967.
63. This is the core thesis of Jerry L. Mashaw and David L. Harfst's influential book, *The Struggle for Auto Safety*. In updating their account in 2017, Mashaw and Harfst argued that auto safety regulation since their book was first published in 1990 has become *even weaker and grimmer*. Mashaw and Harfst, "From Command and Control to Collaboration and Deference."

Chapter 6 • *Discovering and (Not) Controlling Automotive Air Pollution*

1. Dewey, *Don't Breathe the Air*, 38; Krier and Ursin, *Pollution and Policy*, 45.
2. Krier and Ursin, *Pollution and Policy*, 58; Thackray and Myers, *Arnold O. Beckman*, 217.

3. Dyck, *Evolution of Federal Air Pollution Control Policy*, 19.

4. Stradling, *Smokestacks and Progressives*, 163–166.

5. Jacobs and Kelly, *Smogtown*.

6. Jacobs and Kelly, *Smogtown*, chapter 1.

7. Krier and Ursin, *Pollution and Policy*, 57–60.

8. "Text of Report and Conclusions of Smog Expert," *Los Angeles Times*, January 19, 1947.

9. On the founding of the air pollution control district and Louis McCabe's management of it, see Jacobs and Kelly, *Smogtown*, 37–67.

10. Jacobs and Kelly, *Smogtown*, 38.

11. Uekoetter, "The Strange Career of the Ringelmann Smoke Chart."

12. Jacobs and Kelly, *Smogtown*, 39–40.

13. Stanford Research Institute, The Smog Problem in Los Angeles County: Second Interim Report by Stanford Research Institute on Studies to Determine the Nature and Sources of Smog (Los Angeles: Committee on Smoke and Fumes, Western Oil and Gas Association, 1949), 37–45.

14. Stanford Research Institute, The Smog Problem in Los Angeles County, 45.

15. McCabe, *Air Pollution*, v.

16. Raghu Garud and others have highlighted the role that conferences play in configuring organizational fields. Raghu Garud, "Conferences as Venues for the Configuration of Emerging Organizational Fields."

17. Hofflund, "National Aspects of Air Pollution Legislation," 770.

18. Kingsland, "Frits Went's Atomic Age Greenhouse."

19. On Beckman's work on smog, see Thackray and Myers, *Arnold O. Beckman*, 217–228.

20. Thackray and Myers, *Arnold O. Beckman*, 220–221.

21. Haagen-Smit, "Chemistry of Flavor."

22. Dunsby, "Clarifying Smog," 109.

23. Middleton, Kendrick, and Schwalm, "Injury to Herbaceous Plants by Smog or Air Pollution."

24. Thackray and Myers, *Arnold O. Beckman*, 222.

25. Assembly of the State of California, "Study and Analysis of the Facts Pertaining to Air Pollution Control in Los Angeles County," Report of the Subcommittee of the Assembly Interim Committee on Governmental Efficiency and Economy, 1953, 14.

26. Jaffe, "Effects of Photochemical Oxidants on Materials."

27. Haagen-Smit, "Chemistry and Physiology of Los Angeles Smog," 1342.

28. Bradley and Haagen-Smit, "Application of Rubber in the Quantitative Determination of Ozone."

29. Haagen-Smit, "Chemistry and Physiology of Los Angeles Smog," 1343.

30. Haagen-Smit, "The Air Pollution Problem in Los Angeles." On Haagen-Smit's work and the story of air pollution control in California, see also Dewey, *Don't Breathe the Air*.

31. Krier and Ursin, *Pollution and Policy*, 86–87.

32. As quoted in Krier and Ursin, *Pollution and Policy*, 89.

33. Thackray and Myers, *Arnold O. Beckman*, 223.

34. Krier and Ursin, *Pollution and Policy*, 81.

35. As quoted in Thackray and Myers, *Arnold O. Beckman*, 223.

36. Bailey, *Congress and Air Pollution*, 91–93.

37. Sundquist, *Politics and Policy*, 351–353.

38. Bailey, *Congress and Air Pollution*, 94–95.

39. Stahman, Mills, and Korth, *Federal Air Pollution Efforts*, 1. The center had been completed in 1954.

40. Rose, and Brandt, "Environmental Irradiation Test Facility"; Rose, Stahman, and Korth, "Dynamic Irradiation Tests of Automotive Exhaust, Part 1."

41. Krier and Ursin, *Pollution and Policy*, 93–94.

42. Krier and Ursin, *Pollution and Policy*, 117.

43. Sundquist, *Politics and Policy*, 332.

44. Krier and Ursin, *Pollution and Policy*, 110, 169.

45. US Public Health Service, *Motor Vehicles, Air Pollution, and Health*.

46. This and the following two paragraphs rely on Krier and Ursin, *Pollution and Policy*, 137–169.

47. Dewey, "'The Antitrust Case of the Century,'" 351.

48. On the relationship between corporations and external intellectual property, see Usselman, *Regulating Railroad Innovation*.

49. The most controversial technology that the Environmental Protection Agency tested was the so-called LaForce Engine, created by Edward LaForce and his brother, Robert. Congress took interest in the engine, which the LaForce brothers claimed was far more efficient than the average Detroit vehicle. EPA staff lambasted the LaForce engine in a Senate hearing, claiming that it contained no new or real contributions and that it had severely failed federal emissions tests. See Environmental Protection Agency, "Testing of the LaForce Engine"; Environmental Protection Agency, "Evaluation of the LaForce-Modified AMC Hornet"; "EPA Tests Refute LaForce Engine Claim." The Securities and Exchange Commission later found Edward LaForce guilty of criminal contempt after he continued issuing stock to fund his efforts after the SEC ordered him to stop doing so. See "Edward P. LaForce Convicted."

50. Muskie, *Journeys*, 83.

51. Sundquist, *Politics and Policy*, 353.

52. The best treatment of the NAPCA's abatement efforts is Esposito, *Vanishing Air*, chapter 6.

53. For discussion of Muskie's role in the hearings process and his turn toward pollution control, see Krier and Ursin, *Pollution and Policy*, 173–174.

54. Krier and Ursin, *Pollution and Policy*, 174.

55. Scheberle, *Federalism and Environmental Policy*, 4.

56. Krier and Ursin, *Pollution and Policy*, 173–175.

57. Stahman, Mills, and Korth, "Federal Air Pollution Efforts," 4.

58. Esposito, *Vanishing Air*, 23.

59. Esposito, *Vanishing Air*, 23.

60. Esposito, *Vanishing Air*, 24.

61. Jones, "The Limits of Public Support," 504.

62. Jones, "The Limits of Public Support," 503.

63. Jones, "The Limits of Public Support," 503.

64. Esposito, *Vanishing Air*, 52.

65. For a painfully detailed account of the creation of the Federal Test Procedure, see Vinsel, "Federal Regulatory Management of the Automobile," 212–219.

66. Kruse and Huls, "Development of the Federal Urban Driving Schedule," 2.

67. Kruse and Huls, "Development of the Federal Urban Driving Schedule," 2–4.

68. Coordinating Research Council, "A Survey of Average Driving Patterns in the Six Urban Areas of the United States: Summary," CRC Study CAPE-10 (copy in author's personal collection). Soon after this study was finished, the EPA ceased working with the Coordinating Research Council, because high-level officials worried that this collaborative research might lead to collusion. Eric Stork, interview with the author, April 10, 2010.

Chapter 7 • Command and Control

1. Flippen, *Nixon and the Environment*.

2. Muskie, *Journeys*, 82.

3. Lippman and Hansen, *Muskie*, 143.

4. This and the following quotations in this paragraph are from Muskie, *Journeys*, 84.

5. Arlington, "Defining Gasoline," 49–51.

6. "Exhaust Gas Made Safe," *Popular Mechanics* (September 1951): 134.

7. Campbell, "Luncheon Address," 19.

8. Campbell, "Luncheon Address," 21.

9. Muskie, "Banquet Address," 1.

10. Muskie, "Banquet Address," 2.

11. Dewey, "'The Antitrust Case of the Century.'"

12. Dewey, "'The Antitrust Case of the Century,'" 348; Esposito, *Vanishing Air*, 41.

13. Esposito, *Vanishing Air*, 42.

14. Dewey, "'The Antitrust Case of the Century,'" 353.

15. Esposito, *Vanishing Air*, 25. On Muskie's reaction to these criticisms, see also Krier and Ursin, *Pollution and Policy*, 203.

16. Lippman and Hansen, *Muskie*, 153.

17. The paper was eventually published as Barth, "Federal Motor Vehicle Emission Goals for CO, HC, and NO_x Based on Desired Air Quality Levels." Much of the following biographical description comes "Delbert S. Barth," Commissioned Corps of the U.S. Public Health Service, https://dcp.psc.gov/OSG/engineer/history -dbarth.aspx. For other biographical information on Barth, see *Who's Who in Science and Engineering, 2000–2001*, 5th ed. (New Providence, N.J.: Marquis Who's Who, 1999).

18. One person who worked for both the National Air Pollution Control Administration and the EPA later recounted that he had always heard that Barth had posited the reductions needed to make the automobile insignificant in ambient air pollution and that these numbers had gotten picked up in the Clean Air Act Amendments of 1970. Eric Stork, interview with author, Arlington, Va., April 10, 2010. But another person, who also worked for both agencies, told me that he was sure Leon Billings, Edmund Muskie's chief of staff and a key author of the Clean Air Act Amendments, had in fact set the reduction levels. Edwin Tuerk, phone interview with author, April 22,

2010. On Billings, see *Who's Who in American Politics*, 22nd ed. (New Providence, N.J.: Marquis Who's Who, 2009).

19. Agis Salpukas, "Court Fines Ford $7-Million in Suit on Pollution Test," *New York Times*, February 14, 1973.

20. Colburn, "Washington's RX," 72.

21. Colburn, "Washington's RX," 72.

22. Like previous federal automotive emissions laws, the Clean Air Act Amendments of 1970 allowed for the state of California to set automotive emissions standards stronger than the national ones. This policy created a "two car system," in which cars manufactured for California would have more effective emission control systems than automobiles made for the rest of the country. Technically, the EPA had to hold public hearings before giving the OK to California's tighter standards. But a former EPA employee told me that the Office of Mobile Sources saw these hearings as a necessary but ultimately empty, even boring, activity. As the law was written, the agency had to have a *very* good reason for denying California's requests, and no such reasons existed. The EPA allowed California's tighter automotive emission control programs to go unchallenged until 2007, when the George W. Bush administration blocked the state's attempt to regulate emissions of greenhouse gasses, particularly carbon dioxide (CO_2), from cars. On the Bush administration's refusal of California's request, see Bryan Walsh, "California's Clean-Air Slapdown," *Time*, December 20, 2007.

23. Colburn, "Washington's RX," 73. Stork often made statements like this one, and news sources enjoyed quoting his philosophical reflections on the "short hair."

24. Colburn, "Washington's RX," 73.

25. Jacobs, *Panic at the Pump*, 27.

26. Lippman and Hansen, *Muskie*, 143.

27. As quoted in Jacobs, *Panic at the Pump*, 27.

28. Flippen, *Nixon and the Environment*, 29.

29. Flippen, *Nixon and the Environment*, 84–89.

30. Esposito, *Vanishing Air*, 60–62.

31. Eric Stork, email to author, December 31, 2009.

32. A definitive history of the technology assessment idea has yet to be written. Bruce Allen Bimber's *The Politics of Expertise in Congress: The Rise and Fall of the Office of Technology Assessment* (1996) is helpful for the history of that office, but Bimber does not really deal with the intellectual origins of the idea, or with its early history. Some of the early thinking about technology assessment can be gleaned from Kaspar, *Technology Assessment*.

33. Joseph Somers, former EPA staff member, email to author, February 18, 2016.

34. Many of these reports are extremely difficult to locate and belong to what some archivists and researchers call the "gray literature," ephemeral publications rarely collected in "official" depositories such as libraries and archives. For one such technology assessment, see Emission Control Technology Division, Office of Mobile Source Air Pollution Control, "Automobile Emission Control—The Development, Status, Trends, and Outlook as of December 1976."

35. Stork, interview, April 10, 2010.

36. The advisory circulars are now available on the EPA's website, https://iaspub .epa.gov/otaqpub/publist_gl.jsp?guideyear=All%20Years. Business historian Joann

Yates has written about how corporations managed communication to increase coordination and efficiency. While we may interpret Stork's advisory circulars as a form of adjudication, we may also view it as a way of managing regulatory processes. In this case, the seemingly strict line between the state and private businesses breaks down, and both sides can be seen as part of a single process that requires coordination. In another example of management via issuance, Stork created papers called fact sheets to manage public inquiries. Federal agencies typically receive huge numbers of inquiries either sent directly from the public or, more often, forwarded by legislators on behalf of their constituents. These inquiries can easily overwhelm the agency's office space. In an effort to stem the tide, Stork and his subordinates wrote the fact sheets, which summarized the agency's position on certain topics and the current state of knowledge on technical and scientific issues. The fact sheets have been collected in a binder housed at the EPA's National Vehicle and Fuel Emissions Laboratory library in Ann Arbor, Michigan. A complete set of the documents is in the author's possession.

37. Existing advisory circulars were updated through the early 1980s, but the agency issued no new ones after Stork left in 1978.

38. Eric Stork, phone interview with author, February 25, 2011.

39. Gray and Alson, *Moving America to Methanol.*

40. Somers, "Conformational Properties of Some Phosphorinane Derivatives."

41. Hellman, "A Kinetic Theory Analysis of the Evolution of Shock, Thermal, and Expansion Waves."

42. Joseph Merenda, interview with author, Rockville, Maryland, August 16, 2008.

43. Stork, phone interview, February 25, 2011.

44. Karl Hellman, interview with author, August 19, 2008.

45. Stork, interview, April 10, 2010. Government Accounting Office (GAO), "Examination into the Adequacy."

46. GAO, "Examination into the Adequacy," 4.

47. GAO, "Examination into the Adequacy," 4.

48. GAO, "Examination into the Adequacy," 4.

49. William Ruckelshaus, phone interview with author, February 20, 2009.

50. GAO, "Examination into the Adequacy," 14.

51. Salpukas, "Court Fines Ford $7-Million."

52. Charles Camp and Walter Mossberg, "Sniffing at the Tailpipe: Fuss at Ford Shows How Weak Enforcers Are," *Wall Street Journal,* June 6, 1972.

53. Stork, interview, April 10, 2010.

54. Mondt, *Cleaner Cars,* 69–74.

55. The staff members constantly bumped into what Herbert A. Simon called "bounded rationality," the limits of their own knowledge and ability to process information. Simon, *Administrative Behavior.*

56. Memorandum, E. J. Brune to Director, MSPCP, EPA, OAP, "Characterization and Use of Emission Control Systems That Operate Only under Specific Ambient Conditions or Vehicle Operational Modes," July 7, 1972, p. 1, personal papers of Eric Stork.

57. Memorandum, E. J. Brune to Director, 3.

58. On the notion of "interpretive flexibility" in technology, see Pinch and Bijker, "The Social Construction of Facts and Artifacts."

59. MSPC [Mobile Source Pollution Control] Advisory Circular No. 24, "Prohibition of Use of Emission Control Defeat Devices," December 11, 1972, and Ruckelshaus announcement in press release, Environmental Protection Agency, January 3, 1973. This advisory circular and the press release are in the "Defeat Control" binder and, thus, in the author's possession. But the complete collection of advisory circulars is also available at https://iaspub.epa.gov/otaqpub/publist_gl.jsp?guideyear=All%20Years.

60. Stork, interview, April 10, 2010.

61. Emission Control Technology Division, Office of Mobile Sources, "Automobile Sulfuric Acid Emission Control—The Development Status as of December 1975: A Report to the Administrator, EPA," p. 2-1. Eric Stork provided the author a copy of this report, but it is also available online at "New Titles List," National Service Center for Environmental Publications, EPA, http://nepis.epa.gov/EPA/html/newtitles.html. See also, Eric Stork, "The United States Experience with Imposing Automobile Emission Standards," paper presented to the Australian Society of Automotive Engineers, Perth, Australia, September 22, 1976 (given to the author by Eric Stork). In this paper, Stork described the controversy regarding catalytic converters and sulfate particulates as the "sulfate flap," an unnecessary state of worry created by his adversaries' imprudent claims.

62. Emission Control Technology Division, Office of Mobile Sources, "Automobile Sulfuric Acid Emission Control," p. 2-2; Joe Merenda, interview with author, Rockville, Maryland, July 16, 2008.

63. The historian Jongmin Lee has begun some fascinating research that views the sulfate issue from the viewpoint of the EPA's Office of Research and Monitoring. See, for instance, "Regulatory Engineering in Pollution and Health Research: U.S. Environmental Protection Agency's CHESS, 1969–1977," paper presented at the annual meeting of the Society for the History of Technology, Tacoma, Wash., October 2, 2010.

64. For the Office of Research and Monitoring's work on sulfur oxides, see Environmental Protection Agency, "Health Consequences of Sulfur Oxides."

65. Eric Stork, "United States Experience with Imposing Automobile Emission Standards." See also "Do Auto Catalysts Meet the Safety Test?" *Business Week* (October 27, 1973), 116.

66. Merenda, interview, August 16, 2008.

67. "Do Auto Catalysts Meet the Safety Test?" 116.

68. "Do Auto Catalysts Meet the Safety Test?" 116. On Greenfield, see *Who's Who in America, 1994* (New Providence, N.J.: Marquis Who's Who, 1994). On Moran, see *Who's Who in America, 2008* (New Providence, N.J.: Marquis Who's Who, 2008).

69. Stork, phone interview, February 25, 2011.

70. This and the following quotations in this paragraph are from Lawrence Mosher, "Catalytic Converter: Next Street Hazard?" *National Observer*, undated, from Eric Stork's clipping files.

71. Mosher, "Catalytic Converter."

72. Eric Stork to Representative James F. Hastings, November 14, 1975, in Eric Stork's personal collection; Ned MacKay, "The EPA Sniffs Sulfuric Acid," *Palo Alto Times*, March 12, 1975.

73. "EPA Would Rather Switch than Fight, Substitutes Stork for Moran," *Environmental Health Letter*, July 15, 1975.

74. Stork, "United States Experience with Imposing Automobile Emission Standards."

75. "Do Auto Catalysts Meet the Safety Test?" 116.

76. Roger Strelow, Assistant Administrator of Air and Waste Management, to Ernest R. Starkman, Vice President, Environmental Activities Staff, General Motors Corporation, June 197, 1975; Strelow to Starkman, July 7, 1975, both letters in Eric Stork's personal collection. Strelow replaced Robert Sansom as the assistant administrator of air and waste management and as Stork's direct supervisor.

77. Wilson et al., "General Motors Sulfate Dispersion Experiment."

78. Papers from this effort were eventually collected into the volume Stevens et al., eds., *General Motors/Environmental Protection Agency Sulfate Dispersion Experiment.*

79. Stork, email, February 26, 2011.

Chapter 8 • Establishing the State of the Art

1. Salpukas, "Court Fines Ford $7-Million."

2. The EPA held a third set of hearings in early 1975, but by that time the debate had changed considerably, as I discuss briefly at the end of this chapter. The issue of whether the standards should be suspended eventually returned to Congress once the EPA no longer had legislative authority to grant extensions to the automakers. Congress later passed the Clean Air Act Amendments of 1977, which froze emissions standards at their interim levels for five years.

3. Margaret Taylor and her coauthors have written about how regulation fosters knowledge production and information sharing. See Taylor, "The Influence of Government Actions on Innovative Activities in the Development of Environmental Technologies"; Taylor, Rubin, and Hounshell, "Regulation as the Mother of Innovation."

4. Eric Stork, in National Research Council, "Transcript from Meeting on Resolution of Scientific Issues and the Judicial Process: *International Harvester* v. *Ruckelshaus*," May 14, 1976, p. 4, Stork personal papers (hereafter cited as NRC Transcript with speaker's name). Stork attributes this phrase to Chrysler engineer Charles Heinen.

5. National Academy of Sciences (NAS) and US Environmental Protection Agency (EPA), *Semiannual Report by the Committee on Motor Vehicle Emissions*, January 1, 1972, p. 12.

6. NAS and EPA, *Semiannual Report by the Committee on Motor Vehicle Emissions*, January 1, 1972, p. 12.

7. As described in the previous chapter, the manufacturers' use of air pumps was a major factor in pushing the EPA to change its testing procedures, since the additional air effectively "watered down" the controlled emissions' concentration.

8. *International Harvester v. Ruckelshaus*, 478 F.2d 615 (D.C. Cir. 1973), 623.

9. Hustead in Environmental Protection Agency (EPA), *Auto Emissions Extension*, 1469.

10. Sansom, in EPA, *Auto Emissions Extension*, 1471.

11. On how engineers frame problems around flaws or "reverse salients," see Hughes, *Networks of Power*, 14.

12. Stempel, in EPA, *Auto Emissions Extension*, 1445.

13. Stempel, in EPA, *Auto Emissions Extension*, 1446.

14. Eric Stork eventually took part in litigation over a case of fire supposedly caused by a catalytic converter. He testified for the auto companies and maintains that catalytic converters were not really a fire hazard. Stork, author interview, April 10, 2010.

15. Stempel, in EPA, *Auto Emissions Extension*, 1445.

16. Hustead, in EPA, *Auto Emissions Extension*, 1403–1404.

17. Unleaded gasoline, however, provided a rich example of managing complexity in systems: if you change one thing in order to solve a problem, you are likely to create new problems. In the 1970s, for reasons related to both emissions control and fuel efficiency, manufacturers began adopting electronic fuel injection. Auto engineers had used mechanical fuel injection in diesel engines for many years, but the oily nature of diesel fuel gave it lubricating properties that unleaded gasoline lacked. Engineers found that unleaded fuel more quickly deteriorated fuel injection systems. For instance, Volvo learned that unleaded gasoline created deposits on the injection needle, the part of the fuel injection system that projected into the cylinder. In one of its test vehicles, the fuel injection system failed from using unleaded gasoline before the catalytic converter did. Unleaded gasoline was necessary for catalytic converters, but it was by no means a trouble-free proposition. Volvo staff member Elljbaer, in EPA, *Auto Emissions Extension*, 91. Clogged fuel injectors continued to be a problem until 1994, when the federal government required petroleum companies to put detergent in gasoline to stave off the problem.

18. On Cole, see McCarthy, *Auto Mania*, 179.

19. Letter from William Ruckelshaus describing the Second Decision, in US Senate, Committee on Public Works, *Decision of the Administrator of the Environmental Protection Agency Regarding Suspension of the 1975 Auto Emissions Standards*, p. 3 (hereafter cited as *Second Decision*).

20. George Allen, in NRC Transcript, 7.

21. Historians and sociologists have long noted that states have formed bureaucracies, professionalized civil services, and adopted scientific and statistical methods to downplay the political, contested nature of their decisions. See, for instance, Skowronek, *Building a New American State*; Porter, *Trust in Numbers*; and Kahn, *Budgeting Democracy*.

22. Allen, in NRC Transcript, 6.

23. Allen, in NRC Transcript, 8.

24. Ruckelshaus, in EPA, *Auto Emissions Extension*, 21.

25. Allen, in NRC Transcript, 4.

26. William Ruckelshaus, in EPA, *Auto Emissions Extension*, vol. 1, p. 4.

27. Ruckelshaus, in EPA, *Auto Emissions Extension*, vol. 1, p. 4.

28. Ruckelshaus, in EPA, *Auto Emissions Extension*, vol. 1, p. 6.

29. William Ruckelshaus, interview with author, February 20, 2009.

30. Ruckelshaus, in EPA, *Auto Emissions Extensions*, 8.

31. On the "division of intellectual (or 'innovative') labor," see Arora, Fosfuri, and Gambardella, *Markets for Technology*. For an examination of how central firms and

suppliers partition knowledge to foster innovation, see Lee and Veloso, "Inter-firm Innovation under Uncertainty."

32. For a popular recounting of AMC's financial woes, see Bradsher, *High and Mighty*.

33. Doyle, *Taken for a Ride*, 122–123.

34. The consumer advocates' and environmentalists' unwillingness to flout liberal norms is striking. Although one community organizer did suggest that the ideal was that people would stop driving, most other progressive witnesses steered well clear of such talk. In his testimony, David Hawkins, who was then a lawyer at the National Resources Defense Council and who later served as the EPA's assistant administrator of air, noise, and radiation under President Jimmy Carter, felt compelled to say, "A halt to manufacturing operations in Detroit in 1975 is neither a real nor desirable possibility. Accordingly, such a result should not be hypothesized in calculating whether a suspension is essential to the public interest." Hawkins, in EPA, *Auto Emissions Extension*, 2572. If nothing else, Hawkins's statement demonstrates the degree to which debates over emission control in the United States took place within the framework of a liberal consensus.

35. Ruckelshaus proposed this rule change on October 31, 1972. See 37 Fed. Reg. 23,778–23,779 (October 31, 1972). The proposed rule contained a stipulation that the consumer's failure to have the maintenance done could not improve the performance of the vehicle. Many of the automakers were building bypass systems into their emission control plans to address cases of overheating and the like. If a system failed permanently and was bypassed, this could hypothetically improve vehicle performance as raw exhaust was vented into the air. The rule probably sought to avoid giving consumers the incentive *not* to get their emission controls repaired. As I explain in chapter 7, the EPA had already dealt with automakers' intentionally using defeat devices to disable emissions controls and improve drivability. The EPA did not want a replay of that controversy.

36. Eric Stork, in NRC Transcript, 8.

37. Stork, in NRC Transcript, 5.

38. T. E. Hustead in EPA, *Auto Emissions Extension*, 1356. Hustead was the Head Engineer of GM's AC Spark Plug Division.

39. Hustead, in EPA, *Auto Emissions Extension*, vol. 6, p. 1358.

40. R. W. Bowditch and Hustead, in EPA, *Auto Emissions Extension*, vol. 6, pp. 1360–1361. Bowditch was the director of GM's Automotive Emissions Control Office.

41. Bowditch and Hustead, in EPA, *Auto Emissions Extension*, vol. 6, p. 1361.

42. Hustead, in EPA, *Auto Emissions Extension*, vol. 6, p. 1457. Hustead called it the IC: "I have 128 people at IC now."

43. Stork, interview, January 5, 2010.

44. Ruckelshaus, in EPA, *Auto Emissions Extension*, vol. 6, p. 1342.

45. E. S. Starkman and Hustead, in EPA, *Auto Emissions Extension*, vol. 6, p. 1343. Starkman was the vice-president of GM's Environmental Activities Staff.

46. Starkman and Hustead, in EPA, *Auto Emissions Extension*, vol. 6, p. 1343.

47. R. C. Stempel, in EPA, *Auto Emissions Extension*, vol. 6, p. 1345. Stempel was a motor engineer in GM's Oldsmobile Division and future president of General Motors.

48. Stempel, in EPA, *Auto Emissions Extension*, vol. 6, p. 1345.

49. Ruckelshaus, interview with author, February 20, 2009.

50. Stempel, in EPA, *Auto Emissions Extension*, vol. 6, p. 1346.

51. Ruckelshaus, in EPA, *Auto Emissions Extension*, vol. 6, p. 1346.

52. Starkman, in EPA, *Auto Emissions Extension*, vol. 6, p. 1347.

53. Norman Shutler, in EPA, *Auto Emissions Extension*, vol. 6, p. 1359. Shutler was the director of the EPA's Division of Mobile Source Enforcement.

54. Stempel, in EPA, *Auto Emissions Extension*, vol. 6, p. 1371.

55. Sansom, in EPA, *Auto Emissions Extension*, vol. 6, p. 1372.

56. Stempel, in EPA, *Auto Emissions Extension*, vol. 6, p. 1372–1373.

57. Stempel, in EPA, *Auto Emissions Extension*, vol. 6, p. 1373.

58. Stempel, in EPA, *Auto Emissions Extension*, vol. 6, p. 1373.

59. Allen, in EPA, *Auto Emissions Extension*, vol. 12, p. 2751.

60. Starkman, in EPA, *Auto Emissions Extension*, vol. 12, p. 2754.

61. Allen, in EPA, *Auto Emissions Extension*, vol. 12, p. 2758.

62. Fri, in EPA, *Auto Emissions Extension*, vol. 12, p. 2759.

63. Starkman, in EPA, *Auto Emissions Extension*, vol. 12, pp. 2760–2761.

64. Joel Mokyr lays out the helpful explanatory dichotomy of prescriptive versus propositional knowledge in *The Gifts of Athena*, esp. chapter 1.

65. Stork, interview, April 10, 2010.

66. Syd Terry, in EPA, *Auto Emissions Extension*, 2065–2066.

67. Terry, in EPA, *Auto Emissions Extension*, 2066.

68. Huebner, in EPA, *Auto Emissions Extension*, 2101.

69. Terry, in EPA, *Auto Emissions Extension*, 2098–2099.

70. Charles Heinen, in NRC Transcript, 4.

71. Stork: "I hope we can figure out how to make it as tough as it should be." Stork, in EPA, *Auto Emissions Extension*, 2124.

72. *Second Decision*, 3.

73. From Ruckelshaus's First Decision, as quoted in *International Harvester v. Ruckelshaus*, 478 F.2d 615, at 625.

74. Wald, "Thirty Years of Administrative Law in the D.C. Circuit Court."

75. In environmental regulation, the "hard look" approach originated in earlier cases, such as *Citizens to Preserve Overton Park, Inc. v. Volpe* (1971).

76. NAS and EPA, *Semiannual Report to the Environmental Protection Agency*, January 1, 1972; idem, *Interim Standards Report*, April 26, 1972.

77. NAS and EPA, *Semiannual Report by the Committee on Motor Vehicle Emissions*, February 12, 1973, p. 7 (hereafter cited as NAS Report).

78. NAS Report, 8.

79. Colburn, "Washington's RX."

80. Stork, quoted in Colburn, "Washington's RX."

81. Colburn, "Washington's RX."

82. NAS Report, 2.

83. NAS Report, 3. The NAS also insisted that tests results should be averaged "within a vehicle and engine class."

84. The dynamic here strongly resembles the one that Theodore Porter describes in *Trust in Numbers*, whereby mechanical mathematical devices are substituted for interpersonal confidence.

85. Stork, in NRC Transcript, 12.

86. *Second Decision*, 23. Anecdotally, the core of the methodology may have been based on General Motor's production "matrix" formulas, discussed earlier in this chapter. A few former and present EPA employees suggested this to me in passing.

87. *Second Decision*, 23.

88. Bill Pedersen, in NRC Transcript, 17. Pedersen was an employee in the EPA General Counsel's Office during the hearings.

89. Clarence Ditlow, in NRC Transcript, 17.

90. *Second Decision*, 23.

91. The historian of science Peter Galison has called Monte Carlo simulations an "artificial reality." See Galison, *Image and Logic*, chapter 8.

92. *Second Decision*, 52.

93. Allen, in EPA, *Auto Emissions Extension*, 1528.

94. The hearings and the EPA's technology assessments were by no means the only ways that automakers and suppliers were learning about the current state of the art. The publications and conferences of the Society of Automotive Engineers were also important venues, though some anecdotal evidence suggests that SAE members were chary about not violating the consent decree at SAE meetings.

95. Eric Stork, in NRC Transcript, 4.

96. George Allen, in NRC Transcript, 7.

97. "Do Auto Catalysts Meet the Safety Test?" 116.

Chapter 9 • The Bureaucratic Struggle over Fuel Economy

1. Peter B. Clark and James Q. Wilson, "Incentive Systems: A Theory of Organizations," *Administrative Science Quarterly* 6, no. 2 (September 1961), 157, as quoted in Down, *Inside Bureaucracy*, 8. See also Down's discussion of "bureau territoriality" and struggles over turf in *Inside Bureaucracy*, chapter 17.

2. This finding fits with Daniel Carpenter's argument that a great deal of government work is aimed at enhancing and preserving *reputation*. See Carpenter, *Reputation and Power*.

3. Eric Stork, "The Genesis of the EPA Fuel Economy Program: An Exercise in Bureaucratic Gamesmanship," Stork personal papers. In reflecting in this seminar presentation on his past experience, Stork focused on the EPA's involvement in and battles over measuring and publishing data on automotive fuel economy. His written text for this seminar provides a unique lens through which to observe one such battle. This text, from which Stork worked as he shared his reflections at several seminars, is in Eric Stork's personal collection.

4. Stork, phone interview, January 5, 2010.

5. Organizations' practice of finding and assimilating external technologies has long been one of the practical aspects of R&D. Instead of developing the technologies on their own, organizations often learn how to use technologies developed elsewhere. Cohen and Levinthal, "Absorptive Capacity." See also Miller and Arikan, "Technology Search Investments"; and Brenner, "Technology Intelligence and Technology Scouting."

6. Stivender, "Development of a Fuel-Based Mass Emission Measurement Procedure." This paper was delivered at the SAE's midyear meeting held in Montreal,

Canada, June 7–11, 1971. Stivender was a forward-thinking engineer who worked in General Motors' research laboratories. He specialized in emission controls, electronic controls (which were directly related to controlling emissions), and alternative power systems, especially gas turbines. On Stivender's biography, see *Who's Who in America, 2007*, 61st ed. (New Providence, N.J.: Marquis Who's Who, 2006).

7. Here, following the language of "research and development," I'm suggesting that Stivender's paper constituted the actual "research," while EPA staff members "developed" the procedure by putting it into practice and simultaneously transforming it from its original conception. Society of Automotive Engineers, *Automotive Fuel Economy*. Of the twenty-two technical papers in a Society of Automotive Engineers volume that collected the most important studies of fuel economy, Austin, Hellman, or both men together wrote four of them, or nearly one-fifth. In my count of twenty-two papers, I am neglecting one essay, which recounts an SAE development of fuel economy standards but does not generate new knowledge in the same way the other papers do.

8. Mobile Source Pollution Control Program, *Fuel Economy and Emission Control*.

9. Environmental Protection Agency, "Fact Sheet 13: Automobile Fuel Economy," library of EPA's Ann Arbor office. There are two versions of this fact sheet in the library; one is an updated version of the other. Both are undated, but they were written after 1978 as they make reference to changes in the EPA's fuel economy program that occurred in that year.

10. The historian Kevin Borg has written about the phenomenon of mechanics removing emissions controls in *Auto Mechanics*. This phenomenon connects deeply with historians' and economists' examinations of how "users" shape technology. On users, see Kline and Pinch, "Users as Agents of Technological Change"; Oudshoorn and Pinch, *How Users Matter*; Kline, *Consumers in the Country*; Yates, *Structuring the Information Age*; Yates, "How Business Enterprises Use Technology"; von Hippel, *Democratizing Innovation*.

11. "Junk Anti-smog Devices, Save Gas, G.M. Head Urges," *Seattle Times*, November 15, 1973. On Ed Cole, see McCarthy, *Auto Mania*, chapter 9.

12. All of the foregoing quotations are from "Junk Anti-smog Devices, Save Gas, G.M. Head Urges."

13. The account and quotations in this paragraph come from Environmental Protection Agency (EPA), "A Study of Fuel Economy Changes Resulting from Tampering with Emission Control Devices," January 1974, pp. 1–5, library of EPA's Ann Arbor office.

14. EPA, "A Study of Fuel Economy Changes," 4.

15. EPA, "A Study of Fuel Economy Changes," 4.

16. Robert W. Irvin, "U.S. Urges Keeping Anti-smog Systems: You Won't Save on Gas by Tinkering, Drivers Told," *Detroit News*, January 4, 1974.

17. EPA, "A Study of Fuel Economy Changes," 5.

18. Stork, quoted in Douglas Williams, "Foiling Smog Gear Can Hurt Mileage, EPA Test Confirms," *Detroit Free Press*, January 31, 1974. Stork's statement also speaks to his faith in expertise at a time when, as a number of scholars have argued, many people in the United States were losing faith in expertise. On the decline of faith in experts, see Balogh, *Chain Reaction*, 17.

19. Douglas Williams, "Unhook Smog Gear? EPA Tests the Idea," *Detroit Free Press*, January 18, 1974.

20. For a clear, broad survey of the "energy crises" of the 1970s, see Yergin, *The Prize*, esp. chapters 31–33. Another helpful volume, this one focusing primarily on the late-1970s but offering perspectives on the entire period, is Merrill, *The Oil Crisis of 1973–1974*. See also Horowitz, *Jimmy Carter and the Energy Crisis of the 1970s*.

21. Nixon, "Special Message to the Congress on Energy Policy," April 18, 1973, American Presidency Project, University of California Santa Barbara, http://www .presidency.ucsb.edu/ws/index.php?pid=3817&st=&stl=. Also see "Environment: At Last, The Energy Message," *Time*, April 30, 1973.

22. Rothschild, "What Is the 'Energy Crisis'?"

23. Eric Stork, phone interview, January 5, 2010.

24. Nixon, "Special Message to the Congress on Energy Policy."

25. 38 Fed. Reg. 10,867–10,908 (May 2, 1973).

26. 38 Fed. Reg. 10,868 (May 2, 1973). The term *cold-start* refers to running a car when the engine has not had time to warm up. During these periods the car produced many more emissions and had lower fuel efficiency. For this reason, the Federal Test Procedure required periods known as cold soaks in which a car would be left to sit overnight to ensure that the test conducted the next day would capture this cold-start phase.

27. Stork, "The Genesis of the EPA Fuel Economy Program." In a Senate hearing discussed below, it was clear from the testimony of members of the Office of Management and Budget—Roy Ash, director; Frank Zarb, associate director of energy, science, and natural resources; and James Tozzi, chief of the environmental branch, all of whom wanted to centralize energy efficiency labeling under the Department of Commerce's National Bureau of Standards (NBS)—that the NBS had no hand in developing the EPA's fuel economy labeling or information publication program.

28. 38 Fed. Reg. 22,944–22,947 (August 27, 1973). In accord with the EPA's exclusive control of the automotive labeling program, the agency made the announcement solely under its own name; the contact information included an agency address; and the announcement declared, "The Environmental Protection Agency is particularly interested in receiving comments on the effectiveness of the recommended vehicle label." The notice also highlighted the centrality on the labels of the agency's fuel economy data.

29. See, for example, "Gasoline Guzzlers Unmasked by EPA Tests," *Chicago Tribune*, September 19, 1973; and "Honda Civic Tops Lists in '74 Auto Mileage Test," *New York Times*, September 19, 1973.

30. Williams, "The Mazda Fuel Furor." The *Review of Southern California Journalism*, in which this article appeared, was a small, student-run publication out of California State University, Long Beach, that ran from 1971 to 1975. Douglas Williams, "'Gas Hog' Tag Cost $100 Million—Mazda," *Detroit Free Press*, February 25, 1974.

31. Miranda Lorraine, "EPA Issues New Mazda Data: Rotary Engine Still Rates Low," *Washington Post*, April 12, 1974.

32. Williams, "The Mazda Fuel Furor."

33. Stork, phone interview, January 5, 2010.

34. Douglas Williams, "EPA Admits Blunder on Mileage Figures,' *Detroit Free Press*, May 30, 1974; Agis Salpukas, "E.P.A. to Change Auto Fuel Data," *New York Times*, May 30, 1974.

35. In response to the OPEC Oil Embargo, President Nixon proposed reorganizing the bulk of federal energy functions into one central authority. He created this organization, the Federal Energy Office, through an executive order in December 1973. When Congress authorized the organization in May 1974, it was named the Federal Energy Agency. The FEA handled many federal energy programs during the "crises" of the 1970s, until President Jimmy Carter created the Department of Energy in 1977.

36. As my colleague Eric Hounshell put it to me, leaders in business and government have always known instinctively what Bruno Latour later described: success depends on forming networks, or associations, of strength. Fittingly, the Office of Mobile Sources' attempts to preserve its fuel economy testing program frequently involved convincing other agencies and organizations that the EPA programs benefited their interests.

37. Stork, "The Genesis of the EPA Fuel Economy Program."

38. Tom Kleene, "SAE to Design EPA Mileage Test: Engineers Face June Deadline," *Detroit Free Press*, March 8, 1974.

39. Stork, "The Genesis of the EPA Fuel Economy Program."

40. Rowland Evans and Robert Novak, "Fuel Economy Testing: Undercutting the EPA," *Washington Post*, May 4, 1974.

41. Stork, "The Genesis of the EPA Fuel Economy Program."

42. Stork, phone interview, January 5, 2010.

43. As quoted in Williams, "EPA Admits Blunder on Mileage Figures."

44. Dan Fisher and Paul E. Steiger, "Gas Mileage Tests by U.S. May Be Changed," *Los Angeles Times*, April 25, 1974.

45. Douglas Williams, "Mileage Test Jurisdiction Up for Grabs," *Detroit Free Press*, April 26, 1974.

46. Evans and Novak, "Fuel Economy Testing."

47. Evans and Novak, "Fuel Economy Testing."

48. Evans and Novak, "Fuel Economy Testing."

49. Stork, phone interview, January 5, 2010. On Train's close relationship with Rowland Evans, whom Train called "Rowly," see Flippen, *Conservative Conservationist*, esp. 177, which mentions that Train regularly dined at Evans's house.

50. In an oral history videotaped for the National Archives in March 2009, Jim Tozzi, who was in charge of the EPA's budget at OMB, said that because the OMB was so close to the president, it felt considerable pressure during the Watergate period. The video is archived at Center for Regulatory Effectiveness, http://www.thecre.com/video/National_Archive.html.

51. US Senate, Special Subcommittee on Science, Technology, and Commerce, "Hearings on the National Fuel Economy Testing Act of 1974," 98.

52. Stork, "The Genesis of the EPA Fuel Economy Program."

53. George C. Wilson, "EPA Wins Its Fight to Keep Car-Mileage Tests in Agency," *Washington Post*, May 18, 1974.

54. "EPA Still Holds Reins in Mileage Labeling," *Automotive News*, May 27, 1974.

55. Comptroller General of the United States, "Review of the Automobile Fuel Economy Testing and Labeling Program."

56. Comptroller General, "Review of the Automobile Fuel Economy Testing and Labeling Program," 3.

57. Comptroller General, "Review of the Automobile Fuel Economy Testing and Labeling Program," 17.

58. Comptroller General, "Review of the Automobile Fuel Economy Testing and Labeling Program," 19.

59. Opening statement by Russell E. Train, Administrator, U.S. Environmental Protection Agency, "1975 Fuel Economy Data," September 20, 1974, in Stork personal collection. The EPA knew about—and early in the year, publicly predicted—these fuel economy gains. Because the companies were relying on catalytic converters, they could move away from systems and procedures, such as spark retard and "riching-up" the fuel mix, that increased fuel consumption and decreased performance. "EPA's Chief Asserts '75 Autos Will Use Less Fuel than '74s," *Wall Street Journal*, March 26, 1974.

60. Stork, "The Genesis of the EPA Fuel Economy Program."

61. Goodson's résumé is available online at Faculty Directory, Ross School of Business, University of Michigan, www.bus.umich.edu/FacultyBios/CV/ggoodsn.pdf. After completing his DOT stint, Goodson returned to Purdue until the early 1980s when he moved to industry. During this period he was also a member of the Policy Board of the US-Japan Study on the Future of the Automobile, coadministered by the Massachusetts Institute of Technology and the University of Michigan.

62. Stork, "The Genesis of the EPA Fuel Economy Program."

63. Memo, Study Task Force Co-chairmen [Goodson and Stork] to Distribution, "120-Day Fuel Economy Study—Panel Workshop, August 28, 1974, Office of Mobile Source Pollution Control, Air Standards and Regulations, 1966–1977, Environmental Protection Agency, Record Group 412, National Archives at College Park (hereafter cited as OMSPC EPA Records).

64. "Outline: 120-Day Study; Chapter 3," undated draft outline, OMSPC EPA Records.

65. For instance, one draft study, "Improved Fuel Economy for New Automobiles and Small Trucks," mentions "free-market response" as one solution to the issue of fuel economy (document unpaginated). Another paper, which was considering strengths of "voluntary fuel economy goals and voluntary fuel economy labeling," noted that one of the "positive aspects" of such a program was that it "relies on free market (e.g. consumers demand more efficient vehicles and goals to provide producer incentive), rather than federal intervention." "Improvement of New Car Fuel Economy: Discussion of Four Policy Options," August 19, 1974, p. 3. Both studies in OMSPC EPA Records.

66. Department of Transportation (DOT) and Environmental Protection Agency, "Potential for Motor Vehicle Fuel Economy Improvement," Report to Congress, October 24, 1974, p. 3.

67. DOT and EPA, "Potential for Motor Vehicle Fuel Economy Improvement.".

68. See "Potential for Improvement of Energy Efficiency of Automobiles and Other Major Appliances," August 29, 1975, a report (in Eric Stork's persona papers) prepared by the staff of the House Energy and Power Subcommittee, which used the 120-Day Study and a report by the RAND Corporation to justify a 28 MPG level.

69. For a history of the Energy Policy and Conservation Act, see Jacobs, *Panic at the Pump*, 144–152. Jacobs's account focuses primarily on energy price controls, which were extremely controversial, particularly with the emerging New Right.

70. US House of Representatives, Committee on Interstate and Foreign Commerce, *Summary of Major Provisions of Energy Conservation and Oil Policy Act of 1975 (H.R. 7014) as Passed by the House*, 5. "Standards are required to be adjusted downward to reflect any decrease in fuel economy in excess of ½ MPG which results from more stringent motor vehicle emission standards" (ibid.).

71. Dingell renamed the Committee on Interstate and Foreign Commerce the "Committee on Energy and Commerce" in 1981 in order to highlight its role in rationalizing and planning the nation's energy policy. Ironically, Dingell made this name change at the same time Ronald Reagan assumed the presidency. Having campaigned on a platform that included abolishing the Department of Energy (as well as the Department of Education), Reagan believed that the best energy policy was no federal energy policy. The energy market would tend to its own needs. And so the very impetus for the committee's new name—the desire to use federal power to "plan" the nation's energy use—withered on the vine.

72. Dingell will one day be the subject of a fascinating biography, but no such work yet exists. Several works, however, describe the representative's legislative prowess at taking control of any issue that he has desired in the course of congressional "turf wars." Any law that affects the auto industry clearly affects Dingell's interests. See, for instance, King, *Turf Wars*.

73. Douglas Williams, "EPA Jumps in on Gas Issue," *Detroit Free Press*, January 18, 1974.

74. Stigler "Theory of Economic Regulation"; McCormick, "A Review of the Economics of Regulation."

Chapter 10 • Deregulation and Its Limits

1. Myers, *A Report on the Closing of the NASA Electronics Research Center*.

2. Murphy, *Science, Geopolitics, and Federal Spending*, 387–388.

3. Richard John, phone interview with author, March 1, 2016.

4. Reich and Donahue, *New Deals*.

5. The trade-off between innovation and productivity is William J. Abernathy's central argument in *The Productivity Dilemma*. For an earlier thesis that bureaucracy, rigidity, and "demoralization" were leading to decline in the auto industry, see Rothschild, *Paradise Lost*.

6. Deregulation efforts are more accurately described as "regulatory restructuring." No industry went from being governed by a slew of rules to having none whatsoever.

7. This and the quotations that follow in this paragraph are from McCormick, "Review of the Economics of Regulation," 16, 18.

8. Scholars often refer to the economic philosophy that came to reign during this period as "neoliberalism," an acknowledgment that its laissez-faire vision harkened back to classical the liberal political economy of the eighteenth and nineteenth centuries. There is much to learn from the literature on this topic. Yet, largely through overuse and sloppy application, the term *neoliberalism* has become vague, often obscuring more than it unveils. Moreover, many left-leaning scholars use it as a

dysphemism to slander individuals and trends that they abhor. Few, if any, individuals call themselves neoliberals. For all of these reasons, I avoid using the term and, as much as possible, see the individuals described in this chapter through their own self-understanding. Two primers on neoliberalism are Harvey, *A Brief History of Neoliberalism*; and Steger and Roy, *Neoliberalism*. I cite other works on neoliberalism in the substantive discussion below.

9. US House of Representatives, Committee on Interstate and Foreign Commerce, *Federal Regulation and Regulatory Reform*, 160.

10. US House of Representatives, Committee on Interstate and Foreign Commerce, *Federal Regulation and Regulatory Reform*, 167–168.

11. "Auto Safety Program of the Administration Apparently Has a Chief," *Wall Street Journal*, September 12, 1969.

12. Graham, *Auto Safety*, 45. See also Wetmore, "Systems of Restraint."

13. Graham, *Auto Safety*, 37–38.

14. International Conference on Passive Restraints, North Atlantic Treaty Organization, General Motors Proving Grounds, *International Conference on Passive Restraints*.

15. Graham, *Auto Safety*, 63–64.

16. Graham, *Auto Safety*, 71.

17. Graham, *Auto Safety*, 63–64.

18. Graham, *Auto Safety*, 80.

19. The Department of Education was carved out of the Department of Health, Education, and Welfare, the remaining portions of which became the Department of Health and Human Services. The Department of Energy was created out of several preexisting agencies, including the Federal Energy Agency, the Federal Power Commission, and the Energy Research and Development Administration.

20. On the history of the character and stereotype of the Dragon Lady, see Prasso, *The Asian Mystique*.

21. Lemov, *Car Safety Wars*, 135.

22. Joan Claybrook, "The Airbag Issue: Whether and When," *Washington Post*, September 1, 1976.

23. On the development of this consumer information program, see Finkelstein, "Consumer Safety Information as a Government Policy Tool.".

24. Lemov, *Car Safety Wars*, 143.

25. Both Nader and Claybrook discuss their falling out in the documentary *An Unreasonable Man* (2006).

26. The seminar papers were eventually collected in Ginsberg and Abernathy, *Government, Technology, and the Future of the Automobile*. For this quotation see, Editor's Note, "The Visible Hand," ibid., 3.

27. John, Coonley, Ricci, and Rebinger, "Mandated Fuel Economy Standards as a Strategy for Improving Motor Vehicle Fuel Economy."

28. White, "Automobile Emissions Control Policy."

29. Abernathy, *The Productivity Dilemma*. Abernathy was only fifty years old when he died in 1983.

30. For a popular treatment of how Japanese cars got a foothold in the United States, see David Halberstam, *The Reckoning* (New York: William Morrow, 1986).

31. Hayes and Abernathy, "Managing Our Way to Economic Decline."

32. Abernathy, Clark, and Kantrow, *Industrial Renaissance*.

33. Vogel, *Japan as Number One*.

34. Womack, Jones, and Roos, *The Machine That Changed the World*.

35. John et al., "Mandated Fuel Economy Standards as a Strategy for Improving Motor Vehicle Fuel Economy."

36. US House of Representatives, Committee on Government Operations, *Administration's Proposals to Help the U.S. Auto Industry*, 16.

37. Luger, *Corporate Power, American Democracy, and the Automobile Industry*, 98–104; Reich and Donahue, *New Deals*.

38. On Kahn's deregulation of the airline industry, see McCraw, *Prophets of Regulation*, chapter 7.

39. Horowitz, *Jimmy Carter and the Energy Crisis of the 1970s*, 109.

40. Horowitz, *Jimmy Carter and the Energy Crisis of the 1970s*, 115.

41. See, for example, Reagan, "A Vision for America," speech delivered November 3, 1980, in Horowitz, *Jimmy Carter and the Energy Crisis of the 1970s*, 167–171.

42. Skinner, Anderson, and Anderson, *Reagan's Path to Victory*, 237–238 (emphasis in the original). Many of Reagan's radio speeches were written by others, so it is difficult, often impossible, to attribute authorship.

43. Bruce-Briggs, *The War against the Automobile*.

44. Lasch, "Same Old New Class."

45. Bell, "The New Class."

46. For a popular, highly critical, if weakly researched, portrait of the Chicago School, see Klein, *The Shock Doctrine*. For a more balanced but also critical account of the Chicago School and the rise of neoliberalism, see Mirowski and Plehwe, *Road from Mont Pèlerin*.

47. Hayek, *The Road to Serfdom*; Friedman, *Capitalism and Freedom*.

48. McCormick, "Review of the Economics of Regulation."

49. Stigler "The Theory of Economic Regulation."

50. Yandle, "Cost-Benefit Analysis of the 1981–1984 MPG Standard," 300n11.

51. Clarkson, Kadlec, and Laffer, "Regulating Chrysler Out of Business?

52. Yandle, "Cost-Benefit Analysis of the 1981–1984 MPG Standard," 303.

53. Peltzman, "Effects of Automobile Safety Regulation."

54. University of Chicago professors Ronald Coase and Milton Friedman were on the advisory board of the American Enterprise Institute. Peltzman, *Regulation of Automobile Safety*.

55. Kemp, *An American Renaissance*, 37–39.

56. House Committee on Government Operations, *Administration's Proposals to Help the U.S. Auto Industry*, 2.

57. U.S. Department of Transportation, *The Auto Industry, 1980*.

58. House Committee on Government Operations, *Administration's Proposals to Help the U.S. Auto Industry*, 18.

59. House Committee on Government Operations, *Administration's Proposals to Help the U.S. Auto Industry*, 1. "Gipper gap" is a reference to the film *Knute Rockne, All American* (1940), in which Reagan played George Gipp, aka the Gipper.

60. McGarity, *Reinventing Rationality*, 17–25. The literature on regulatory reform in the Reagan administration is truly *enormous*. A few of the books that have informed the account here and that are not otherwise cited include Eads and Fix, *Relief or Reform?*; Friedman, *Regulations in the Reagan-Bush Era*; Eisner, *Regulatory Politics in Transition*.

61. Stork, interview, April 10, 2010.

62. McGarity, *Reinventing Rationality*, 18.

63. The Center for Regulatory Effectiveness (CRE) has helpfully collected the executive orders and other presidential and executive branch papers that document the rise of "centralized regulatory review" from the Johnson through the Obama administration. The CRE's website is at http://www.thecre.com/ombpapers/centralrev.html

64. Luger, *Corporate Power, American Democracy, and the Automobile Industry*, 117.

65. On the earlier history of cost-benefit analysis in government, see Porter, *Trust in Numbers*.

66. Miller, "Early Days of Reagan Regulatory Relief and Suggestions for OIRA's Future"; Tozzi, "OIRA's Formative Years"; Copeland, "Federal Rulemaking."

67. West and Cooper, "The Rise of Administrative Clearance," 197.

68. On Stockman's reign at the Office of Management and Budget, see Stockman, *Triumph of Politics*; and Tomkin, *Inside OMB*.

69. McGarity, *Reinventing Rationality*, 22.

70. The political scientist Stan Luger titles his chapter on this period "The Triumph of Corporate Power" (chapter 6 in *Corporate Power, American Democracy, and the Automobile Industry*).

71. Mayer, *Dark Money*, esp. chapters 1–3.

72. House Committee on Government Operations, *Administration's Proposals to Help the U.S. Auto Industry*, 5, 15–17.

73. Tucker, "R&D Consortia."

74. House Committee on Government Operations, *Administration's Proposals to Help the U.S. Auto Industry*, 16.

75. House Committee on Government Operations, *Administration's Proposals to Help the U.S. Auto Industry*, 17.

76. House Committee on Government Operations, *Administration's Proposals to Help the U.S. Auto Industry*, 110.

77. Luger, *Corporate Power, American Democracy, and the Automobile Industry*, 141–142.

78. House Committee on Government Operations, *Administration's Proposals to Help the U.S. Auto Industry*, 6.

79. House Committee on Government Operations, *Administration's Proposals to Help the U.S. Auto Industry*, 39.

80. Golden. *What Motivates Bureaucrats?* 118.

81. Golden. *What Motivates Bureaucrats?* 110.

82. Heymann, *The Politics of Public Management*, 43.

83. Mooney, *Republican War on Science*, 40; Leary, *Ethics of Dissent*, 53.

84. Collin, *Environmental Protection Agency*, 281.

85. Joseph Somers (former EPA staff member), phone conversation with author, May 10, 2016.

86. Charles Gray, interview with author, August 19, 2008, Ann Arbor, Mich.; Karl Hellman, interview, August 19, 2008. Charles Gray's earlier career is discussed in chapter 7.

87. Gray and Alson, *Moving America to Methanol*.

88. Heymann, *The Politics of Public Management*, 44.

89. Charles A. Radin and Benjamin Taylor, "Burford Resigns from EPA," *New York Times*, March 10, 1983.

90. "Chronology of Events in the E.P.A. Dispute," *New York Times*, March 10, 1983.

91. Philip Shabecoff, "Mrs. Burford Says She Decided to Quit before President Yielded to Congress," *New York Times*, March 11, 1983.

92. Mooney, *Republican War on Science*, 41.

93. Golden, *What Motivates Bureaucrats?* 109.

94. Claybrook, Gillan, and Strainchamps, *Reagan on the Road*, ii.

95. Claybrook, Gillan, and Strainchamps, *Reagan on the Road*, ii.

96. Claybrook, Gillan, and Strainchamps, *Reagan on the Road*, iv.

97. Golden, *What Motivates Bureaucrats?* 44.

98. Golden, *What Motivates Bureaucrats?* 56.

99. On Nader and the Cornell-Liberty Mutual Safety Vehicle, see chapter 3.

100. Statement of Ben Kelley, Senior Vice President of the Insurance Institute for Highway Safety, in US Senate, Committee on Commerce, Science, and Transportation, *Motor Vehicle Safety and the Marketplace*, 58. The company Calspan—a descendent of the Cornell Aeronautical Laboratory, which had expertise in crash studies going back to Hugh DeHaven and the Cornell–Liberty Mutual Safety Car—conducted research in science and technology. Simca, a French automaker, was founded in 1934, became a part of Chrysler in 1970, and disappeared completely in 1978.

101. Friedman, "The Minicars RSV—Still a Car for the Future."

102. Kelley, in Senate Committee on Commerce, Science, and Transportation, *Motor Vehicle Safety and the Marketplace*, 58.

103. Kelley, in Senate Committee on Commerce, Science, and Transportation, *Motor Vehicle Safety and the Marketplace*, 59.

104. Allan Parachini, "The Auto Safety War: What Will Happen to Supercar?" *Los Angeles Times*, September 13, 1981.

105. Parachini, "The Auto Safety War."

106. Claybrook, Gillan, and Strainchamps, *Reagan on the Road*, 5.

107. The former staff member made this statement to me off the record while I was preparing interviews for my dissertation on automobile regulation.

108. I owe this insight to Evan Hepler-Smith.

109. Dole and Dole, *The Doles*, 225.

110. Dole and Dole, *The Doles*, 226; Graham, *Auto Safety*, 180–181.

111. The following figures are from National Highway Traffic Safety Administration, *Motor Vehicle Safety, 1990*.

112. Dole and Dole, *The Doles*, 226.

113. Graham, *Auto Safety*, 228.

114. Moreover, Graham fails to account for the general increased perception of risk and safety (which is extremely difficult to quantify) that some call the "risk society." See Beck, *Risk Society*.

115. Of course, the fields of psychology, marketing, and, more recently, behavioral economics have carried out laboratory experiments on the issue (mostly on captive undergraduate students).

116. Stockman, *The Triumph of Politics*.

117. Leone, "Regulatory Relief and the Auto Industry," 104–105.

Chapter 11 • Indecision, Regulatory Uncertainty, and the Politics of Partisanship

1. Sam Roe, "Political Obstacle Course Proves Fatal for Supercar," *Chicago Tribune*, December 10, 2002.

2. Matthew Wald, "U.S. Drops Research into Fuel Cells for Cars," *New York Times*, May 7, 2009.

3. Bullis, "Q & A: Steven Chu.".

4. Christopher Hare, Keith T. Poole, and Howard Rosenthal, "Polarization in Congress Has Risen Sharply; Where Is It Going Next?" *Washington Post*, Monkey Cage blog, February 13, 2014, https://www.washingtonpost.com/news/monkey-cage/. A June 2014 panel at the Policy History Conference in Columbus, Ohio, brought together a group of esteemed historians who had written about various periods in North American history going back to the colonial period. The historians were given the task of deciding whether the current moment was the most partisan period in the history of the United States. They unanimously answered no. For instance, Americans once killed one another in a civil war. My argument here is not that the United States is more partisan now than in previous periods, but that positions on regulation, the environment and climate change, and other important issues became more strongly tied to party affiliation than previously.

5. I emphasize *elected* party members—or I could say party elites—because there is considerable disagreement about whether the general American public has become more polarized, while there is broad agreement that the US *Congress* has become more polarized.

6. Morris Fiorina, "Americans Have Not Become More Politically Polarized," *Washington Post*, Monkey Cage blog, June 23, 2014: https://www.washingtonpost.com/news/monkey-cage/wp/.

7. "Wide Differences between Conservative Republicans and Liberal Democrats on Likely Effects of Climate Change and Effectiveness of Ways to Address It," Pew Research Center, September 30, 2016, http://www.pewinternet.org/2016/10/04/the-politics-of-climate/ps_2016-.10-04_politics-of-climate_0-02/

8. This is also the finding of Mashaw and Harfst, "From Command and Control to Collaboration and Deference."

9. Conniff, "The Political History of Cap and Trade"; Taylor, Rubin, and Hounshell, "Regulation as the Mother of Innovation."

10. Waxman, "An Overview of the Clean Air Act Amendments of 1990."

11. Luger, *Corporate Power, American Democracy, and the Automobile Industry*, 173.

12. From the book jacket for Doody and Bingaman, *Reinventing the Wheels*.

13. From the book jacket for Ingrassia, *Crash Course*.

14. Luger, *Corporate Power, American Democracy, and the Automobile Industry*, 165.

15. Luger, *Corporate Power, American Democracy, and the Automobile Industry*, 167.

16. Graham and Wiener, *Risk vs. Risk.*

17. Graham, "Saving Gasoline and Lives."

18. Graham, "Saving Gasoline and Lives," 95.

19. Graham, "Saving Gasoline and Lives," 100.

20. Dunn and Johnson, "Chasing Molecules."

21. For a historical overview of the SUV, see Bradsher, *High and Mighty.*

22. Bradsher, *High and Mighty,* chapter 3.

23. Joe Mathews, "The Hummer and Schwarzenegger: They Probably Won't Be Back," *Washington Post,* February 28, 2010.

24. Bradsher, *High and Mighty,* 105.

25. Keith Bradsher provided the best coverage of the problems with SUVs in the *New York Times* during the 1990s. See Bradsher, "A Deadly Highway Mismatch Ignored," *New York Times,* September 24, 1997; Bradsher, "Trucks, Darlings of Drivers, Are Favored by the Law, Too," *New York Times,* September 30, 1997; Bradsher, "Further Problems of Safety Found for Light Trucks," *New York Times,* December 12, 1997.

26. Laurie McGinley, "Agency Issues Auto Safety Standards to Reduce Injuries from Side Crashes," *Wall Street Journal,* October 25, 1990.

27. The Standard 201 revisions, which focused on "upper interior components," became effective September 18, 1995. US Department of Transportation, National Highway Traffic Safety Administration, Federal Motor Vehicle Safety Standards and Regulations, Standard No. 201, Occupant Protection in Interior Impact.

28. Luger, *Corporate Power, American Democracy, and the Automobile Industry,* 175.

29. Doyle, *Taken for a Ride,* chapter 17.

30. Doyle, *Taken for a Ride,* 345.

31. Cheney, "Speech," April 30, 2001, reprinted in Horowitz, *Jimmy Carter and the Energy Crisis,* 174.

32. Cheney, "Speech," 175.

33. Cheney, "Speech," 176.

34. Ralph Nader, "Dick Cheney and Conservation," May 1, 2001, in Horowitz, *Jimmy Carter and the Energy Crisis,* 177–179.

35. The list is viewable by year at "Rewriting the Rules: The Bush Administration's First Term Environmental Record," National Resources Defense Council, http://www.nrdc.org/resources/rewriting-rules-bush-administrations-first-term -environmental-record.

36. Martin Kettle, "Cheney Tells US to Carry on Guzzling," *The Guardian,* May 10, 2001.

37. As quoted in Katharine Q. Seelye, "President Distances Himself from Global Warming Report," *New York Times,* June 5, 2002.

38. Merrill, *The Oil Crisis of 1973–1974,* 149.

39. Neela Banerjee, "The High, Hidden Cost of Saudi Arabian Oil," *New York Times,* October 21, 2001, reprinted in Merrill, *The Oil Crisis of 1973–1974,* 149–152.

40. Energy Information Administration, "STEO Supplement: Summer 2006 Motor Gasoline Prices," https://www.eia.gov/forecasts/steo/special/pdf/gasoline2006.pdf.

41. Al Gore, Davis Guggenheim, Laurie David, Lawrence Bender, Scott Z. Burns, Jeff Skoll, Lesley Chilcott, et al., *An Inconvenient Truth* (Hollywood, Calif.: Paramount, 2006).

42. Limbaugh, *See, I Told You So.*

43. Oreskes and Conway, *Merchants of Doubt.*

44. Judith Lewis, "EPA's California Waiver," *LATimes.com*, January 29, 2009, http://www.latimes.com/opinion/la-oe-lewis29-2009jan29-story.html.

45. Office of the Press Secretary, "Executive Order 13563—Improving Regulation and Regulatory Review."

46. Office of the Press Secretary, "Obama Administration Finalizes Historic 54.5 MPG Fuel Economy Standards."

47. Mayer, *Dark Money.*

48. Environmental Protection Agency, "Clean Power Plan for Existing Power Plants.", https://www.epa.gov/cleanpowerplan/clean-power-plan-existing-power -plants.

49. US Department of Transportation, "Secretary Ray LaHood Distracted Driving Press Conference U.S. Department of Transportation," updated April 18, 2012.

50. Yang et al., "Detecting Driver Phone Use Leveraging Car Speakers."

51. See the "About Us" page at KidsandCars.org.

52. Laura Lane, "12 Years after Son's Tragic Death Father Wins Battle for Rearview Cameras," *Long Island Herald/Oyster Bay Guardian*, April 3, 2014, http:// liherald.com/oysterbay/stories/12-years-after-sons-tragic-death-Father-wins-battle -for-rearview-cameras,90564.

53. "How the Rear Visibility Standard Was Won," KidsandCars.org, https://www .kidsandcars.org/resources/release-the-rear-visibility-standard/.

54. Coral Davenport, "Senate Confirms Scott Pruitt as E.P.A. Head," *New York Times*, February 17, 2017.

55. March for Science, "The Science behind the March for Science Crowd Estimates," *Medium*, May 15, 2017, https://medium.com/marchforscience-blog/the -science-behind-the-march-for-science-crowd-estimates-f337adf2d665.

56. Lisa Friedman and Brad Plumer, "E.P.A. Announces Repeal of Major Obama-Era Carbon Emissions Rules," *New York Times*, October 9, 2017 https://www.nytimes .com/2017/10/09/climate/clean-power-plan.html; Sonari Glinton, "Trump Adminis-tration Takes Key Step to Rolling Back Auto Fuel Standards," National Public Radio, August 14, 2017: https://www.npr.org/2017/08/14/543474251/trump-administration -takes-key-step-to-rolling-back-auto-fuel-standards; Jonathan M. Gitlin, "Trump Administration Reportedly Kills Vehicle-to-Vehicle Safety Mandate," *Ars Technica*, November 1, 2017, https://arstechnica.com/cars/2017/11/trump-administration -reportedly-kills-vehicle-to-vehicle-safety-mandate/.

57. Tucker, "R&D Consortia."

58. Anderson, "Introduction."

59. Alpert and Holt, "Inter-industry Cooperative Research and the Government."

60. Tucker, "R&D Consortia."

61. Tucker, "R&D Consortia," 46.

62. US House of Representatives, Committee on Government Operations, *Administration's Proposals to Help the U.S. Auto Industry.*

63. Teles, *The Rise of the Conservative Legal Movement*, esp. 94–98; Van Horn, "Reinventing Monopoly and the Role of Corporations."

64. Wright, "The National Cooperative Research Act of 1984."

65. Inkpen, "Knowledge Transfer and International Joint Ventures."

66. Luger, *Corporate Power, American Democracy, and the Automobile Industry*, 162.

67. Luger, *Corporate Power, American Democracy, and the Automobile Industry*, 158; Doyle, *Taken for a Ride*, chapter 14; Chris Paine, dir., *Who Killed the Electric Car?* (2006, DVD).

68. Doyle, *Taken for a Ride*, 308.

69. AAMA memo, quoted in Doyle, *Taken for a Ride*, 321.

70. Paine, *Who Killed the Electric Car?*

71. Oscar Sums, "Continental Divide: Californians Collide with Folks in Detroit over the Electric Car," *Wall Street Journal*, January 24, 1994.

72. National Research Council, *Effectiveness of the United States Advanced Battery Consortium as a Government-Industry Partnership.*

73. This series of events is the focus of Paine, *Who Killed the Electric Car?*

74. "Supercar: The Tanking of an American Dream," *Chicago Tribune*, December 7, 2002, http://www.chicagotribune.com/chi-super1part3-story.html.

75. National Research Council, *Review of the Research Program of the Partnership for a New Generation of Vehicles: Fifth Report.*

76. Elizabeth Kolbert, "Running on Fumes," *New Yorker*, November 5, 2007.

77. Roe, "Political Obstacle Course Proves Fatal for Supercar."

78. Roe, "Political Obstacle Course Proves Fatal for Supercar."

79. "Bush Touts Benefits of Hydrogen Fuel," *CNN.com*, February 6, 2003, http://www.cnn.com/2003/ALLPOLITICS/02/06/bush-energy/.

80. "Bush Touts Benefits of Hydrogen Fuel."

81. Amanda Little, "What Can We Learn from Bush's FreedomCAR Program?" *The Grist*, February 26, 2003, http://grist.org/article/tough/.

82. Wald, "U.S. Drops Research into Fuel Cells for Cars."

83. David Biello, "R.I.P. Hydrogen Economy? Obama Cuts Hydrogen Car Funding," *Scientific American*, May 8, 2009, http://blogs.scientificamerican.com/news-blog/rip-hydrogen-economy-obama-cuts-hyd-2009-05-08/.

84. Jurgen, "Introduction," ii.

85. Ann Johnson, "Projects of Auto Emissions Control—Putting a Brain in the Car?"

86. Jurgen, "Introduction."

87. Russell, *Open Standards and the Digital Age.*

88. Jones, "The Challenge of Automotive Electronics in the USA."

89. Jones, "The Challenge of Automotive Electronics in the USA."

90. Jones, "The Challenge of Automotive Electronics in the USA."

91. Borg, *Auto Mechanics*, 163; Johnson, "Routines for Innovation."

92. Denton, *Advanced Automotive Fault Diagnosis*, 127; Luger, *Corporate Power, American Democracy, and the Automobile Industry*, 157.

93. Borg, *Auto Mechanics*, chapter 7.

94. Johnson, *Hitting the Brakes*, 114–115.

95. "U.S. Department of Transportation Issues Advance Notice of Proposed Rulemaking to Begin Implementation of Vehicle-to-Vehicle Communications Technology," August 18, 2014, NHTSA.gov, http://www.nhtsa.gov/About+NHTSA/Press+Releases/NHTSA-issues-advanced-notice-of-proposed-rulemaking-on-V2V-communications.

96. Danny Hakim and Hiroko Tabuchi, "Volkswagen Test Rigging Follows a Long Auto Industry Pattern," *New York Times*, September 23, 2015.

97. Guilbert Gates, Jack Ewing, Karl Russell, and Derek Watkins, "Explaining Volkswagen's Emissions Scandal," *New York Times*, June 1, 2016.

98. Jack Ewing, "Volkswagen Not Alone in Flouting Pollution Limits," *New York Times*, June 9, 2016.

99. Andy Greenberg, "Hackers Remotely Kill a Jeep on the Highway—With Me in It," *Wired*, July 21, 2015.

100. "Robotic Road Trip on Military Mission," *New York Times*, October 9, 2003.

101. "Grand Challenge Overview" (2004), DARPA, archived at http://archive .darpa.mil/grandchallenge04/overview.htm.

102. "Robotic Road Trip on Military Mission."

103. David Szondy, "Autonomous Vehicles to Join the US Army," *Gizmag*, July 4, 2014, http://www.gizmag.com/us-army-autonomous-vehicles/32796/.

104. Thrun, "What We're Driving At."

105. Thrun, "What We're Driving At."

106. Bix, *Inventing Ourselves Out of Jobs?*

107. Derek Thompson, "The End of Work," *The Atlantic*, July–August 2015.

108. For instance, in "The End of Work," Derek Thompson writes, "After 300 years of people crying wolf [about technological unemployment], there are now three broad reasons to take seriously the argument that the beast is at the door: the ongoing triumph of capital over labor, the quiet demise of the working man, and the impressive dexterity of information technology."

109. Mui and Carroll, *Driverless Cars*. Mui wrote a seven-piece series on driverless cars that appeared on Forbes.com. The first part, "Fasten Your Seatbelts: Google's Driverless Car Is Worth Trillions" (January 22, 2013) is available at http://www.forbes.com/sites/chunkamui/2013/01/22/fasten-your-seatbelts-googles -driverless-car-is-worth-trillions/. The quotation from Mui comes from a comment on my blog post, "Searching for the Limits of Innovation Speak," leevinsel.com, December 19, 2013, http://leevinsel.com/blog/2013/12/19/searching-for-the-limits -of-innovation-speak.

110. Mindell, *Our Robots, Ourselves*.

111. Danielle Muoio, "19 Companies Racing to Put Self-Driving Cars on the Road by 2021," *Business Insider*, October 17, 2016, http://www.businessinsider.com /companies-making-driverless-cars-by-2020-2016-10/#tesla-is-aiming-to-have-its -driverless-technology-ready-by-2018-1.

112. Department of Transportation, "U.S. Department of Transportation Releases Policy on Automated Vehicle Development," Transportation.gov, updated May 30, 2013, https://www.transportation.gov/briefing-room/us-department-transportation -releases-policy-automated-vehicle-development.

113. Hope Reese, "Autonomous Driving Levels 0 to 5: Understanding the Differ-ence," TechRepublic.com, November 1, 2016, https://www.techrepublic.com/article /autonomous-driving-levels-0-to-5-understanding-the-differences/.

114. Ryan Beene, "NHTSA Moves to Overhaul Rules for Self-Driving Cars," *Automotive News*, October 27, 2017, http://www.autonews.com/article/20171027 /MOBILITY/171029753/autonomous-cars-regulation-nhtsa.

115. "Automated Driving Systems 2.0: A Vision for Safety," National Highway Traffic Safety Administration, September 12, 2017, p. 16, https://www.nhtsa.gov/sites /nhtsa.dot.gov/files/documents/13069a-ads2.0_090617_v9a_tag.pdf.

116. "Automated Driving Systems 2.0," 12, 15.

117. Sam Frank, "Come With Us if You Want to Live: Among the Apocalyptic Libertarians of Silicon Valley," *Harper's* (January 2015).

Conclusion

1. Peltzman, "Regulation and the Natural Progress of Opulence."

2. Mazzucato, *The Entrepreneurial State*. Mazzucato's findings on the iPhone are consistent with historians' understandings of how the Cold War national security state influenced developments in microelectronics. See, for instance, McCray, "From Lab to iPod."

3. Nelson and Winter, *An Evolutionary Theory of Economic Change*.

4. The literature on experts and expertise is enormous. This book builds on a specific historiographical tradition about expertise known as the organizational synthesis. For the most recent overall statement of the synthesis, see Galambos, *The Creative Society*.

5. New Oxford American Dictionary (Second Edition).

6. The process by which experts take up problems has been a central topic in the history and sociology of science and technology since the very beginning of these fields. A few works that address the issue include Hessen, "The Social and Economic Roots of Newton's *Principia*"; Merton, "Science, Technology, and Society in Seventeenth Century England"; Kuhn, *The Structure of Scientific Revolutions*; Hughes, *Networks of Power*; Constant, *Origins of the Turbojet Revolution*; Vincenti, *What Engineers Know and How They Know It*; and Johnson, *Hitting the Brakes*.

7. Collins, *Weberian Sociological Theory*, 77–79; Mokyr, *Gifts of Athena*.

8. Hounshell, *From the American System to Mass Production*.

9. Sinclair, *Philadelphia's Philosopher Mechanics*.

10. Ann Johnson calls problems "attractors," in *Hitting the Brakes*, 5. Not only do certain problems become sexy within specific communities, but whole fields become sexy and attractive. In his book *Power Loss*, the historian Richard Hirsh shows that before World War II, the electric power industry was a cutting-edge sector that many talented young people found attractive, but after the war, it came to be seen as a backwater. Talented hotshots moved into young exciting fields like electronics and eventually computing. Hirsh argues further that this brain drain away from electric power production came to pose real problems.

11. Usselman, *Regulating Railroad Innovation*; Hounshell and Smith, *Science and Corporate Strategy*, chapter 15.

12. Leslie, *Boss Kettering*.

13. On the creation of the Society of Automotive Engineers, see chapter 1. See also Sinclair and Hull, *A Centennial History of the American Society of Mechanical Engineers*; Post, *The SAE Story*.

14. Johnson, *Hitting the Brakes*.

15. See chapter 2. On dominant design and closed-body cars, see Utterback and Abernathy, "A Dynamic Model of Process and Product Innovation." On riding

comfort, see Mom, "Orchestrating Automobile Technology"; and Vinsel, "Virtue via Association."

16. Mom, *The Evolution of Automotive Technology*, chapters 6, 7, and 11.

17. Flink, *The Automobile Age*, 235–240.

18. The idea that problems become public through social processes emerged with the sociological literature on deviance in the 1950s and 1960s and probably extends back to the pragmatist writings of the late nineteenth and early twentieth centuries, especially the works of John Dewey and Arthur F. Bentley. For the deviance literature, see Abbott, *Chaos of Disciplines*, chapter 3. The best treatment of public problems as social process may still be Gusfield, *The Culture of Public Problems*.

19. Downs, "Up and Down with Ecology."

20. I take the term *focusing device* from Nathan Rosenberg, who borrowed it from Albert Hirschman. Rosenberg, "The Direction of Technological Change." At some level, this idea of focus, in the sense of biologically rooted human attention, is probably more than metaphorical. Ethology and other animal studies suggest that problem solving is deeply rooted in biology. While problem solving is a complex phenomenon, the cognitive capacity for sustained attention appears to be a necessary component. It is probably true that regulations work by getting experts to pay attention. Ultimately, however, such considerations lie beyond the scope of this book.

21. This argument builds in part on and extends the ideas of earlier studies conducted at Carnegie Mellon University, including Taylor, Rubin, and Hounshell, "Regulation as the Mother of Innovation"; Lee, Veloso, and Hounshell, "Linking Induced Technological Change and Environmental Regulation"; Lee, Veloso, Hounshell, and Rubin, "Forcing Technological Change"; and Lee, Veloso, Hounshell, and Rubin, "Innovation in Automotive Emission Control Technologies."

22. Carpenter, "Confidence Games."

23. "Motor Vehicle Accidents, Death Rates, and Deaths, by Type of Accident, 1913–1996," *Historical Statistics of the United States*, table Df448–456.

24. Balogh, *The Associational State*.

25. Vinsel, "Virtue via Association."

26. Part of the divergence between scientific developments in crash safety and those in air pollution during this period may have attributable to technical complexity. Atmospheric chemistry was and is extremely complicated and has involved constant improvements, uncertainties, and revisions to this day, whereas, at its most basic level, impact biomechanics involves limiting human exposure to physical force.

27. Such "inter-sectoral knowledge flows," as Nathan Rosenberg and David Mowery call them, are well known, but the existing literature does not emphasize that regulation often leads to such movement.

28. See Constant, *Origins of the Turbojet Revolution*.

29. Klepper, *Experimental Capitalism*, 106.

30. For a recent examination of regulatory capture in the Environmental Protection Agency, see Margaret Talbot, "Scott Pruitt's Dirty Politics," *New Yorker*, April 2, 2018.

31. US House of Representatives, Committee on Interstate and Foreign Commerce, *Traffic Safety: Hearings on H.R. 13228*.

32. Lee Vinsel and Constantine Samaras, "Some of the Best Parts of Autonomous Vehicles Are Already Here," *The Conversation*, September 19, 2017,

https://theconversation.com/some-of-the-best-parts-of-autonomous-vehicles-are
-already-here-84029.

33. Ayres and Kapczynski, "Innovation Sticks."

34. David Roberts, "The World's Largest Car Market Just Announced an Imminent End to Gas and Diesel Cars," *Vox*, September 13, 2017, https://www.vox.com
/energy-and-environment/2017/9/13/16293258/ev-revolution.

Archives and Personal Papers

Division of Work and Industry, National Museum of American History. Smithsonian Institution, Washington, D.C.

Records of the Committee on the Psychology of the Highway. National Research Council. Archives of the National Academy of Sciences, Washington, D.C.

Records of the Crash Injury Research Project. Cornell Medical College Archives, New York City.

Records of the Council of National Defense, National Archives and Records Administration, College Park, Md.

Records of the General Services Administration. National Archives and Records Administration, College Park, Md.

Hellman, Karl. Personal papers. Now in the Hagley Museum and Library, Wilmington, Del.)

Records of the National Bureau of Standards. National Archives and Records Administration, College Park, Md.

Records of the National Highway Traffic Safety Administration. National Archives and Records Administration, College Park, Md.

Stork, Eric. Personal papers.

Warshaw Collection, National Museum of American History, Smithsonian Institution, Washington, DC

Wayne State University Archives. Clipping files.

Periodicals Cited

Atlantic, The	*Gizmag*
Automobile, The	*Grist, The*
Automobile Magazine	*Guardian, The*
Automotive News	*Harper's*
Business Insider	*Life* Magazine
Business Week	*Motor Age*
Chicago Daily Tribune	*Motor Truck News*
Detroit Free Press	*New Yorker, The*
Detroit News	*New York Times*

Palo Alto Times
Popular Mechanics
Salon.com
Scientific American
Seattle Times
Star Beacon (Ashtabula, Ohio)
Time

Town & Country
Transactions of the Illuminating
   Engineering Society
Wall Street Journal
Washington Post
Wired

Published Sources, Unpublished Theses, and Government Documents

Abbott, Andrew Delano. *Chaos of Disciplines.* Chicago: University of Chicago Press, 2001.

Abernathy, William J. *The Productivity Dilemma: Roadblock to Innovation in the Automobile Industry.* Baltimore: Johns Hopkins University Press, 1978.

Abernathy, William J., Kim B. Clark, and Alan M. Kantrow. *Industrial Renaissance: Producing a Competitive Future for America.* New York: Basic Books, 1984.

Akins, James E. "The Oil Crisis: This Time the Wolf Is Here," *Foreign Affairs,* April 1973.

Albert, Daniel M. "The Nut behind the Wheel: Shifting Responsibilities for Traffic Safety since 1895." In *Silent Victories: The History and Practice of Public Health in Twentieth-Century America,* ed. John W. Ward and Christian Warren, 363–378. Oxford: Oxford University Press, 2007.

Alpert, Norman, and Eugene L. Holt, "Inter-industry Cooperative Research and the Government: Two Case Studies." In *Government, Technology, and the Future of the Automobile,* ed. Douglas H. Ginsburg and William J. Abernathy, 334–346. New York: McGraw-Hill, 1980.

Ammon, S. Von. "Developing a Method for Testing Brake-Linings." SAE Technical Paper 220014 (1922).

Aldrich, Mark. *Safety First: Technology, Labor, and Business in the Building of American Work Safety, 1870–1939.* Baltimore: Johns Hopkins University Press, 1997.

Anderson, Cyrus V. "Introduction," in "Symposium: Antitrust in Times of Economic, Technological, and Ecological Crisis." Special issue, *University of Pittsburgh Law Review* 3, no. 36 (Spring 1975): 591–596.

Arlington, Shane Quinlan. "Defining Gasoline: The Evolution of Internal Combustion Fuel Standards and Dominant Designs." BS thesis, Stevens Institute of Technology, 2015.

Armand, Harry. "Make the Automobile Safer!" *Safety Engineering* (March 1938): 35.

Armand, Harry. "Safety and the New Cars." *Safety Engineering* (December 1938): 32–34.

Arora, Ashish, Andrea Fosfuri, and Alfonso Gambardella. *Markets for Technology: The Economics of Innovation and Corporate Strategy.* Cambridge, MA: MIT Press, 2001.

Assembly of the State of California. "Study and Analysis of the Facts Pertaining to Air Pollution Control in Los Angeles County," Report of the Subcommittee of the Assembly Interim Committee on Governmental Efficiency and Economy. 1953.

"Automobile Headlights." Report No. 11 of the Committee on Glare. *Transactions of the Illuminating Engineering Society* 11 (1916): 29–35.

"Automobile Legislation in the United States." *Horseless Age* 10, no. 19 (November 5, 1902), 505–511.

Ayres, Ian, and Amy Kapczynski. "Innovation Sticks: The Limited Case for Penalizing Failures to Innovate." *University of Chicago Law Review* (2015): 1781–1852.

Bailey, Christopher J. *Congress and Air Pollution: Environmental Policies in the USA*. Manchester: Manchester University Press, 1998.

Bailey, Roy Rutherford. *Sure Pop and the Safety Scouts*. Yonkers-on-Hudson, N.Y.: World Book Co., 1917.

Ballou-Wright Automobile Supplies Catalog. 1906. Reprinted, Portland: Oregon Historical Society, 1971.

Balogh, Brian. *The Associational State: American Governance in the Twentieth Century*. Philadelphia: University of Pennsylvania Press, 2015.

Balogh, Brian. *Chain Reaction: Expert Debate and Public Participation in American Commercial Nuclear Power, 1945–1975*. Cambridge: Cambridge University Press, 1991.

Barth, Delbert S. "Federal Motor Vehicle Emission Goals for CO, HC, and NO_x Based on Desired Air Quality Levels." *Air Pollution Control Administration Journal* 20, no. 8 (August 1970): 519–523.

Bazzell, R. "The SAE Manikin Interior Space Evaluator." SAE Technical Paper 630176 (1963).

Beck, Ulrich. *Risk Society: Towards a New Modernity*. Trans. Mark Ritter. London: Sage, 1992.

Becker, Gary S., and Kevin M. Murphy, "A Theory of Rational Addiction." *Journal of Political Economy* 96 (August 1988): 675–700.

Belden, E. H. "The Future Passenger Car." *Journal of the Society of Automotive Engineers* 5, no. 2 (August 1919), 147.

Bell, Daniel. "The New Class: A Muddled Concept." *Society* 16, no. 2 (January 1979): 15–23.

Bierman, Howard R., and Victor R. Larsen. "Reactions of the Human to Impact Forces Revealed by High Speed Motion Picture Technique," *Journal of Aviation Medicine* 17, no. 5 (1946): 407–412.

Bimber, Bruce Allen. *The Politics of Expertise in Congress: The Rise and Fall of the Office of Technology Assessment*. Albany: State University of New York Press, 1996.

Bix, Amy Sue. *Inventing Ourselves Out of Jobs? America's Debate over Technological Unemployment, 1929–1981*. Baltimore: Johns Hopkins University Press, 2001.

Blatter, Jeremy Todd. "The Psychotechnics of Everyday Life: Hugo Münsterberg and the Politics of Applied Psychology, 1887–1917." Ph.D. diss., Harvard University, 2014.

Bowen, Nancy. *Ralph Nader: Man with a Mission*. Brookfield, Conn.: Twenty-First Century Books, 2002.

Borg, Kevin. *Auto Mechanics: Technology and Expertise in Twentieth-Century America*. Baltimore: Johns Hopkins University Press, 2007.

Bowden, Ernest J. *Our Destiny and Other Poems*. Toronto: William Briggs, 1911.

Bradley, C. E., and A. J. Haagen-Smit. "The Application of Rubber in the Quantitative Determination of Ozone." *Rubber Chemistry and Technology* 24, no. 4 (December 1951): 750–755.

Bradsher, Keith. *High and Mighty: SUVs—The World's Most Dangerous Vehicles and How They Got That Way*. New York: PublicAffairs, 2002.

Brenner, M. S. "Technology Intelligence and Technology Scouting." *Competitive Intelligence Review* 7, no. 3 (1996): 20–27.

Brenner, Robert. "Analysis of Traffic Flow Using 'Time' as the Dependent Value." PhD diss., University of California Los Angeles, 1962.

Brenner, Robert. "The Future of Highway Research." *Highway Research Record*, no. 79 (1965): 58–59.

Bromley, Michael L. *William Howard Taft and the First Motoring Presidency, 1909–1913*. Jefferson, NC: McFarland, 2003.

Brown, Herbert Billings. "The Status of the Automobile." *Yale Law Journal* 27, no. 4 (February 1908), 223–231.

Browne, A. B. and E. H. Lockwood. "The Practical Testing of Motor Vehicles." *S.A.E. Transactions* 10, no. 1 (1915): 68–91.

Bruce-Briggs, B. *The War against the Automobile*. New York: E. P. Dutton, 1977.

Buckhorn, Robert F. *Nader: The People's Lawyer*. Englewood Cliffs, N.J.: Prentice-Hall, 1972.

Buerglener, Robert. "Creating the American Automobile Driver, 1898–1918." PhD diss., University of Chicago, 2006.

Bullis, Kevin. "Q & A: Steven Chu." *MIT Technology Review*, May 14, 2009. https://www.technologyreview.com/s/413475/q-a-steven-chu/.

Bureau of Public Roads, US Department of Agriculture. *Motor Vehicle Traffic Conditions in the United States. Part 1: Uniformity of State Motor-Vehicle Traffic Laws*. Washington, D.C.: Government Printing Office, 1938.

Bureau of Public Roads, US Department of Agriculture. *Motor Vehicle Traffic Conditions in the United States. Part 3: Inadequacy of State Motor-Vehicle Accident Reporting*. Washington, D.C.: Government Printing Office, 1938.

Bureau of Public Roads, US Department of Agriculture. *Motor-Vehicle Traffic Conditions in the United States. Part 4: Official Inspection of Vehicles*. Washington, D.C.: Government Printing Office, 1938.

Burke, John G. "Bursting Boilers and the Federal Power." *Technology and Culture* 7, no. 1 (1966): 1–23.

Burnham, John C. *Accident Prone: A History of Technology, Psychology, and Misfits of the Machine Age*. Chicago: University of Chicago Press, 2009.

Bushnell, David. "Administrative History of the Aeromedical Field Laboratory at the Air Force Missile Development Center, Halloman Air Force Base, New Mexico, 1951–1958." Historical Branch, United States Air Force, 1958.

Byrne, John A. *The Whiz Kids: Ten Founding Fathers of American Business—And the Legacy They Left Us*. New York: Doubleday, 1993.

Cain, Louis P. "Fatalities and Injuries in Motor Vehicle Accidents, by Functional Category of Roadway: 1980–1995." In *Historical Statistics of the United States, Earliest Times to the Present: Millennial Edition*, ed. Susan B. Carter, Scott Sigmund Gartner, Michael R. Haines, Alan L. Olmstead, Richard Sutch, and Gavin Wright, table Df434–447. New York: Cambridge University Press, 2006.

California Department of Motor Vehicles. *Table of Approved Headlight Devices, Candlepowers, and Adjustments, with Photographs of Devices and Light Distributions*. Sacramento: California State Printing Office, 1922.

Campbell, J. M. "Luncheon Address." In *Proceedings of the First National Symposium on Heterogeneous Catalysis for Control of Air Pollution*, ed. Bani R. Banerjee, 17–23. Washington, D.C.: National Air Pollution Control Administration, 1968.

Carlos, James P., Alan M. Gittelsohn, and William Haddon Jr. "Caries in Deciduous Teeth in Relation to Maternal Ingestion of Fluoride." *Public Health Reports* 77, no. 8 (1962): 658.

Carpenter, Daniel. "Confidence Games: How Does Regulation Constitute Markets?" In *Government and Markets: Toward a New Theory of Regulation*, ed. Edward J. Balleisen and David A. Moss, 164–191. Cambridge: Cambridge University Press, 2010.

Carpenter, Daniel. *Reputation and Power: Organizational Image and Pharmaceutical Regulation at the FDA*. Princeton: Princeton University Press, 2010.

Carson, John. "The Science of Merit and the Merit of Science: Mental Order and Social Order in Early Twentieth-Century France and America." In *States of Knowledge: The Co-production of Science and the Social Order*, ed. Sheila Jasanoff, 181–205. New York: Routledge, 2004.

Clarke, Sally H. *Trust and Power: Consumers, the Modern Corporation, and the Making of the United States Automobile Market*. Cambridge: Cambridge University Press, 2007.

Clarkson, Kenneth W., Charles W. Kadlec, and Arthur B. Laffer. "Regulating Chrysler out of Business?" *Regulation* 3 (September–October 1979): 44–49.

Claybrook, Joan, Jacqueline Gillan, and Anne Strainchamps. *Reagan on the Road: The Crash of the U.S. Auto Safety Program*. Washington, D.C.: Public Citizen, 1982.

Cohen, Wesley M., and Daniel Levinthal. "Absorptive Capacity: A New Perspective on Learning and Innovation." *Administrative Science Quarterly* 35, no. 1 (March 1990): 128–152.

Cohen, Wesley M., and Steven Klepper. "A Reprise of Size and R & D." *Economic Journal* 106 (July 1996): 925–951.

Cohn, David L. *Combustion on Wheels: An Informal History of the Automobile Age*. Boston: Houghton Mifflin, 1944.

Collin, Robert W. *The Environmental Protection Agency: Cleaning Up America's Act*. Westport, Ct.: Greenwood, 2006.

Collins, Randall. *Weberian Sociological Theory*. Cambridge: Cambridge University Press, 1986.

Comptroller General of the United States. "Review of the Automobile Fuel Economy Testing and Labeling Program." Report to the Subcommittee on Conservation and Natural Resources, Committee on Government Operations, US House of Representatives, August 15, 1974.

Conniff, Richard. "The Political History of Cap and Trade." *Smithsonian Magazine*, August 2009.

Constant, Edward W. *The Origins of the Turbojet Revolution*. Baltimore: Johns Hopkins University Press, 1980.

Copeland, Curtis W. "Federal Rulemaking: The Role of the Office of Information and Regulatory Affairs." Congressional Research Service, June 9, 2009. https://fas.org/sgp/crs/misc/RL32397.pdf.

Corn, Joseph J. *User Unfriendly: Consumer Struggles with Personal Technologies, from Clocks and Sewing Machines to Cars and Computers*. Baltimore: Johns Hopkins University Press, 2011.

Cravath, J. R. "The Headlight Glare Problem." *Transactions of the Illuminating Engineering Society* 12 (1917): 188–190.

Cruickshank, Barbara. "Revolutions Within: Self-Government and Self-Esteem." In *Foucault and Political Reason: Liberalism, Neo-liberalism, and Rationalities of Government*, ed. Andrew Barry, Thomas Osborne, and Nikolas Rose, 231–252. Chicago: University of Chicago Press, 1996.

Cschowski, W. G., P. C. Skeels, and W. R. Hawkins. "Appraisal of Guardrail Installations by Car Impact and Laboratory Tests." *Highway Research Board Proceedings* 40 (1961): 137–149.

Cutlip, Scott M. *The Unseen Power: Public Relations; A History*. New York: Routledge, 2013.

De Haven, Hugh. "Mechanical Analysis of Survival in Falls from Heights of Fifty to One Hundred and Fifty Feet." *War Medicine* 2 (July 1942): 586–596. Reprinted in *Accident Research: Methods and Approaches*, ed. William Haddon Jr., Edward A. Suchman, and David Klein, 539–547. New York: Harper and Row, 1964.

Denton, Tom. *Advanced Automotive Fault Diagnosis*. New York: Routledge, 2012.

Dewey, John. *Human Nature and Conduct*. New York: Henry Holt and Company, 1922.

Dewey, John. *The Quest for Certainty*. New York: Minton, Balch, 1929.

Dewey, Scott Hamilton. *Don't Breathe the Air: Air Pollution and U.S. Environmental Politics, 1945–1970*. College Station: Texas A&M University Press, 2000.

Dewey, Scott Hamilton. "'The Antitrust Case of the Century': Kenneth F. Hahn and the Fight against Smog." *Southern California Quarterly* 81, no. 3 (1999): 341–376.

Dole, Bob and Elizabeth. *The Doles: Unlimited Partners*. New York: Simon and Schuster, 1988.

Donovan, Frank R. *Wheels for a Nation*. New York: Crowell, 1965.

Doody, Alton F., and Ron Bingaman. *Reinventing the Wheels: Ford's Spectacular Comeback*. New York: Harper and Row, 1990.

Down, Anthony. *Inside Bureaucracy*. Prospect Heights, Ill.: Waveland Press, 1994. Reprinted, Boston: Little, Brown, 2000.

Downs, Anthony. "Up and Down with Ecology: The 'Issue-Attention Cycle.'" *National Affairs* (Summer 1972): 39–50.

Doyle, Jack. *Taken for a Ride: Detroit's Big Three and the Politics of Pollution*. New York: Four Walls Eight Windows, 2000.

Dunsby, Joshua William. "Clarifying Smog: Expert Knowledge, Health, and the Politics of Air Pollution." PhD diss., University of California, San Diego, 2001.

Dunn, Richard Chase, and Ann Johnson. "Chasing Molecules: Chemistry and Technology for Automotive Emissions Control." In *Toxic Airs: Body, Place, Planet in Historical Perspective*, ed. James Rodger Fleming and Ann Johnson, 109–126. Pittsburgh: University of Pittsburgh Press, 2014.

Dyck, Robert G. *Evolution of Federal Air Pollution Control Policy, 1948–1967*. Pittsburgh: Publications Office, Graduate School of Public and International Affairs, University of Pittsburgh, 1971.

Eads, George C., and Michael Fix. *Relief or Reform? Reagan's Regulatory Dilemma*. Washington, D.C.: Urban Institute Press, 1984.

"Eastern Motor Vehicle Department Heads Organize for Co-operation," *Motor Truck News* 11 (December 1921): 16.

Eastman, Joel W. *Styling vs. Safety: The American Automobile Industry and the Development of Automotive Safety, 1900–1966.* Lanham, Md.: University Press of America, 1984.

"Edward P. LaForce Convicted." *SEC News Digest,* November 28, 1978.

Eisner, Marc Allen. *Regulatory Politics in Transition,* 2nd ed. Baltimore: Johns Hopkins University Press, 2000.

Ellison, A. E., R. E. Carroll, W. Haddon Jr., and M. Wolf. "Skiing Injuries: Clinical Study." *Public Health Reports (1896–1970)* (1962): 985–991.

Emerson, Edwin, Jr. "Newport in the Lead." *Automobile Magazine* 1, no. 1 (October 1899), 5–16.

Emission Control Technology Division, Office of Mobile Source Air Pollution Control. "Automobile Emission Control—The Development, Status, Trends, and Outlook as of December 1976: A Report to the Administrator, U.S. Environmental Protection Agency." Environmental Protection Agency, April 1977.

Eno, William Phelps. *The Science of Highway Traffic Regulation, 1899–1920.* Washington, D.C.: Brentano's, 1920.

Eno, William Phelps. *The Story of Highway Traffic Control, 1899–1939.* Saugatuck, Conn.: Eno Foundation for Highway Traffic Control, Inc., 1939.

"Eno Says Welsh Bill Is Bad Traffic Measure." *New York Times,* April 1, 1917.

Environmental Protection Agency. *Auto Emissions Extension. Transcript of Proceedings of Public Hearings on Requests by Five Automobile Manufacturers for a One-Year Suspension of the Effective Date of Emission Limitations Applicable to 1975 Model Year Automobiles.* Washington, D.C.: ACE-Federal Reporters, 1972.

Environmental Protection Agency. "Clean Power Plan for Existing Power Plants." Archived at https://19january2017snapshot.epa.gov/cleanpowerplan/clean-power -plan-existing-power-plants_.html.

Environmental Protection Agency. "Health Consequences of Sulfur Oxides: A Report from Chess, 1970–1971." May 1974. Archived at https://nepis.epa.gov/Exe /ZyNET.exe.

Environmental Protection Agency. "Testing of the LaForce Engine by the Environmental Protection Agency," Staff Report Prepared for the Committee on Commerce. US Senate, March 1975.

Environmental Protection Agency, "Evaluation of the LaForce-Modified AMC Hornet." Washington, D.C.: Office of Mobile Source Air Pollution Control, Emission Control Technology Division, December 1974.

"EPA Tests Refute LaForce Engine Claim." *Chemical Engineering News* 52, no. 51 (1974): 5.

"EPA Would Rather Switch than Fight, Substitutes Stork for Moran." *Environmental Health Letter,* July 15, 1975.

"Environmental Currents." *Environmental Science and Technology* 4, no. 2 (February 1970): 91–95.

Esposito, John C. *Vanishing Air: The Ralph Nader Study Group Report on Air Pollution.* New York: Grossman, 1970.

Finkelstein, Michael M. "Consumer Safety Information as a Government Policy Tool." In *Government, Technology, and the Future of the Automobile,* ed. Douglas H. Ginsberg and William J. Abernathy, 424–430. New York: McGraw-Hill, 1980.

Flink, James J. *America Adopts the Automobile, 1895–1910*. Cambridge, Mass.: MIT Press, 1970.

Flink, James J. *The Automobile Age*. Cambridge, Mass.: MIT Press, 1988.

Flink, James J. *The Car Culture*. Cambridge, Mass: MIT Press, 1975.

Flippen, J. Brooks. *Conservative Conservationist: Russell E. Train and the Emergence of American Environmentalism*. Baton Rouge: Louisiana State University Press, 2006.

Flippen, J. Brooks. *Nixon and the Environment*. Albuquerque: University of New Mexico Press, 2000.

Follows, George H. "Safety and Welfare Work in the Engineer's Education." *Journal of the American Society of Mechanical Engineers* 40 (July 1918): 545–548.

Ford Motor Company. *Freedom of the American Road*. Dearborn, Mich.: Ford Motor Company, 1956.

Foucault, Michel. *The Birth of Biopolitics: Lectures at the College de France, 1978–1979*. Trans. Graham Burchell. New York: Palgrave Macmillan, 2008.

Francis, Devon. "Cars Can Be Safer." *Popular Science* (April 1947): 82–87.

Franz, Kathleen. *Tinkering: Consumers Reinvent the Early Automobile*. Philadelphia: University of Pennsylvania Press, 2005.

Friedman, Barry D. *Regulations in the Reagan-Bush Era: The Eruptions of Presidential Influence*. Pittsburgh: University of Pittsburgh Press, 1995.

Friedman, Donald. "The Minicars RSV—Still a Car for the Future." Center for Injury Research, Paper 09-0480. http://www.centerforinjuryresearch.org/publications /symposiums/.

Friedman, Lawrence M. *American Law in the Twentieth Century*. New Haven: Yale University Press, 2002.

Friedman, Milton. *Capitalism and Freedom*. Chicago: University of Chicago Press, 1962.

Galambos, Louis. *The Creative Society—And the Price Americans Paid for It*. Cambridge: Cambridge University Press, 2012.

Galison, Peter. *Image and Logic: A Material Culture of Microphysics*. Chicago: University of Chicago Press, 1997.

Gandelot, Howard K. "Engineering Safety into Automobile Bodies." SAE Technical Paper 510041 (1951).

Gangloff, Amy Beth. "Medicalizing the Automobile: Public Health, Safety, and American Culture, 1920–1967." PhD diss., Stony Brook University, 2006.

Garud, Raghu. "Conferences as Venues for the Configuration of Emerging Organizational Fields: The Case of Cochlear Implants." *Journal of Management Studies* 45, no.6 (September 2008): 1061–1088.

"General Report on Glare." *Transactions of the Illuminating Engineering Society* 10 (1915): 987–999.

Ginsberg, Douglas H., and William J. Abernathy, eds. *Government, Technology, and the Future of the Automobile*. New York: McGraw-Hill, 1980.

Godin, Benoît. "The Linear Model of Innovation: The Historical Construction of an Analytical Framework." *Science, Technology, and Human Values* 31, no. 6 (November 2006): 639–667.

Goggin, Joyce E., William Haddon Jr., George S. Hambly, and Janet R. Hoveland. "Incidence of Femoral Fractures in Postmenopausal Women;. Before and after Water Fluoridation." *Public Health Reports* 80, no. 11 (1965): 1005.

Golden, Marissa Martino. *What Motivates Bureaucrats? Politics and Administration during the Reagan Years*. New York: Columbia University Press, 2000.

Government Accounting Office. "Examination into the Adequacy of the Environmental Protection Agency's Motor Vehicle Certification Activities." June 12, 1972.

Graham, John D. *Auto Safety: Assessing America's Performance*. Dover, Mass.: Auburn House, 1989.

Graham, John D., and Jonathan Beart Wiener, eds. *Risk vs. Risk: Tradeoffs in Protecting Health and the Environment*. Cambridge, Mass.: Harvard University Press, 1995.

Graham, John D. "Saving Gasoline and Lives." In *Risk vs. Risk: Tradeoffs in Protecting Health and the Environment*, ed. John D. Graham and Jonathan Beart Wiener, 87–103. Cambridge, Mass.: Harvard University Press, 1995.

Gray, Charles L., and Jeffrey A. Alson. "The Case for Methanol." *Scientific American*, November 1989.

Gray, Charles L., and Jeffrey A. Alson. *Moving America to Methanol: A Plan to Replace Oil Imports, Reduce Acid Rain, and Revitalize Our Domestic Economy*. Ann Arbor: University of Michigan Press, 1985.

Gurdjian, E. S., and H. R. Lissner, "Mechanism of Head Injury as Studied by the Cathode Ray Oscilloscope Preliminary Report," *Journal of Neurosurgery* 1, no. 6 (1944): 393–399.

Gusfield, James R. *The Culture of Public Problems: Drinking-Driving and the Symbolic Order*. Chicago: University of Chicago Press, 1981.

Haagen-Smit, Arie. "The Air Pollution Problem in Los Angeles." *Engineering and Science* 14 (December 1950): 1–7.

Haagen-Smit, Arie. "Chemistry and Physiology of Los Angeles Smog." *Industrial and Engineering Chemistry* 44, no. 6 (June 1952): 1342–1346.

Haagen-Smit, Arie. "The Chemistry of Flavor." *Engineering and Science Monthly* 12, no. 4 (January 1949): 3–7.

Habermas, Jurgen. *Legitimation Crisis*. Trans. Thomas McCarthy. Boston: Beacon Press, 1975.

Haddon, William, Jr. "Research with Respect to Fatal Accident Causes: Implications for Vehicle Design." SAE Technical Paper 610255 (1961).

Haddon, William, Jr. "A Note concerning Accident Theory and Research with Special Reference to Motor Vehicle Accidents." *Annals of the New York Academy of Sciences* 107 (1963): 635–646.

Haddon, William, Jr. "The Changing Approach to the Epidemiology, Prevention, and Amelioration of Trauma: The Transition to Approaches Etiologically Rather than Descriptively Based." *American Journal of Public Health* 58, no. 8 (August 1968): 1431–1438.

Haddon, William, Jr., P. Valien, J. R. McCarroll, and C. J. Umberger. "A Controlled Investigation of the Characteristics of Adult Pedestrians Fatally Injured by Motor Vehicles in Manhattan." *Journal of Chronic Diseases* 14, no. 6 (1961): 655–678.

Halberstam, David. *The Reckoning*. New York: William Morrow, 1986.

Hansen, A. M. "SAE Test Procedure for Instrument Panels: An Impact Pendulum for Testing Safety Characteristics of Padded Instrument Panels and Other Passenger Compartment Components." SAE Paper 650963, 1965.

Hanson, Harlan L. "Energy Absorbing Steering Column: A Case History." *Proceedings of the Tenth Stapp Car Crash Conference* (November 8–9, 1966).

Harriss, John A. "Signal Control of Traffic." *Transactions of the Illuminating Engineering Society* 17, no. 5 (May 1922): 245–247.

Harvey, David *A Brief History of Neoliberalism*. Oxford: Oxford University Press, 2005.

Hass, G. C., M. P. Sweeney, and J. N. Pattison. "Laboratory Simulation of Driving Conditions in the Los Angeles Area." SAE Paper 660546, 1966.

Hawley, Ellis W. "Herbert Hoover, the Commerce Secretariat, and the Vision of an Associative State, 1921–1928." *Journal of American History* (1974): 116–140.

Hayek, Friedrich. *The Road to Serfdom*. Chicago: University of Chicago Press, 1944.

Hayes, Robert H., and William J. Abernathy. "Managing Our Way to Economic Decline." *Harvard Business Review* (July–August 1980): 67–77.

Hellman, Karl Harry. "A Kinetic Theory Analysis of the Evolution of Shock, Thermal, and Expansion Waves Generated by an Impulsively Moved, Fractionally Accommodating Piston." PhD diss., Northwestern University, 1970.

Hessen, Boris. "The Social and Economic Roots of Newton's *Principia*." In *The Social and Economic Roots of the Scientific Revolution*, 41–101. New York: Springer, 2009.

Heymann, Philip B. *The Politics of Public Management*. New Haven: Yale University Press, 1989.

Hilgartner, Stephen. *Science on Stage: Expert Advice as Public Drama*. Stanford, Calif.: Stanford University Press, 2000.

Hirsh, Richard F. *Power Loss: The Origins of Deregulation and Restructuring in the America Electric Utility System*. Cambridge, Mass.: MIT Press, 2001.

"History of Automotive Standardization." *SAE Journal* 26, no. 6 (June 1930): 698–704.

Hobbs, Caswell O. "Legal Issues in FTC Trade Regulation Rules." *Food, Drug, and Cosmetic Law Journal* 32 (1977): 414–422.

Hodgson, Godfrey. *The Gentleman from New York: Daniel Patrick Moynihan; A Biography*. Boston: Houghton Mifflin, 2000

Hofflund, John L. "National Aspects of Air Pollution Legislation." In *Air Pollution: Proceedings of the United States Technical Conference on Air Pollution*, ed., Louis McCabe, 765–772. New York: McGraw-Hill, 1952.

Hoffman, Paul G. *Seven Roads to Safety: A Program to Reduce Automobile Accidents*. New York: Harper and Brothers, 1939.

Hofstadter, Richard. *Anti-intellectualism in American Life*. New York: Alfred A. Knopf, 1963.

Holsworth, Robert D. *Public Interest Liberalism and the Crisis of Affluence: Reflections on Nader, Environmentalism, and the Politics of a Sustainable Society*. Boston: G. K. Hall, 1980.

Hoover, Herbert. *The Memoirs of Herbert Hoover: The Cabinet and the Presidency, 1920–1933*. New York: Macmillan, 1952.

Hord, Eugene F. *History and Organization of Automobile Insurance*. New York: Insurance Society of New York, 1919.

Horowitz, Daniel. *Jimmy Carter and the Energy Crisis of the 1970s: The "Crisis of Confidence" Speech of July 15, 1979; A Brief History with Documents*. Boston: Bedford/St. Martin's, 2005.

Horwitz, Morton J. *The Transformation of American Law, 1870–1960: The Crisis of Legal Orthodoxy.* Cambridge, Mass.: Harvard University Press, 1977.

Hounshell, David A. "The Cold War, RAND, and the Generation of Knowledge, 1946–1962." *Historical Studies in the Physical and Biological Sciences* 27, no. 2 (1997): 237–267.

Hounshell, David A. *From the American System to Mass Production, 1800–1932: The Development of Manufacturing Technology in the United States.* Baltimore: Johns Hopkins University Press, 1984.

Hounshell, David A., and John Kenly Smith Jr. *Science and Corporate Strategy: Du Pont R&D, 1902–1980.* Cambridge: Cambridge University Press, 1988.

Huff, Russell. "Factors of Safety." *S.A.E. Transactions* 11, part 2 (1917): 70–87.

Hughes, Thomas Parke. *American Genesis: A Century of Invention and Technological Enthusiasm, 1870–1970.* New York: Viking, 1989.

Hughes, Thomas Parke. *Networks of Power: Electrification in Western Society, 1880–1930.* Baltimore: Johns Hopkins University Press, 1983.

Huls, Thomas A. "Evolution of Federal Light-Duty Mass Emission Regulations." SAE Paper 730554, 1973.

Illuminating Engineering Society. *Specifications of Laboratory Tests for Approval of Electric Headlighting Devices for Motor Vehicles, Approved November 11, 1922.* Washington, D.C.: Government Printing Office, 1924.

Ingrassia, Paul. *Crash Course: The Automobile Industry's Road from Glory to Disaster.* New York: Random House, 2010.

Inkpen, Andrew C. "Knowledge Transfer and International Joint Ventures: The Case of NUMMI and General Motors." *Strategic Management Journal* 29, no. 4 (2008): 447–453.

International Conference on Passive Restraints, North Atlantic Treaty Organization, General Motors Proving Grounds. *International Conference on Passive Restraints, May 11–12, 1970.* Washington, D.C.: Government Printing Office, 1970.

Jacobs, Chip, and William J. Kelly. *Smogtown: The Lung-Burning History of Pollution in Los Angeles.* Woodstock, N.Y.: Overlook Press, 2008.

Jacobs, Meg. *Panic at the Pump: The Energy Crisis and the Transformation of American Politics in the 1970s.* New York: Hill and Wang, 2016.

Jaffe, Louis S. "The Effects of Photochemical Oxidants on Materials." *Journal of Air Pollution Control Association* 17, no. 6 (1967): 375–378.

Jardini, David R. "Out of the Blue Yonder: The RAND Corporation's Diversification into Social Welfare Research, 1946–1968." PhD diss., Carnegie Mellon University, 1996.

Jasanoff, Sheila. *The Fifth Branch: Science Advisers as Policymakers.* Cambridge, Mass.: Harvard University Press, 1990.

John, Richard R., Philip S. Coonley, Robert C. Ricci, and Bruce Rebinger. "Mandated Fuel Economy Standards as a Strategy for Improving Motor Vehicle Fuel Economy." In *Government, Technology, and the Future of the Automobile,* ed. Douglas H. Ginsberg and William J. Abernathy, 118–143. New York: McGraw-Hill, 1980.

Johnson, Ann. *Hitting the Brakes: Engineering Design and the Production of Knowledge.* Durham, N.C.: Duke University Press, 2009.

Johnson, Ann. "Routines for Innovation: Problem-Oriented Knowledge Communities and the Production of Clean Combustion." In *The Emergence of Routines: Entrepreneurship, Organization, and Business History*, ed. Daniel M. G. Raff and Philip Scranton, 314–336. Oxford: Oxford University Press, 2017.

Johnson, H. M. Review of *Psychology and the Motorist*, by Herbert A. Toops and S. Edson Haven. *Psychological Bulletin* 35, no. 8 (October 1938): 561–564.

Jones, Charles O. "The Limits of Public Support: Air Pollution Agency Development." *Public Administration Review* 32, no. 5 (1972): 502–508.

Jones, Trevor O. "The Challenge of Automotive Electronics in the USA." SAE Technical Paper No. 741221 (1974).

Jurgen, Ronald K. "Introduction" to *History of Automotive Electronics: The Early Years*. Warrendale, Pa.: Society of Automotive Engineers, 1998.

Kagan, Robert A. *Adversarial Legalism: The American Way of Law*. Cambridge, Mass.: Harvard University Press, 2003.

Kahn, Jonathan. *Budgeting Democracy: State Building and Citizenship in America, 1890–1928*. Ithaca, N.Y.: Cornell University Press, 1997.

Kaptur, Vincent D., and Michael C. Myal. "The General Motors Comfort Dimensioning System." SAE Paper 610174, 1961.

Kaspar, Raphael G. *Technology Assessment: The Proceedings of a Seminar Series at the Program of Policy Studies in Science and Technology at George Washington University, January–April, 1969*. Washington, D.C.: Government Printing Office, 1969.

Katz, Harold A. "Liability of Automobile Manufacturers for Unsafe Design of Passenger Cars." *Harvard Law Review* 69 (1955): 863.

Kearney, Paul W. "A Safety Expert Looks at the '59 Cars." *Popular Science* (January 1959): 89–92, 230–232.

Keats, John. *The Insolent Chariots*. Philadelphia: Lippincott, 1958.

Keats, John. *Schools without Scholars*. New York: Houghton Mifflin, 1958.

Kemp, Jack. *An American Renaissance: A Strategy for the 1980s*. New York: Harper and Row, 1979.

Kintner, Earl W., and Christopher Smith. "The Emergence of the Federal Trade Commission as a Formidable Consumer Protection Agency." *Mercer Law Review* 26 (1974–1975): 651.

Kevles, Daniel J. "Testing the Army's Intelligence: Psychologists and the Military in World War I." *Journal of American History* 55, no. 3 (December 1968): 565–581.

King, David C. *Turf Wars: How Congressional Committees Claim Jurisdiction*. Chicago: University of Chicago Press, 1997.

Kingsland, Sharon E. "Frits Went's Atomic Age Greenhouse: The Changing Labscape on the Lab-Field Border." *Journal of the History of Biology* 42, no. 2 (Summer 2009): 289–324.

Kisiel, Ralph. "Soichi Kawazoe." *Automotive News*, May 19, 2008.

Klein, Naomi. *The Shock Doctrine: The Rise of Disaster Capitalism*. New York: Picador, 2007.

Klepper, Steven. *Experimental Capitalism: The Nanoeconomics of American High-Tech Industries*. Princeton: Princeton University Press, 2016.

Kline, Ronald R. *Consumers in the Country: Technology and Social Change in Rural America.* Baltimore: Johns Hopkins University Press, 2000.

Kline, Ronald R., and Trevor Pinch. "Users as Agents of Technological Change: The Social Construction of the Automobile in the Rural United States." *Technology and Culture* 37 (October 1996): 763–795.

Knowles, Scott Gabriel. *The Disaster Experts: Mastering Risk in Modern America.* Philadelphia: University of Pennsylvania Press, 2012.

Krier, James E., and Edmund Ursin. *Pollution and Policy: A Case Essay on California and Federal Experience with Motor Vehicle Air Pollution, 1940–1975.* Berkeley: University of California Press, 1977.

Kruse, Ronald E., and Thomas A. Huls. "Development of the Federal Urban Driving Schedule." SAE Paper 730553, 1973.

Kuhn, Thomas S. *The Structure of Scientific Revolutions.* Chicago: University of Chicago press, 2012.

Lasch, Christopher. "Same Old New Class." *New York Review of Books*, September 28, 1967.

Leary, Rosemary. *The Ethics of Dissent: Managing Guerilla Government.* Thousand Oaks, Calif.: Congressional Quarterly Press, 2013.

Lee, Jaegul, and Francisco Veloso. "Inter-firm Innovation under Uncertainty: Empirical Evidence for Strategic Knowledge-Partitioning." *Journal of Product Innovation Management* 25, no. 5 (2006): 418–435.

Lee, Jaegul, Francisco M. Veloso, and David A. Hounshell. "Linking Induced Technological Change and Environmental Regulation: Evidence from Patenting in the US Auto Industry." *Research Policy* 40, no. 9 (2011): 1240–1252.

Lee, Jaegul, Francisco M. Veloso, David A. Hounshell, and Edward S. Rubin. "Forcing Technological Change: A Case of Automobile Emissions Control Technology Development in the US." *Technovation* 30, no. 4 (2010): 249–264.

Lee, Jaegul, Francisco M. Veloso, David A. Hounshell, and Edward S. Rubin. "Innovation in Automotive Emission Control Technologies: Government Actions and Inventive Activities." In *Creating Actionable Knowledge: The Academy of Management Conference Annual Meeting, New Orleans, Louisiana, August 6–11, 2004; Conference Proceedings.* Briarcliff, N.Y.: Academy of Management, 2004.

Lee, Jongmin. "Regulatory Engineering in Pollution and Health Research: U.S. Environmental Protection Agency's CHESS, 1969–1977." Paper presented at annual meeting of Society for the History of Technology, Tacoma, Wash., October 2, 2010.

"The Legislative Era," *The Horseless Age* 7, no. 22 (February 27, 1901): 1.

Lemov, Michael R. *Car Safety Wars: One Hundred Years of Technology, Politics, and Death.* Madison, Wisc.: Farleigh Dickinson University Press, 2015.

Leone, Robert A. "Regulatory Relief and the Auto Industry." In *The Reagan Regulatory Strategy*, ed. George C. Eads and Michael Fix, 87–105. Washington, D.C.: Urban Institute Press, 1984.

Lerner, Barron H. *One for the Road: Drunk Driving since 1900.* Baltimore: Johns Hopkins University Press, 2011.

Leslie, Stuart W. *Boss Kettering.* New York: Columbia University Press, 1983.

Lev, Peter. *Transforming the Screen, 1950–1959*. Berkeley: University of California Press, 2003.

Levison, Mathew E., and William Haddon Jr. "The Area Adjusted Map: An Epidemiological Device." *Public Health Report* 80, no. 1 (January 1965): 55–59.

Light, Jennifer S. *From Warfare to Welfare: Defense Intellectuals and Urban Problems in Cold War America*. Baltimore: Johns Hopkins University Press, 2003.

"Light Weight." *Ford Times* 2, no. 3 (November 1, 1908): 16–17.

Limbaugh, Rush H., III. *See, I Told You So*. New York: Pocket Books, 1993.

Lippman, Theo, Jr., and Donald C. Hansen. *Muskie*. New York: W. W. Norton, 1971.

Lissner, Hebert J. "Human and Animal Impact Studies in U.S. Universities." In National Academy of Sciences, Space Science Board, Man in Space Committee. *Impact Acceleration Stress*, 61–68. Washington, D.C.: National Research Council, 1962.

"A Live and Progressive Organization" *Automobile Topics* 16, no. 2 (October 19, 1907).

Luce, R. Duncan, and Howard Raiffa. *Games and Decisions: Introduction and Critical Survey*. New York: John Wiley and Sons, 1957.

Luger, Stan. *Corporate Power, American Democracy, and the Automobile Industry*. Cambridge: Cambridge University Press, 2000.

Lundstrom, L. C., and P. C. Skeels. "Full-Scale Appraisals of Guardrail Installations by Car Impact Tests." *Highway Research Board Proceedings* 38 (1959): 353–355.

Marks, L. B. "Tenth Anniversary of the Illuminating Engineering Society." *Transactions of the Illuminating Engineering Society* 11 (1916): 593–600.

Mashaw, Jerry L., and David L. Harfst. "From Command and Control to Collaboration and Deference: The Transformation of Auto Safety Regulation." *Yale Journal on Regulation* 34 (2017).

Mashaw, Jerry L., and David L. Harfst. *The Struggle for Auto Safety*. Cambridge, Mass.: Harvard University Press, 1990.

Mazzucato, Mariana. *The Entrepreneurial State: Debunking Public vs. Private Sector Myths*. Revised ed. London: Anthem Press, 2015.

Mayer, Jane. *Dark Money: The Hidden history of the Billionaires behind the Rise of the Radical Right*. New York: Doubleday, 2016.

McCabe, Louis, ed. *Air Pollution: Proceedings of the United States Technical Conference on Air Pollution*. New York: McGraw-Hill, 1952.

McCarthy, Tom. *Auto Mania: Cars, Consumers, and the Environment*. New Haven: Yale University Press, 2007.

McCarroll, James R., and William Haddon. "A Controlled Study of Fatal Automobile Accidents in New York City." *Journal of Chronic Diseases* 15, no. 8 (1962): 811–826.

McCarry, Charles. *Citizen Nader*. New York: Saturday Review Press, 1972.

McCormick, Robert E. "A Review of the Economics of Regulation: The Political Process." In *Regulation and the Reagan Era: Politics Bureaucracy, and the Public Interest*, ed. Roger E. Meiners and Bruce Yandle, 16–37. New York: Homes and Meier, 1989.

McCraw, Thomas K. *Prophets of Regulation: Charles Francis Adams, Louis D. Brandeis, James M. Landis, Alfred E. Kahn*. Cambridge, Mass.: Belknap Press of Harvard University Press, 1984.

McCray, W. Patrick. "From Lab to iPod: A Story of Discovery and Commercialization in the Post–Cold War Era." *Technology and Culture* 50, no. 1 (2009): 58–81.

McCubbins, Matthew D., Roger G. Noll, and Barry R. Weingast. "The Political Origins of the Administrative Procedure Act." *Journal of Law, Economics, and Organization* 15, no. 1 (Spring 1999), 180–217.

McGarity, Thomas O. *Reinventing Rationality: The Role of Regulatory Analysis in the Federal Bureaucracy*. Cambridge: Cambridge University Press, 1991.

McShane, Clay. *Down the Asphalt Path: the Automobile and the American City*. New York: Columbia University Press, 1994.

McShane, Clay. "The Origins and Globalization of Traffic Control Signals." *Journal of Urban History* 25, no. 3 (1999): 379–404.

"Medical News." *Journal of the American Medical Association* 164, no. 14 (August 3, 1957): 1595–1599.

Mehling, Harold. "Big Three Fight over How Safe to Make Cars." *Bluebook*, October 1955. Reprinted in US House of Representatives, *Traffic Safety: Hearings before a Subcommittee Investigation of Highway Traffic Accidents of the Committee on Interstate and Foreign Commerce*, 84th Congress, 2nd sess. (1956), 40–47.

Menand, Louis. *The Metaphysical Club: A Story of Ideas in America*. New York: Farrar, Straus and Giroux, 2002.

Merrill, Karen R. *The Oil Crisis of 1973–1974: A Brief History with Documents*. Boston: Bedford/St. Martin's, 2007.

Merton, Robert K. "The Normative Structure of Science." In *The Sociology of Science: Theoretical and Empirical Investigations*, 267–280. Chicago: University of Chicago Press, 1973.

Merton, Robert K. "Science, Technology and Society in Seventeenth Century England." *Osiris* 4 (1938): 360–632.

Metzgar, Carl R. "A Note concerning Accident Theory and Research with Special Reference to Motor Vehicle Accidents." *Professional Safety* 50, no. 6 (June 2005): 57–58.

Middleton, John T., J. B. Kendrick Jr., and H. W. Schwalm. "Injury to Herbaceous Plants by Smog or Air Pollution." *Plant Disease Reporter* 34, no. 9 (September 15, 1950): 245–252.

Miller, Douglas T., and Marion Nowak. *The Fifties: The Way We Really Were*. Garden City, N.Y.: Doubleday, 1975.

Miller, James C., III, "The Early Days of Reagan Regulatory Relief and Suggestions for OIRA's Future," *Administrative Law Review* (Special Edition) 63 (2011).

Miller, K. D., and A. T. Arikan. "Technology Search Investments: Evolutionary, Option Reasoning, and Option Pricing Approaches." *Strategic Management Journal* 25, no. 5 (2004): 473–486.

Miller, Matthew S. "'Death Car' Reckoning: Responses to the Automobile Slaughter in Chicago, 1920–1938." PhD. diss., Northwestern University, 2012.

Mindell, David. *Our Robots, Ourselves: Robotics and the Myths of Autonomy*. New York: Penguin, 2015.

Mirowski, Philip and Dieter Plehwe, eds. *The Road from Mont Pèlerin: The Making of the Neoliberal Thought Collective*. Cambridge, Mass.: Harvard University Press, 2009.

Mobile Source Pollution Control Program. *Fuel Economy and Emission Control*. Washington, D.C.: Environmental Protection Agency, Office of Air and Water Programs, November 1972.

Mohun, Arwen P. *Risk: Negotiating Safety in American Society*. Baltimore: Johns Hopkins University Press, 2013.

Mokyr, Joel. *The Gifts of Athena: Historical Origins of the Knowledge Economy*. Princeton: Princeton University Press, 2002.

Mom, Gijs. *The Evolution of Automotive Technology: A Handbook*. Warrendale, Pa.: SAE International, 2014.

Mom, Gijs. "Orchestrating Automobile Technology: Comfort, Mobility Culture, and the Construction of the 'Family Touring Car,' 1917–1940." *Technology and Culture* 55, no. 2 (2014): 299–325.

Mondt, Robert. *Cleaner Cars: The History and Technology of Emission Controls since the 1960s*. Warrendale, Pa.: SAE International, 2000.

Mooney, Chris. *The Republican War on Science*. New York: Basic Books, 2006.

Morris, Edmund. *Theodore Rex*. New York: Random House, 2001.

Moss, Fred A. "Standardized Tests for Automobile Drivers." *Public Personnel Studies* 3, no. 5 (1925): 147–165.

Moynihan, Daniel Patrick. "Epidemic on the Highways." *The Reporter* 20, no. 9 (April 1959): 16–23.

Moynihan, Daniel Patrick. "The Legal Regulation of Automobile Design." In *Passenger Car Design and Highway Safety*, 265–285. New York: Association for the Aid of Crippled Children; Consumers Union, 1962.

Moynihan, Daniel Patrick. *Maximum Feasible Misunderstanding: Community Action in the War on Poverty*. New York: Free Press, 1969.

Mui, Chunka, and Peter B. Carroll. *Driverless Cars: Trillions Are Up for Grabs*. New York: Cornerloft Press, 2013.

Murphy, Thomas P. *Science, Geopolitics, and Federal Spending*. Lexington, Mass.: Heath Lexington Books, 1971.

Muskie, Edmund S. "Banquet Address." In *Proceedings of the First National Symposium on Heterogeneous Catalysis for Control of Air Pollution*, ed. Bani R. Banerjee, 1–6. Washington, D.C.: National Air Pollution Control Administration, 1968.

Muskie, Edmund S. *Journeys*. New York: Doubleday, 1972.

Myers, Boyd C. *A Report on the Closing of the NASA Electronics Research Center*. Washington, D.C.: National Aeronautics and Space Administration, 1970.

Nader, Ralph. *The Ralph Nader Reader*. New York: Seven Stories Press, 2000.

Nader, Ralph. "The Safe Car You Can't Buy." *The Nation*, April 11, 1959. Reprinted in *The Ralph Nader Reader*, 266–272. New York: Seven Stories Press, 2000.

Nader, Ralph. "Taming the Corporate Tiger." In *The Ralph Nader Reader*, 133–144. New York: Seven Stories Press, 2000.

Nader, Ralph. *Unsafe at Any Speed*. New York: Grossman, 1965.

National Academy of Sciences and US Environmental Protection Agency. *Interim Standards Report*. April 26, 1972.

National Academy of Sciences and US Environmental Protection Agency. *Semiannual Report by the Committee on Motor Vehicle Emissions of the National Academy of Sciences to the Environmental Protection Agency*. February 12, 1973.

National Academy of Sciences and US Environmental Protection Agency. *Semiannual Report by the Committee on Motor Vehicle Emissions of the National Academy of Sciences to the Environmental Protection Agency*. January 1, 1972.

National Bureau of Standards. *National Directory of Commodity Specifications.* Washington, D.C.: Government Printing Office, 1925.

National Bureau of Standards. *Standards Yearbook.* Washington, D.C.: Government Printing Office, 1928.

National Center for Health Statistics. *Weight, Height, and Selected Body Dimensions of Adults, United States, 1960–1962.* Washington, D.C.: Public Health Service, US Department of Health, Education and Welfare, 1965.

National Conference on Street and Highway Safety. *Report of the Committee on the Causes of Accidents.* Washington, D.C.: Government Printing Office, 1926.

National Highway Traffic Safety Administration. *Motor Vehicle Safety, 1990.* Washington, D.C.: Department of Transportation, 1992.

National Research Council. *Effectiveness of the United States Advanced Battery Consortium as a Government-Industry Partnership.* Washington, D.C.: National Academy Press, 1998.

National Research Council. *Review of the Research Program of the Partnership for a New Generation of Vehicles: Fifth Report.* Washington, D.C.: National Academy Press, 1999.

Nelson, Richard, and Sidney Winter. *An Evolutionary Theory of Economic Change.* Cambridge, Mass.: Belknap Press of Harvard University Press, 1982.

Nielsen, Howard A., Jr. *From Locomotives to Strain Gages: The True Story of a Tortuous Conversion from a Smokestack to High-Technology Industry.* New York: Vantage Press, 1985.

Niquette, Paul. *101 Doozies I've Met* (1996). http://niquette.com/books/doozies /dooz101.html

Noble, David F. *America by Design: Science, Technology, and the Rise of Corporate Capitalism.* New York: Alfred A. Knopf, 1977.

Norton, Peter D. *Fighting Traffic: The Dawn of the Motor Age in the American City.* Cambridge, Mass.: MIT Press, 2011.

Norton, Peter D. "Four Paradigms: Traffic Safety in the Twentieth-Century United States." *Technology and Culture* 56, no. 2 (2015): 319–334.

O'Brien, John. "Flashlight Signal Control of Traffic." *Transactions of the Illuminating Engineering Society* 17, no. 5 (May 1922): 248–251.

O'Connell, Jeffrey, and Arthur B. Myers. *Safety Last: An Indictment of the Auto Industry.* New York: Random House, 1966.

Office of the Press Secretary. "Executive Order 13563—Improving Regulation and Regulatory Review," January 18, 2011. The White House: President Barack Obama. Archived at https://obamawhitehouse.archives.gov/the-press-office/2011/01/18 /executive-order-13563-improving-regulation-and-regulatory-review.

Office of the Press Secretary. "Obama Administration Finalizes Historic 54.5 MPG Fuel Economy Standards," August 28, 2012. The White House: President Barack Obama. Archived at https://obamawhitehouse.archives.gov/the-press-office/2012 /08/28/obama-administration-finalizes-historic-545-mpg-fuel-efficiency -standard.

Ogburn, William. *Social Change with Respect to Culture and Original Nature.* 1922. Reprinted, New York: Viking Press, 1950.

Olds Motor Works. *Goop Talk.* Detroit: Olds Motor Works, 1905.

Oreskes, Naomi, and Erik M. Conway. *Merchants of Doubt: How a Handful of Scientists Obscured the Truth on Issues from Tobacco Smoke to Global Warming.* New York: Bloomsbury Press, 2010.

Oudshoorn, Nelly, and Trevor Pinch, eds. *How Users Matter: The Co-construction of Users and Technology.* Cambridge, Mass.: MIT Press, 2003.

Packer, Jeremy. *Mobility without Mayhem: Safety, Cars, and Citizenship.* Durham, N.C.: Duke University Press, 2008.

Patrick, Lawrence M. "A Look Back on My Stapp Conferences through the Years." *Stapp Car Crash Journal* 44 (November 2000): xi–xv.

Patrick, Lawrence M., Herbert R. Lissner, and Elisha S. Gurdjian. "Survival by Design: Head Protection." In *Proceedings: American Association for the Advancement of Automotive Medicine Annual Conference* 7 (1963): 483–499.

Peltzman, Sam. "The Effects of Automobile Safety Regulation." *Journal of Political Economy* 83, no. 4 (August 1975): 677–726.

Peltzman, Sam. "Regulation and the Natural Progress of Opulence." *Economic Affairs* 30, no. 2 (June 2010): 33–39.

Peltzman, Sam. *Regulation of Automobile Safety.* Washington, D.C.: American Enterprise Institute for Public Policy Research, 1975.

Perry, C. C., and Herbert R. Lissner. *The Strain Gage Primer.* New York: McGraw-Hill, 1955.

Pinch, Trevor J., and Wiebe E. Bijker. "The Social Construction of Facts and Artifacts; Or, How the Sociology of Science and the Sociology of Technology Might Benefit Each Other." *Social Studies of Science* 14, no. 3 (1984): 399–441. doi:10.1177/030631284014003004.

Pitofsky, Robert. "Beyond Nader: Consumer Protection and the Regulation of Advertising." *Harvard Law Review* 90, no. 4 (February 1977): 661–701.

Porter, Theodore M. *Trust in Numbers: The Pursuit of Objectivity in Science and Public Life.* Princeton: Princeton University Press, 1996.

Post, Robert C. *The SAE Story: One Hundred Years of Mobility.* Warrendale, Pa.: SAE International, 2005.

Powell, Mark R. *Science at EPA: Information in the Regulatory Process.* Washington, D.C.: Resources for the Future, 1999.

Prasad, Monica. *The Politics of Free Markets: The Rise of Neoliberal Economic Policies in Britain, France, Germany, and the United States.* Chicago: University of Chicago Press, 2006.

Prasso, Sheridan. *The Asian Mystique: Dragon Ladies, Geisha Girls, and Our Fantasies of the Exotic Orient.* New York: PublicAffairs, 2009.

Proceedings of Tenth Stapp Car Crash Conference, November 8–9, 1966. New York: Society of Automotive Engineers, 1967.

"Proposed Standardization of Colored Signal Lights." *Transactions of the Illuminating Engineering Society* 17, no. 5 (May 1922): 205–207.

Pursell, Carroll. "The Rise and Fall of the Appropriate Technology Movement in the United States, 1965–1985." *Technology and Culture* 34, no. 3 (July 1993): 629–637.

Reich, Robert B., and John D. Donahue. *New Deals: The Chrysler Revival and the American System.* New York: Times Books, 1985.

Roach, Mary. *Stiff: The Curious Lives of Human Cadavers.* New York: W. W. Norton, 2003.

Roberts, E. F. "General Inspection." *Automotive Industries* 28 (June 26, 1913), 1306–1308.

Rose, Andrew H., Jr., and C. Stafford Brandt. "Environmental Irradiation Test Facility." *Journal of Air Pollution Control Association* 10, no. 4 (1960): 331–335.

Rose, Andrew H., Jr., Ralph C. Stahman, and Merrill W. Korth. "Dynamic Irradiation Tests of Automotive Exhaust, Part 1." *Journal of Air Pollution Control Association* 12, no. 10 (1962): 468–478.

Rose, Mark H., and Raymond A. Mohl. *Interstate: Highway Politics since 1939.* 3rd ed. Knoxville: University of Tennessee Press, 2012.

Rosenberg, Nathan. "The Direction of Technological Change: Inducement Mechanisms and Focusing Devices." *Economic Development and Cultural Change* 18, no. 1, part 1 (1969): 1–24.

Rothschild, Emma. *Paradise Lost: The Decline of the Auto-Industrial Age.* New York: Vintage Books, 1974.

Rothschild, Emma. "What Is the 'Energy Crisis'?" *New York Review of Books,* July 19, 1973.

"Rules of the Road." *New York Times,* January 20, 1920.

Russell, Andrew L. *Open Standards and the Digital Age: History, Ideology, and Networks.* Cambridge: Cambridge University Press, 2014.

Russell, Andrew L., and Lee Vinsel. "The Dynamic Interplay between Standards and Routines: Lessons from Industry and Government." In *The Emergence of Routines,* ed. Daniel Raff and Philip Scranton. Oxford: Oxford University Press, 2017.

Ryan, Craig. *Sonic Wind: The Story of John Paul Stapp and How a Renegade Doctor Became the Fastest Man on Earth.* New York: W. W. Norton, 2015.

Scheberle, Denise. *Federalism and Environmental Policy: Trust and the Politics of Implementation.* Washington, D.C.: Georgetown University Press, 1997.

Schumpeter, Joseph A. *Capitalism, Socialism, and Democracy.* New York: Harper and Brothers, 1942.

Sears, David O., and Jack Citrin. *Tax Revolt: Something for Nothing in California.* Cambridge, Mass.: Harvard University Press, 1982.

Seely, Bruce Edsall. *Building the American Highway System: Engineers as Policy Makers.* Philadelphia: Temple University Press, 1987.

Severy, Derwyn M., Arnold Siegel, and John H. Mathewson. *Statement on Crashworthiness of Automotive Seatbelts.* Los Angeles: Department of Engineering, University of California, 1957.

Shapin, Steven, and Simon Schaffer. *Leviathan and the Air Pump: Hobbes, Boyle, and the Experimental Life.* Princeton: Princeton University Press, 1985.

Sherry, Ralph H. "Metallurgy in the Automotive Industry." *S.A.E. Transactions* 12, no. 2 (1918), 323–350.

Skowronek, Stephen. *Building a New American State: The Expansion of National Administrative Capacities, 1877–1920.* Cambridge: Cambridge University Press, 1982.

Siegel, Greg. *Forensic Media: Reconstructing Accidents in Accelerated Modernity.* Durham, N.C.: Duke University Press, 2014.

Simon, Herbert A. *Administrative Behavior: A Study of Decision-Making Processes in Administrative Organization.* New York: Macmillan, 1947.

"Simmons and the Strain Gage." *Engineering & Science* (September 1986): 19–23.

Simpson, Andrew T. "Transporting Lazarus: Physicians, the State, and the Creation of the Modern Paramedic and Ambulance, 1955–73." *Journal of the History of Medicine and Allied Sciences* 68, no. 2 (2013): 163–197.

Sinclair, Bruce. *Philadelphia's Philosopher Mechanics: A History of the Franklin Institute, 1824–1865.* Baltimore, Johns Hopkins University Press, 1974.

Sinclair, Bruce, and James P. Hull. 1980. *A Centennial History of the American Society of Mechanical Engineers, 1880–1980.* Toronto: University of Toronto Press Published for American Society of Mechanical Engineers, 1980.

"Single Effort by Hartford Dealers Makes Better Show." *Motor Age* 39 (March 3, 1921): 31.

Skinner, Kiron K., Annelise Anderson, and Martin Anderson, eds. *Reagan's Path to Victory: The Shaping of Ronald Reagan's Vision; Selected Writings.* New York: Free Press, 2004.

Slaton, Amy E. *Reinforced Concrete and the Modernization of American Building, 1900–1930.* Baltimore: Johns Hopkins University Press, 2001.

Snow, Adolph Judah. "Tests for Chauffeurs." *Industrial Psychology* 2, no. 1 (1926): 30–45.

Society of Automotive Engineers. *Automotive Fuel Economy.* Warrendale, Pa., 1976.

Somers, Joseph Henry. "Conformational Properties of Some Phosphorinane Derivatives." PhD diss., Duke University, 1970.

"Some Leading Automobile Suits." *Horseless Age* 10, no. 19 (November 5, 1902), 512.

Stack, Herbert J. *Safety for Greater Adventures: The Contributions of Albert Wurts Whitney.* New York: Center for Safety Education, New York University, 1953.

Stahman, R. C., K. D. Mills, and M. W. Korth. *Federal Air Pollution Efforts.* No. CONF-890240. Warrendale, Pa.: Society of Automotive Engineers, 1989.

Stanford Research Institute. *The Smog Problem in Los Angeles County: Second Interim Report by Stanford Research Institute on Studies to Determine the Nature and Sources of Smog.* Los Angeles: Committee on Smoke and Fumes, Western Oil and Gas Association, 1949.

"State Crusades Inspire Sales of Headlights." *Automotive News*, October 19, 1925.

Steger, Manfred B., and Ravi K. Roy *Neoliberalism: A Very Short Introduction.* Oxford: Oxford University Press, 2010.

Stevens, R. K., P. J. Lamothe, W. E. Wilson, J. L. Durham, and T. G. Dzubay, eds. *The General Motors/Environmental Protection Agency Sulfate Dispersion Experiment.* Research Triangle Park, N.C.: Environmental Protection Agency, 1976.

Stevens, Thomas Wood. *Adventure: A Pageant of Life and Chance.* New York: National Bureau of Casualty and Surety Underwriters, 1923.

Stigler, George J. "The Economics of Information." *Journal of Political Economy* 69, no. 3 (June 1961): 213–225.

Stigler, George J. "The Theory of Economic Regulation." *Bell Journal of Economics and Management Science* 2, no. 1 (1971): 3–21.

Stivender, D. L. "Development of a Fuel-Based Mass Emission Measurement Procedure." SAE Paper 710604, 1971.

Stockman, David Alan. *The Triumph of Politics: Why the Reagan Revolution Failed.* New York: Harper and Row, 1986.

Stokes, Donald. *Pasteur's Quadrant: Basic Science and Technological Innovation.* Washington, D.C.: Brookings Institution Press, 1997.

Stonex, Kenneth A. "How Roadside Ditches and Slopes Can Be Designed for Safety." *General Motors Engineering Journal* 8, no. 3 (July–September 1961): 31–36.

Stonex, Kenneth A. "Roadside Design for Safety." *Highway Research Board Proceedings* 39 (1960): 120–156.

Stonex, Kenneth A., and Paul C. Skeels. "A Summary of Crash Research Techniques Developed by the General Motors Proving Grounds." *General Motors Engineering Journal* 10, no. 4 (1963): 7–11.

Stradling, David. *Smokestacks and Progressives: Environmentalists, Engineers, and Air Quality in America, 1881–1951.* Baltimore: Johns Hopkins University Press, 1999.

Sundquist, James L. *Politics and Policy: The Eisenhower, Kennedy, and Johnson Years.* Washington, D.C.: Brookings Institution Press, 1968.

Taylor, Margaret R. "The Influence of Government Actions on Innovative Activities in the Development of Environmental Technologies to Control Sulfur Dioxide Emissions from Stationary Sources." PhD diss., Carnegie Mellon University, 2001.

Taylor, Margaret R., Edward S. Rubin, and David A. Hounshell. "Regulation as the Mother of Innovation: The Case of SO_2 Control." *Law & Policy* 27, no. 2 (April 2005): 348–378.

Teague, D. M. "Los Angeles Traffic Pattern Survey." Paper no. 171, presented to the Society of Automotive Engineers National West Coast Meeting, August 1957. Collected in SAE. *Vehicle Emissions (Selected SAE Papers).* New York: SAE, 1964.

Teles, Steven M. *The Rise of the Conservative Legal Movement: The Battle for the Control of the Law.* Princeton: Princeton University Press, 2008.

Tenney, Edward A. *The Highway Jungle.* New York: Exposition Press, 1962.

"Test of Speedometers," *Motor Age* 17, no. 23 (June 9, 1910), 30–31.

"Texas Headlight Laws Ridding Roads of Glare." *Automotive News,* October 27, 1925.

Thackray, Arnold, and Minor Myers Jr., *Arnold O. Beckman: One Hundred Years of Excellence.* Philadelphia: Chemical Heritage Foundation, 2000.

Thompson, George V. "Intercompany Technical Standardization in the Early American Automobile Industry." *Journal of Economic History* 14, no. 1 (1954): 1–20.

Thrun, Sebastian. "What We're Driving At." *Google Official Blog,* October 9, 2010. http://googleblog.blogspot.com/2010/10/what-were-driving-at.html.

Tomkin, Shelley Lynne. *Inside OMB: Politics and Process in the President's Budget Office.* Armonk, N.Y.: M. E. Sharpe, 1998.

Tozzi, Jim. "OIRA's Formative Years: The Historical Record of Centralized Regulatory Review Preceding OIRA's Founding." *Administrative Law Review* (Special Edition) 63 (2011).

Tucker, Jonathan B. "R&D Consortia: Can U.S. Industry Beat the Japanese at Their Own Game?" *High Technology* 4 (1984): 46–52.

Uekoetter, Frank. "The Strange Career of the Ringelmann Smoke Chart." *Environmental Monitoring and Assessment* 106, nos. 1–3 (July 2005): 11–26.

[Updike, John]. "Packaged." *New Yorker,* September 29, 1956.

Ullmann, John E. "Some Economic Aspects of Automobile Safety Devices." In *Passenger Car Design and Highway Safety,* 248–264. New York: Association for the Aid of Crippled Children and the Consumers Union, 1962.

US Department of Health, Education, and Welfare. *Report of the Secretary's Advisory Committee on Traffic Safety, February 29, 1968*. Washington, D.C.: Government Printing Office, 1968).

US Department of Transportation. *The Auto Industry, 1980*. Washington, D.C.: Government Printing Office, 1981.

US Department of Transportation. "Secretary Ray LaHood Distracted Driving Press Conference U.S. Department of Transportation," updated April 18, 2012. Trans portation.gov. https://www.transportation.gov/briefing-room/secretary-ray -lahood-distracted-driving-press-conference-us-department-transportation.

US Department of Transportation and Environmental Protection Agency. "Potential for Motor Vehicle Fuel Economy Improvement." Report to Congress, October 24, 1974.

US Department of Transportation, National Highway Traffic Safety Administration. Federal Motor Vehicle Safety Standards and Regulations, Standard No. 201, Occupant Protection in Interior Impact. http://www.nhtsa.gov/cars/rules/import /FMVSS/.

US House of Representatives. *Automobile Seat Belts: Report of the Special Subcommittee on Traffic Safety of the Committee on Interstate and Foreign Commerce*. 85th Congress, 1st sess. (1957).

US House of Representatives. *Motor Vehicle Safety: Hearings before a Subcommittee of the Committee on Interstate and Foreign Commerce*. 86th Congress, 1st sess. (1959).

US House of Representatives. *Research Needs in Traffic Safety: Hearing before a Subcommittee of the Committee on Interstate and Foreign Commerce*. 85th Congress, 2nd sess. (1958).

US House of Representatives. *Traffic Safety: Hearings before a Subcommittee Investigation of Highway Traffic Accidents of the Committee on Interstate and Foreign Commerce*. 84th Congress, 2nd sess. (1956)

US House of Representatives, Committee on Government Operations, *Administration's Proposals to Help the U.S. Auto Industry: Hearings*. 97th Congress, 1st session, 1981. Washington, D.C.: Government Printing Office, 1981.

US House of Representatives, Committee on Interstate and Foreign Commerce. *Federal Regulation and Regulatory Reform: Support by the Subcommittee on Oversight and Investigations*. 94th Congress, 2nd sess., October 1976.

US House of Representatives, Committee on Interstate and Foreign Commerce. *Traffic Safety: Hearings on H.R. 13228 and Other Bills Relating to Traffic Safety*. 89th Congress, 2nd sess. Washington, D.C.: Government Printing Office.

US House of Representatives, Committee on Interstate and Foreign Commerce. *Summary of Major Provisions of Energy Conservation and Oil Policy Act of 1975 (H.R. 7014) as Passed by the House*, September 26, 1975. Washington, D.C.: Government Printing Office, 1975.

US Public Health Service. *Motor Vehicles, Air Pollution, and Health: A Report of the Surgeon General to the U.S. Congress in Compliance with Public Law 86–493, the Schenck Act*. Washington, D.C., 1962.

Usselman, Steven W. *Regulating Railroad Innovation: Business, Technology, and Politics in America, 1840–1920*. Cambridge: Cambridge University Press, 2002.

US Senate. Committee on Commerce, Science, and Transportation. *Motor Vehicle Safety and the Marketplace: Hearings Before the Subcommittee on Surface*

Transportation. 98th Congress, 1st sess., February 17, March 10 and 11, 1983. Washington, D.C.: Government Printing Office, 1983.

US Senate. Committee on Government Operations. *Federal Role in Traffic Safety: Hearings before the Subcommittee on Executive Reorganization.* 89th Congress, 2nd sess. (1965).

US Senate. Committee on Labor and Public Welfare. "Motor Vehicle Safety Responsibility Act of District of Columbia." July 10, 1953. PL 83–365.

US Senate. Committee on Public Works. *Highway Safety Act: Hearings on S. 3052.* 89th Congress, 2nd sess. (1966)

US Senate. Committee on Public Works, Subcommittee on Air and Water Pollution. *Decision of the Administrator of the Environmental Protection Agency regarding Suspension of the 1975 Auto Emissions Standards.* 93rd Congress, 1st sess. April 16–18, 1973.

US Senate. Committee on the Judiciary. "Compensation for Civilian Injuries and Death from Collision of Army Truck and Automobile." July 9, 1953 (Bill no. 83, HR 724).

US Senate. Committee on the Judiciary. "Compensation to Mrs. Cecil N. Broy for Husband's Death in Automobile Accident While in Foreign Service." June 22, 1954 (Bill no. 83, HR 7886).

US Senate. Special Subcommittee on Science, Technology, and Commerce of the Committee on commerce. "Hearings on the National Fuel Economy Testing Act of 1974." 93rd Congress, 2nd sess. May 17, 1974,

Utterback, James M., and William J. Abernathy. "A Dynamic Model of Process and Product Innovation." *Omega* 3, no. 6 (1975): 639–656.

Van Horn, Rob. "Reinventing Monopoly and the Role of Corporations: The Roots of Chicago Law and Economics." In *The Road from Mont Pèlerin: The Making of the Neoliberal Thought Collective*, ed. Philip Mirowski and Dieter Plehwe, 204–238. Cambridge, Mass.: Harvard University Press, 2009.

Vincenti, Walter G. *What Engineers Know and How They Know It: Analytical Studies from Aeronautical Engineering.* Baltimore: John Hopkins University, 1990.

Vinsel, Lee. "The Crusade for Credible Energy Information and Analysis in the United States, 1973–1982." *History and Technology* 28, no. 2 (2012): 149–176.

Vinsel, Lee. "Federal Regulatory Management of the Automobile in the United States, 1966–1988." PhD diss., Carnegie Mellon University, 2011.

Vinsel, Lee. "'Safe Driving Depends on the Man at the Wheel': Psychologists and the Subject of Auto Safety, 1920–55." *Osiris* 33, no. 1 (2018): 191–209.

Vinsel, Lee. "Virtue via Association: The National Bureau of Standards, Automobiles, and Political Economy, 1919–1940." *Enterprise & Society* 17, no. 4 (2016): 1–30.

Viteles, Morris S. "Transportation Safety by Selection and Training." *Industrial Psychology Monthly* 2 (1927): 119–128.

Vogel, Ezra. *Japan as Number One: Lessons for America.* Cambridge, Mass: Harvard University Press, 1979.

Von Hippel, Eric. *Democratizing Innovation.* Cambridge, Mass.: MIT Press, 2005.

Wald, Patricia M. "Thirty Years of Administrative Law in the D.C. Circuit Court." Harold Leventhal Talk, DC Bar, July 1, 1997. https://www.dcbar.org/communities /administrative-law-and-agency-practice/wald.cfm.

Walsh, Bryan. "California's Clean-Air Slapdown." *Time*, December 20, 2007.

Ward, John W., and Christian Warren. *Silent Victories: The History and Practice of Public Health in Twentieth-Century America.* Oxford: Oxford University Press, 2007.

Waxman, Henry A. "An Overview of the Clean Air Act Amendments of 1990." *Environmental Law* 21 (1991): 1721–1816.

Weart, Spencer R. *The Discovery of Global Warming.* Cambridge, Mass.: Harvard University Press, 2003.

Weingroff, Richard E. "President Dwight D. Eisenhower and the Federal Role in Highway Safety," Federal Highway Administration, http://www.fhwa.dot.gov /infrastructure/safety.cfm.

Weiss, Albert P., and Alvhh R. Lauer. *Psychological Principles in Automotive Driving.* Columbus: Ohio State University Press, 1930.

Weschler, David. "Tests for Taxicab Drivers." *Personnel Journal* 5 (1926): 24–30.

West, William F., and Joseph Cooper. "The Rise of Administrative Clearance." In *The Presidency and Public Policy Making,* ed. George C. Edwards III, Steven A. Shull, and Norman C. Thomas, 192–214. Pittsburgh: University of Pittsburgh Press, 1985.

Wetmore, Jameson Michael. "Systems of Restraint: Redistributing Responsibilities for Automobile Safety in the United States since the 1960s." PhD diss., Cornell University, 2003.

"What Is Class Legislation?" *The Horseless Age* 19, no. 9 (February 27, 1907), 300.

"What Some Missourians Think of Autos," *The Automobile* 15, no. 25 (December 20, 1906).

White, Lawrence J. "Automobile Emissions Control Policy: Success Story or Wrongheaded Regulation?" In *Government, Technology, and the Future of the Automobile,* ed. Douglas H. Ginsberg and William J. Abernathy, 401–420. New York: McGraw-Hill, 1980.

Whitney, Albert W. *How Knowledge Driveth Away Fear: A Morality Play for Children.* New York: National Safety Council, 1927.

Whitney, Albert W. *Man and the Motor Car.* New York: National Bureau of Casualty and Surety Underwriters, 1936.

Whitney, Albert W. *The Place of Standardization in Modern Life.* Washington, D.C.: Government Printing Office, 1924.

Whyte, Kenneth. *Hoover: An Extraordinary Life in Extraordinary Times.* New York: Alfred K. Knopf, 2017.

Wiener, Jonathan B., and Michael D. Rogers. "Comparing Precaution in the United States and Europe." *Journal of Risk Research* 4, no. 4 (2002): 317–349.

Williams, Douglas. "The Mazda Fuel Furor." *Review of Southern California Journalism,* no. 11 (May 1974).

Wilson, William E., et al. "General Motors Sulfate Dispersion Experiment: Summary of EPA Measurements." *Journal of the Air Pollution Control Association* 27, no. 1 (January 1977): 46–51.

Winner, Langdon. "Do Artifacts Have Politics?" In "Modern Technology: Problem or Opportunity?" Special issue, *Daedalus* 109, No. 1 (Winter 1980): 121–136.

Withrow, James R. "The Organizing Conference of the Federated American Engineering Societies." *Ohio State Engineer* 4, no. 1 (November 1920): 12–15, 18.

Woodward, Fletcher D. "Medical Criticism of Modern Automotive Engineering." *Journal of the American Medical Association* 138, no. 9 (1948): 627–631.

Womack, James P., Daniel T. Jones, and Daniel Roos. *The Machine That Changed the World*. New York: Rawson Associates, 1990.

Wright, Christopher O. B. "The National Cooperative Research Act of 1984: A New Antitrust Regime for Joint Research and Development Ventures." *Berkeley Technology Law Journal* 1, no. 1 (January 1986): 133–193.

Yandle, Bruce. "A Cost-Benefit Analysis of the 1981–1984 MPG Standard." *Policy Analysis* 6 (Summer 1980): 291–304.

Yang, Jie, Simon Sidhom, Gayathri Chandrasekaran, Tam Vu, Hongbo Liu, Nicolae Cecan, Yingying Chen, Marco Gruteser, and Richard P. Martin. "Detecting Driver Phone Use Leveraging Car Speakers." In *Proceedings of the 17th Annual International Conference On Mobile Computing And Networking*, 97–108. New York: Association for Computing Machinery, 2011.

Yates, JoAnne. "How Business Enterprises Use Technology: Extending the Demand-Side Turn." *Enterprise and Society* 7, no. 3 (2006): 422–455.

Yates, JoAnne. *Structuring the Information Age: Life Insurance and Technology in the Twentieth Century*. Baltimore: Johns Hopkins University Press, 2005.

Year Book of the American Engineering Standards Committee 1923

Yergin, Daniel. *The Prize: The Epic Quest for Oil, Money, and Power*. New York: Simon and Schuster, 1991.

Yoganandan, Narayan, Alan M. Nahum, and John Melvin. *Accidental Injury: Biomechanics and Prevention*. New York: Springer, 2015.